GeoJournal Library

Volume 113

More information about this series at http://www.springer.com/series/6007

Xueming Chen • Qisheng Pan
Editors

Building Resilient Cities in China: The Nexus between Planning and Science

Selected Papers from the 7th International Association for China Planning Conference, Shanghai, China, June 29 – July 1, 2013

Springer

Editors
Xueming Chen
L. Douglas Wilder School
of Government and Public Affairs
Virginia Commonwealth University
Richmond, VA, USA

Qisheng Pan
Department of Urban Planning
and Environmental Policy
Texas Southern University
Houston, TX, USA

School of Urban Design
Wuhan University
Wuhan, China

ISSN 0924-5499 ISSN 2215-0072 (electronic)
GeoJournal Library
ISBN 978-3-319-14144-2 ISBN 978-3-319-14145-9 (eBook)
DOI 10.1007/978-3-319-14145-9

Library of Congress Control Number: 2015939439

Springer Cham Heidelberg New York Dordrecht London

Printed on acid-free paper

Springer International Publishing AG Switzerland is part of Springer Science+Business Media (www.springer.com)

Preface

Urban resiliency in face of uncertainty is a priority for planning and governance in communities worldwide. In China, which has suffered many of the world's most devastating floods, earthquakes, and typhoons, preparing for the threat of disaster has long been an important planning objective. Recent calamities, such as the 2008 Winter Storms, the 2008 Wenchuan Earthquake, and the 2012 Beijing Floods have only made planning for resiliency more urgent.

As planners work to prepare for such events, interdisciplinary collaboration becomes increasingly important. Planners need the tools and insights offered by other fields, including both the natural and social sciences. At the same time, these interdisciplinary relationships also help to shape the identity of urban-rural planning in its new role as one of China's primary academic disciplines.

This book discusses various planning and management issues related to urban resiliency. It includes 30 peer-reviewed and rigorously edited papers, which were originally presented at the 7th International Association for China Planning (IACP) Conference held from June 29 to July 1, 2013, in Shanghai, China, on the subjects of urban planning, environmental planning, transportation planning, historical preservation, emergency relief and management, geographic information systems (GIS) and other technological applications, and others.

In chapter "Empirical Research on Global Value Chain and Industrial Structure Upgrading of Shanghai Putuo District," Huixia Bian introduces an empirical research on global value chain and industrial structure upgrading of Shanghai Putuo District. She uses a multivariate linear regression model to demonstrate the positive connection between the degree of integration of global value chain and the industry structure. Based on her research, the author suggests that the top-end service industry and the headquarter economy should have a higher priority in the development of Putuo District.

Chapter "Spatial and Social Assessment of Chinese Urban Neighbourhoods Undergoing Change: A Case Study of a Large Scale Resettlement Neighborhood in South Jiangsu Province" by Ying Chang, Jing Lu and Xiaonan Zhang conducts a spatial and social assessment of the undergoing change of Chinese urban neighbourhoods through an empirical case in a large-scale resettlement neighborhood in South Jiangsu Province. They have found that the quality of public open space is a critical factor in the development of a more flourishing neighbourhood life.

In chapter "The Unprecedented Age Wave and Challenged Fixed-Route and Specialized Transportation Services: The Case of Richmond, Virginia," Xueming Chen discusses the severe age wave and challenging transit and paratransit planning, funding and operating issues in Richmond, Virginia. In Chen's opinion, it is necessary to take a holistic view of these issues and tackle them from political, economic, planning and technological perspectives in a concerted manner.

With respect to how to build resilient cities, which is the main theme of chapter "Resilient Planning Frame for Building Resilient Cities," Zhiduan Chen and Baoxing Qiu suggest that there are two types of resilience: urban ecological resilience and city engineering resilience. They recommend a major resilient urban planning initiative, intended to improve the thought and method of urban planning, expand urban systematic research and strengthen cooperation among different disciplines involved in resilient urban planning and development.

Flood control and stormwater drainage are important urban resilient issues. In chapter "The Construction of the Neighborhood Unit System of Flood Control and Drainage Based on Landscape Infrastructure," Guan Shaoping and Zhang Xi specifically examine the issues related to the construction of the neighborhood unit system of flood control and drainage based on landscape infrastructure. Their model shows the effects in flood control and drainage through the integration of engineering technology, nature, and landscape architecture, providing the benefits of improving ecological, economic and societal resiliency to promote the healthy and sustainable development of the city.

Community life amid rapid urbanization is the focus of chapter "Community Life Regression Under the Background of Rapid Urbanization." In that chapter, Liang Guo and Min Li describes the current community life types, standards, facilities and community organizations, based on which they make some constructive suggestions on how to build community life under the background of the rapid urbanization.

In chapter "Addressing Social Revitalization in Conservation of Historic Quarters in China: A Social Capital Initiated Urban Conservation Approach," Jie Han, Liangliang Wang, and Chye Kiang Heng take a social capital-initiated urban conservation approach to address social revitalization in conservation of historic quarters in China. Through a pilot study on the conservation of the Xi'an Muslim District, this paper examines this alternative urban conservation approach and the potential implications for conservation planning in China.

Chapter "The Strategies Research of Open Space Planning Under the Goal of Building Resilient City" reports the research results of open space planning under the goal of building resilient city. In this chapter, Hu Yuwei and Huang Jianyun have found that "Resilient city" and its own internal mechanism can adapt to changes, especially in response to climate change. Open space can improve urban environment and provide disaster refuge, thus helping build a resilient city.

Recognizing the important roles of Geographic Information Systems (GIS), in chapter "A People's Atlas of Muncie: Citizen Representations of Urban Space," Junfeng Jiao, Steven M. Radil, Jenna Harbin, and Yuan Li show how to use mental mapping as a tool to collect data and insight from community members around Muncie, Indiana, and introduce different data collection methods to students.

In chapter "The Ecological Impact of the Korean Saemaul (New Rural Community) Movement, 1970–1979," Chung Ho Kim assesses the ecological impact of the Korean Saemaul (New Rural Community) Movement, or KSM, between 1970 and 1979. Even though the historical reviews of KSM are generally positive, a detailed examination of its economic and social impacts is still lacking. This chapter finds the high correlation between ecological impact of KSM and Korean Reforestation, and concludes that KSM was a spiritual platform for national agendas to be implemented according to changing situations; and KSM caused regime changes of resource management from natural resource-based materials into industrial resource-based materials.

Chapter "The Analysis of Two Increases and Two Reductions Policy of Shanghai" by Liao Yu-qing and Huang Jian-yun examines the implications of "Two Increases and Two Reductions" Policy of Shanghai and comes up with an idea about its future development in three aspects, including urban design, traffic impact assessment and resilient city.

Chenchen Li, Jie Yin and Yang Lei summarize the utilizing ways and façade renovation patterns of the shops along the "shi-glyph" commercial streets in Pingyao ancient city, the world cultural heritage, and show that facade subdivision caused by multiple different utilizing ways has a wide and strong destructive impact on the style and features of the compound facade along the historical commercial streets. Read chapter "Facade Renovation and Utilizing Ways in the Historical Commercial Streets: Rigid Planning at the Compound Level Needed in Pingyao Ancient City in China" for more details.

In chapter "Method Exploration of Sensitive Public Space System Planning in Urban Newly Areas on the Viewpoint of Systematology," Mengchen Li and Ying Lin present a case study of urban design of Bali Lake District in Jiujiang City, Jiangxi, China, which explores the method of sensitive public space system planning in urban newly areas on the viewpoint of systematology. This chapter provides new perspectives and thoughts for exploration of generality integrated urban design methodology.

Pengbo Li, Jun Wu, Yan Jiang and Lei Meng believe that urban natural landscape system plays a very important role in the protection and rebuilding of the urban eco-environment. In chapter "Analysis of Framework and Optimization of the Urban Natural Landscape System (UNLS): Case Research of Eco-efficiency of UNLS in Tianjin, China," these authors construct a framework, the Urban Natural Landscape System (UNLS), within which to understand the improvement of the urban ecoenvironment and optimization of the urban landscape. The Tianjin UNLS serves as an example for this chapter.

Mingxing Liu and Gang Zeng conduct a study on the indicator system of eco-regional development planning for Shanghai Chongming Island, with its results documented in chapter "Study on the Indicator System of Eco-regional

Development Planning for Shanghai Chongming Island." The case of Chongming eco-island can shed light on other underdeveloped areas by bringing fresh air to the discussion of the reorganization of the urban-rural dualistic structure and the improvement of the urban-rural integration.

In chapter "A New Thinking of the Urbanization Route in Forest Ecological Function Regions: The Case of Shen-Nongjia," Liu Yun, Chang Lili and Liu Ying provide a new thinking of the urbanization route in forest ecological function regions. The case of Shen-Nongjia reveals the importance of harmonizing the development among urbanization, industrialization and ecology.

In chapter "Quantitative Study of Housing Price Based on Huff Model and Hedonic Method," Yuan Li, Lang He, Junfeng Jiao and Guoqiang Shen use the Hedonic and Huff model to conduct an empirical study of the ordinary commercial housing price's spatial pattern in Xiamen City, China. They have found that allocation of educational resources and distance between housing and shopping malls and supermarkets based on Huff model have significant impact on housing price.

In chapter "Develop A GIS Based Risk Model to Evaluate the Economic Resilience of Houston Neighborhoods for the Next Oil Bust," Qisheng Pan and Dong Zhai document the process to develop a GIS based risk model to evaluate the economic resilience of Houston neighbourhoods for the next oil bust. The risk model they developed takes into account multiple factors such as the house ownership in the neighbourhood, the availability of public transportation, and the average education level of its residents, etc., which will help to understand the resilience of a neighbourhood in case of another oil bust in Houston.

Chapter "Storm Preparedness: A Case Study of Delaware County, Indiana" by Junfeng Jiao, Timothy R. Phelps and Yuan Li uses ArcGIS to analyze the effectiveness of Delaware County, Indiana, in severe weather warning. Looking at the distribution of emergency equipment, this research shows that Delaware County is well suited to handle emergency storm situations, with only Yorktown being insufficient in the distribution of warning sirens.

Shan Shuang reports the research results on the quality and level of urbanization in the Baoshan District, Shanghai, based on the TOPSIS evaluation method (TOPSIS). Based on the composite score and a comparative analysis, this chapter "Research on the Quality and Level of Urbanization in the Baoshan District, Shanghai" demonstrates the advantages and disadvantages of the urban development process in Baoshan District, and provides a theoretical basis for policy-making in the future.

Chapter "The "Policy-Projects Districts" Model of New Urban Area Development in China and Its Analysis" introduces the "Policy-projects districts" model of new urban area development in China. In this chapter, Wang Jin-Bai argues that the core mechanism of China's new urban area development is the combination of the centralized allocation of resources with decentralized market-oriented development, and its logical relationship model could be constructed by two linear functions and a negative correlation curve function. He also discusses the urban-rural development and transitioning urbanization mechanisms, which could provide guidelines for future local development practice.

Wen Jie and Huang Jianyun present their study on transportation space improvement and adaptation in Changxing County, Huzhou in chapter "A Study on Transportation Space Improvement and Adaptation in Old Downtown Area of Middle and Small Size Cities: A Case Study for Changxing County, Huzhou." Changxing County is a representative model of middle and small size cities.

Jun Wu, Naiwen An, Pengbo Li and Min Zhang believe that Point of Interest (POI) is the place where public likes to perform activities, which can arouse public interest, attract public activities and satisfy public participation. It is necessary to plan and design landscape notes into POI to provide suitable spaces and increase the attractiveness and efficient use.

Based on significant research of landscape notes of Tianjin urban open space, chapter "Constructure and Update of Point of Interest (POI) in Urban Open Space: A Case Study of Assessment and Reform of POI in Water Park Tianjin, China" reveals the three constituent factors of POI and analyzes the principles of the POI construction using Water Park of Tianjin as an empirical example.

Chapter "Comparison of Two Approaches for Evacuation Plan with Multiple Exits" compares two approaches for evacuation plan with multiple exits. In this chapter, Jun Xie, Qing Wan, Chi Zhang and Xiang Li specifically examine both virtual node approach and optimal allocation approach, under different scenarios with the road network of Jing'an District, Shanghai. A simulation program is also developed to visualize the evacuation process. Results illustrate that the optimal allocation approach performs better than the virtual node approach under these scenarios.

In chapter "Study of Urban Patterns Optimization Employing CFD Method: A Case Study of Chenjiazhen Experimental Ecological Community, Chongming, Shanghai," Chao Liu, Peng Xu, Weizhen Chen, Liang Zhang and Weilin Li study the urban patterns optimization employing CFD method with a case study of Chenjiazhen Experimental Ecological Community, Chongming, Shanghai. Their study finds how urban patterns affect wind flow in city level, based on which several optimizing strategies and estimation of corresponding energy saving have been given.

Jiawen Yang and Chuanglin Fang assess the public-sector transportation finance and planning issues in urban China in chapter "Assessing Public-Sector Transportation Finance and Planning in Urban China." In their opinion, as the new economic and transportation environment pushes for a better integrated transportation system, two recent reforms, the national fuel tax and the consolidation of transportation management functionalities within the megacities, provide an opportunity to restructure the supply of urban and regional mobility in China.

In chapter "Rational Traffic Dispersion: Take the Example of No Left Turn in the Rush Hours Along Renmin Road in Suzhou City," Yao Yangyang demonstrates a method of rational traffic dispersion using the example of No Left Turn in the rush hours along Renmin Road in Suzhou City. Based on the data of multiple aspects (the delay of stop, the flow of vehicles, average speed of vehicles and so on), Yao offers a solution to address the related traffic problem.

Using Nanjing's Mai-Gao-Qiao Area as an empirical case, Xin Yi probes into the problems and proposes the redevelopment strategies for this old industrial area in chapter "Problems and Redevelopment Strategy for Old Industrial Area: After an Accident in Nanjing's Mai-Gao-Qiao Area." This chapter examines the functional relationship of this area with adjacent areas, especially a newly developed satellite city to the east. The social and economic profiles of different social groups are also portrayed.

Chapter "Analysis of the Planning Mode of Resilient Urban Space Structure" by Zhou Min, Lin Kaixuan and Huang Yaping analyzes the planning mode of resilient urban space structure. This chapter contends that a resilient urban spatial structure mode should have four basic attributes of ecology, growth, economy and coordination. Based on this study, the chapter proposes a "multi-center, organizing network" within the overall urban spatial structure planning mode.

Finally, chapter "The Impact of Building Control on Urban Planning and Building Management in Hong Kong" by Han Zou and Charlie Q.L. Xue assesses the impact of building control on urban planning and building management in Hong Kong by using the historical research method. It takes the background of high density urban development mode, and sums up the influences of building control on the evolution of urban form by analyzing commercial and private housing cases. It also tries to sum up the historical experiences and explore the scientific method of building control in Hong Kong.

As the proceedings of the 2013 IACP annual conference, this 30-chapter book is targeted at those faculty members, students, practitioners, and the general public interested in the subjects of urban planning, urban design, land use, transportation, housing and community development, infrastructure planning, geographic information system technological developments and applications, climate change and ecological planning, emergency management and disaster relief, risk mitigation and anticipatory practice, and others. The key contribution of the book is to address the nexus between planning and science, which is critically important in building resilient cities in China. The Chinese planning experience and lessons we present in this book can also be learned by scholars, planning practitioners, and policy makers in many other countries.

Richmond, VA, USA Xueming Chen
Houston, TX, USA Qisheng Pan

Contents

Empirical Research on Global Value Chain and Industrial Structure Upgrading of Shanghai Putuo District

Huixia Bian

1 Introduction

In the traditional industrial structure change theory, there are two classic facts: Kuznets established the "Kuznets facts" in 1966, that is, with the development level, the proportion of agriculture in the economy gradually decreases, while the proportion of industry and services gradually increases [1]; Later, Daniel Bell (1973) brought in the concept of "post-industrial fact" [2]. It was believed that when economic development became into a higher level, the proportion of industry began to decline and that of the service industry continued to rise. Maddison (1997) conducted empirical researches on the industrial structure changes in developed countries and OECD countries, which confirmed that those two facts are common phenomenon [3].

Zhang and Liu (2009) proposed that the global value chain (GVC), as the main form of international industrial transfer of the organization, has a fundamental influence on the microscopic foundation of the global economy [4]. Li (2008) has shown that with the economic globalization, the modern metropolis has gone through the space reorganization led by the GVC. Moreover, the integration of elements of the industry, human capital, information in high-tech industry and high-end service industry, promoted the development of transformation and upgrading of the city [5].

Therefore, this research selected Putuo District as the study area, which is the central urban area of Shanghai – the most affected frontier by globalization in the Yangtze River Delta region. By analyzing the developmental evolution of the industrial structure and its integration into the upper side of GVC, this paper

H. Bian (✉)
Department of Urban and Regional Economics, East China Normal University (ECNU), Shanghai 200241, China
e-mail: ibhx@163.com

X. Chen, Q. Pan (eds.), *Building Resilient Cities in China: The Nexus between Planning and Science*, GeoJournal Library 113, DOI 10.1007/978-3-319-14145-9_1

attempts to empirically investigate the reasons, power and intrinsic link of those facts based on 2004–2011 economic and social development data of Putuo District.

2 Background of the Research Area

2.1 The Evolution of Industrial Development Stage of Putuo District

There are five periods of industry development: (a) (before 1990) the traditional manufacturing industry formation period. Putuo District, along the banks of the Suzhou River flowing through downtown Shanghai had developed national industry, which was mainly traditional manufacturing industry. (b) (1991–2000) The traditional commerce industry formation period. The textile industry and other pillar manufacturing industries transferred while traditional service industry experienced a rapid development. First, the various types of agriculture by-product commodity trading markets played supporting function for the city life. Second, with its prosperity and development, the real estate industry became the first pillar industry of the Putuo District. (c) (2001–2010) The consumer-oriented service sector accumulation period. On the one hand, relying on the agglomeration and transportation hub, "Trading Center" has formed. It has a significant industrial clustering effect, business and commercial activities as the main function. On the other hand, based on the residence in a certain geographical area, "Community Center" enhanced the service function to meet residents' consumption and community commercial activities need. (d) (since 2010) The commerce and trade with other industrial integration development expansion period . Commerce industry and other industries showed a good integration of the development trend, and traditional commerce industry achieved industrial upgrading. First of all, Changfeng Ecological Business District agglomerated modern service area and building economy to promote the development of headquarters economy. Secondly, Shanghai Northwest Logistics Park bonded logistics center, Changfeng Financial Harbor as the carrier continue to play a role of platform organized by government.

Briefly, in the past 20 years, the process of industrial development of Putuo District, is a graduate process with a reduce in secondary industry, increase in tertiary industry and services keeping the high-end forward.

2.2 Maintaining the Integrity of the Specifications

The Industries Integration to the High-End of the GVC in Putuo District.

The 1990s was a period when Putuo rapidly increased the level of opening. In this period, the total foreign trade volume increased from 177 million USD in 1997 to 1.52 billion USD in 2011 (data from the "Putuo Statistical Yearbook 1998–2012"), the average increase rate is 16.6 %. Some companies in the region gradually become a part of the global industrial chains, while a number of multinational regional headquarters settled in Putuo as their location, which means interactive high degree of integration. Furthermore, with the rapid increase in cost of Putuo land price and other resources, only the high added value part in GVC, such as the modern service industry, the technology industry and headquarters economy etc., can generate high profits which reduces the input-output ratio.

3 Model Construction and Data Selection

3.1 Model Construction

In order to demonstrate the relationship between the global value chain and industry changing, we established a model (1), expressed as:

$$y_t = \beta_0 + \beta_1 openness + \beta_2 gdp_t + \beta_3 g_t + \varepsilon_t \tag{1}$$

In model (1), y_t represents the proportion of the tertiary industry and regional GDP, which measures the direction of change of industrial structure. *Openness* refers to the import and export trade and regional GDP ratio, that is, external dependence degree, which measures the degree of Putuo into global value chain. gdp_t is the per capita regional GDP as a measure of the level of economic development. Raise the level of regional economic development will lead to the change of its capital, technology, resource endowments and bring the change of the demand for human consumption preferences, which have resulted in the change of the industrial structure of the region. g_t refers to the proportion of government expense and regional gross domestic product, which measures the degree of government intervention in economic development. β_0 refers to regression constant. $\beta_1, \beta_2, \beta_3$ refers to Regression coefficient. ε_t refers to residual.

3.2 Data Selection

Given the availability of data, the sample time period for the years which we study is from 2004 to 2011. Putuo District data derives from "Putuo Statistical Yearbook 2012" and "Putuo District National Economic and Social Development Statistical Bulletin (2004–2011)". Because of the import and export trade data in U.S. dollar, we converted them to RMB with 2004–2011 exchange rate of annual in "China Statistical Yearbook (2012)". In order to eliminate price element, the per Capita Gross Regional Product data are deflating handled.

4 Results

The relevant economic and social development data are analyzed through Eview7.0, the values of variables are listed in Table 1.

The empirical analysis shows the regression coefficient of openness and of GDP to the industrial structure are 0.1262 and 0.2439 respectively, which indicates a positive impact. However, the regression coefficient of the degree of intervention in economic development to the industrial structure is −0.2395, which has a negative impact on the optimization of industrial structure.

Judging from the output results, we can make the following explanations: (a) The degree of openness can effectively promote the upgrading of industrial structure. Possible reasons are that Putuo District has location endowments, which determines the high economic density of this region. By vigorously developing its economic development level in headquarters economy, modern service industry and other high-end industries, Putuo District has gradually upgraded its position in global value chain, and further enhanced the clustering effect of modern service industry in the region which therefore led to the formation of headquarters economic zone and commercial consumption experience areas and promoted the continued optimization of the industrial structure. (b) Significant effect of GDP per capita on the upgrade of industrial structure in Putuo District. Results were mainly due to two aspects: first of all, Putuo had higher level of economic development (In 2011 per capita GDP was more than 10,000 U.S. dollars, which reached a high-income developed country standard), and entered the stage of industrialization. The experience in the developed countries shows that during the period of industrialization, the proportion of tertiary industry has the trend of continuous increase with the improvement of the level of economic development; Secondly, Putuo District is located in the northwest of the central city of Shanghai, with the region's rapid economic development, the residents has the increased demand for housing, consumer and entertainment services, resulting in a change in public policy. Therefore, regional business costs are also rising, and industrial enterprises in the regions gradually move out of the modern service industry. This phenomenon led to the continuous rise in the proportion of the higher investment in unit of output industry. (c) Government intervention in economic development is not significant affected. The reason may be related to the following aspects: firstly, in order to attract high-quality companies, Putuo District has implemented more favorable

Table 1 Regression results

Variable	Regression coefficient	Statistic	P -value
β_1	0.1262*	2.2405	0.0886
β_2	0.2439***	4.6988	0.0093
β_3	−0.2395	−1.4956	0.2091
β_0	−0.8935**	−3.6580	0.0216
$R^2 = 0.9390$		D.W. = 2.6713	

Note: ***, **, * mean that the coefficient is significant in the condition of 1 %, 5 %, 10 %

policies to attract talented people to work in the district. For instant, large state-owned enterprises helped pre-development in planning area, in order to attract other enterprises forming the industrial clusters. For an individual person, government provides them priority to solve the issue of their children's education, etc. All above supports are difficult to measure by government expense. In addition, the infrastructure such as part of park construction, viaduct building are invested by the municipal level of government intervention, which cannot be reflected in district-level financial expenditure data.

5 Suggestion

By empirical result analysis, the process of integration into the global value chain and the rapid escalation process of their leading industries are synchronous in Putuo District. Participating in international competition and gradually climbing to the high-end of GVC is the main driving force to promote the industrial structure upgrading.

Therefore, in the future development we suggest Putuo District to further increase the degree of openness, and focus on the development of modern service industry and headquarters economy, which is the high value-added economic pattern. In order to fully enhance core position in the global economy, Putuo District should actively participate in the competition in the global value chain, and continue to gradually increase the level of regional industrial. Apart from that, government intervention guide should gradually shift from capital-oriented to service oriented platform.

References

1. Kuznets S (1966) Modern economic growth: rate, structure and spread. Yale University Press, New Haven/London, pp 86–101
2. Bell D (1973) The coming of post-industrial society: a venture in social forecasting. Basic Books, New York, pp 32–45
3. Maddison A (1997) Causal influences on productivity performance 1820–1992: a global perspective. J Prod Anal 8:325–359
4. Zhang S, Liu Z (2009) Industry transference of GVC mode-force, influence and inspiration for China's industrial upgrading and balanced development of areas. China Ind Econ 11:5–15
5. Li C (2008) Urban spatial restructuring and the function of new industry value chain. Jianghai Acad J 4:219–224

Spatial and Social Assessment of Chinese Urban Neighbourhoods Undergoing Change

A Case Study of a Large Scale Resettlement Neighborhood in South Jiangsu Province

Ying Chang, Jing Lu, and Xiaonan Zhang

1 Introduction

A significant consequence of China's rapid urbanisation has been the creation of a large group of landless farmers. It is predicted that by 2030 the total population of landless farmers will exceed 78 million [1]. 'Upgrading from village to neighbourhood', a main approach that has contributed significantly to the urbanisation rate in China, mainly includes two outcomes: firstly, to relocate landless farmers into resettlement neighbourhoods; and secondly, to change their *hukou* registration category from agricultural to non-agricultural to enable them to receive a pension and other allowances or subsidies which can be claimed by local citizens. However, it does not mean that they necessarily enjoy the same quality of life as other citizens. This paper has assessed one resettlement neighbourhood that is undergoing significant changes, mainly from two perspectives: firstly, the demographic and social changes and their demand, and secondly, the quality of public open space and how it can be upgraded to accommodate more social activities.

This research is funded by Suzhou Association of Philosophy and Social Sciences (12-B-022012) and Xi'an Jiaotong-Liverpool University Research Development Fund (RDF11-03-11).

Y. Chang (✉) • J. Lu • X. Zhang
Department of Urban Planning and Design, Xi'an Jiaotong-Liverpool University, Suzhou Industrial Park, Suzhou, Jiangsu Province, China
e-mail: ying.chang@xjtlu.edu.cn

X. Chen, Q. Pan (eds.), *Building Resilient Cities in China: The Nexus between Planning and Science*, GeoJournal Library 113, DOI 10.1007/978-3-319-14145-9_2

2 Literature Review

2.1 *Emerging New Poverty of Landless Farmers*

It is a universal phenomenon that the landless farmers who have lost their primary livelihood (the land), are very likely to fall into poverty [2–5]. However, they have not only been deprived of their land, their dwellings and other natural resources, but also human and social capital. Furthermore, they are more likely to lose local economic, social and cultural assets, as well as opportunities and rights [6]. Involuntary urbanisation is a process that normally takes place suddenly, without a comprehensive consideration of the long term perspective, and is also an irreversible process [7]. Although the landless farmers have citizenship, they lack the capability to adapt to city life. For instance, they lack the ability to express their demand, they are disadvantaged in job competition and they have difficulties adjusting to the cash based urban economy [8]. In addition, they are also disadvantaged from the perspective of opportunity because the city is selective [7]. Lack of employment opportunities, or other income generation methods, is associated with high living expenses, which together has led to new poverty. Although the landless farmers share the income from the collective economy, it is reported that many collective economies are struggling to maintain their business [9]. A new group of landless farmers who have no land to farm, no opportunities for employment, and no minimum livelihood security subsidies, have become the new poor and a marginalized groups in cities.

2.2 *The Neighbourhood and Community of Landless Farmers and the Demand for High Quality Public Open Space*

The amount of living space landless farmers have decreased greatly after their relocation and they often cannot adapt to this new situation easily [1, 10–13]. In fact, recent research (sample size 288) [14] has shown that landless farmers still retain the economic and community structure of a rural society and almost all of them would like to retain their original community relationships.

Living spaces play a key role in enhancing social networks and social capital [15]. Green infrastructure is the main means by which urban residents can be close to nature, reduce stress, exercise and participate in other social activities [16–24], all of which are essential to a fulfilling life. Research undertaken in the UK (sample size 366,000) has shown that residents living close to a small green area display much lower rates of heart attacks or strokes [25]. The public space is also more important for disadvantaged groups because it is probably the only place they can afford to have social activities, sports or recreation [26–29]. However, some empirical research has shown that both the quantity and quality of public green space cannot meet the demand of older people [30, 31].

3 Methodology

A systematic sampling method was employed to conduct a household survey. Every habitable room, including the garage on the ground floor and the loft of the fifth floor, was recognised as one unit. There were 113 buildings in the case study neighbourhood. We randomly chose one building as the starting point and then every tenth building was selected for sampling. From each chosen building, we randomly selected one unit as the start and undertook the household questionnaire with every tenth household. Using this method, we successfully avoided selecting the units at the same side of the wing or on the same floor (each wing has five floors and each floor has two households). The survey was undertaken in Nov, 2012. In total, 100 valid questionnaires were collected. Questionnaires requested information with regard to individual situations, household information, housing conditions, local facilities, neighbourhood interactions and sense of community, as well as satisfaction with housing quality and the neighbourhood. Pictures replaced texts to collect households' answers with regard to the use of different facilities.

Data about the use of space was collected through two full-day (one weekday and one weekend in mid Nov) distance sampling observations from 8 AM to 7 PM. Line-transect distance sampling, a widely-used method to estimate the density or abundance of a biological population, was adapted in this study to estimate the public open space events within the neighbourhood. Observers performed a standardized survey along a series of parallel lines. They looked for a variety of activities, photographed the activities taking place, completed the information on the survey and recorded the location on the map.

All the activities, by gender, age, size and purposes were mapped and further classified into three major categories: (1) social activities, (2) optional activities and (3) necessary activities [32]. Social Activities refer to all activities that depend upon the presence of others in neighbourhood public open spaces, including children at play, chatting and greeting. Optional activities are those pursuits that are only participated in if there is a desire to do so. Examples of optional activities include taking a walk, standing around, and sunbathing. Necessary activities are those activities that are more or less compulsory, e.g. waiting for a person, running errands, and doing household work (Fig. 1). In this study, the term '*passive contact*' is used to summarize all individual activities and '*positive contact'* refers to activities participated in by two or more residents. Geographical Information System (GIS) software was then employed to elucidate patterns of social and cultural interactions and uses of public open space in the case study neighbourhood.

In addition, three focus groups were organised with local residents. The participants of the first two focus groups were invited through the key community informants and the participants of the last group were invited by short messages and posters. Forty residents in total attended the focus groups.

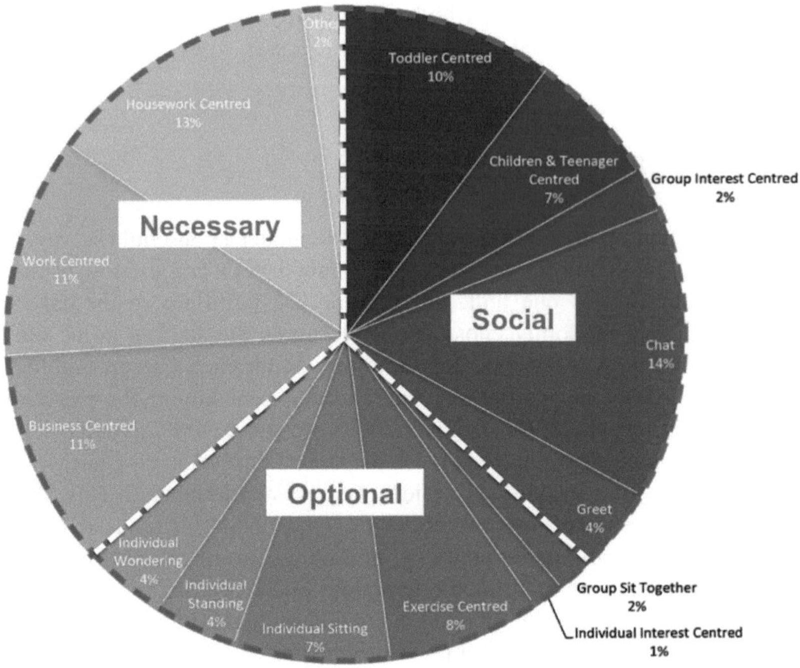

Fig. 1 Public open space activity classification

4 Findings on Social Issues of the Neighbourhood

The number of registered residents (with local *Hukou*) was 5,071 in total. Among them, elderly people (aged 60 or over) accounted for 20.4 % (1,037). The number of children (under age 7) was 402, and the number of youth (8–18) was 407. In addition, the number of migrant labourers (intra and inter-provincial) in the community is estimated to be over 22,000 (from interview with local office). In our survey, migrant households account for 72 % of all households in the community.

4.1 High Percentage of 'Migrant' Households and High Rate Degree of Mobility

The households could be divided into three categories: relocated farmers (original residents), migrant families from other provinces and collective migrant households who share one unit.[1] The second type of household counts for 39.6 % of all the

[1] It is found that normally six collective household share one unit.

surveyed households. The first and third types of household counted for 28.1 % and 31.3 % respectively. The average family size (people who share one room or unit) for these three groups were 3.7, 2.5 and 2.2 respectively.

Twenty one households out of 100 had stayed less than half a year. In these cases no further questions were asked. Thirty seven per cent of the households surveyed had lived in the neighbourhood for 1 year or more, and 12 % households had lived there for more than 5 years. About 23 % of households change their living place frequently (less than every 6 months). Their main reasons for moving were changes of employments (41.4 %), moving to better housing (24.1 %) and increased rent (13.8 %). Nearly half of the surveyed households were planning to stay in the community for at least another three years.

4.2 Satisfaction with Housing and the Neighbourhood

Tenants occupied the majority of the apartments. Only 33 % were owner occupied. In order to have more space to let out, some owners had converted their original garage space into apartments. These apartments very often lack basic sanitary facilities. For example, 9 % of apartments were without a private toilet, only 20 % had their own kitchen and 30 % were without a shower or bathing space.

Overall the satisfaction rate was recorded as being high. Forty-one per cent were satisfied or very satisfied with their apartments. Forty-five per cent felt neutral about their apartments. Among the criteria used for assessing the satisfaction of their current living condition, only the rent criterion scored 2.5, which was below the average (see Fig. 2).

4.3 Social Life and Sense of Belonging to the Community

About 51 % households surveyed have some sense of belonging to the community. The index was created based on previous research conducted by Wu (2012) and the Sense of Community Index II adapted from McMillan and Chavis (1986). This includes criteria such as membership, influence, meeting needs and a shared emotional connection.

The most popular social activities in the neighbourhood were chatting (either in a public space or a neighbour's apartment), having dinner together and singing Karaoke. The participation rate was low for events organized by the committee of the community (people employed by the local government to run events) (Fig. 3).

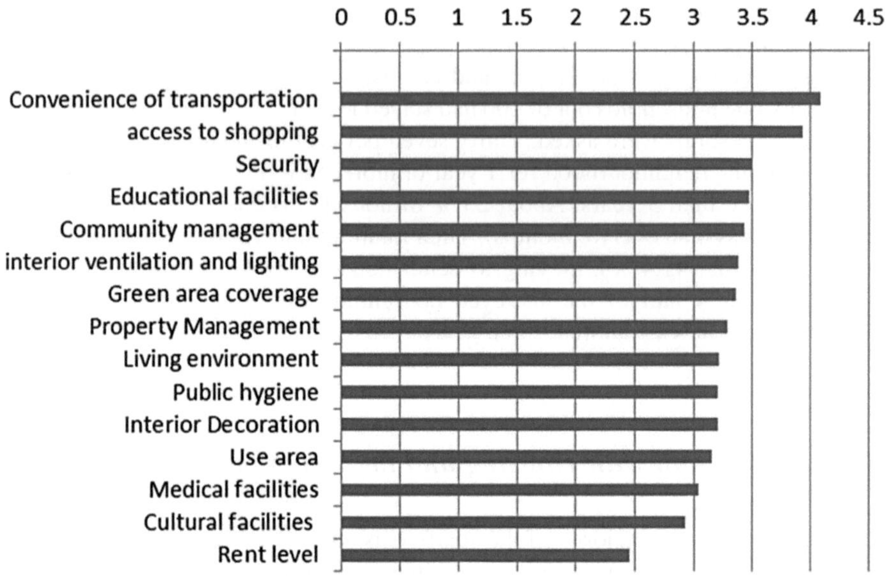

Fig. 2 Residents' satisfaction on different criteria (sample size: 79)

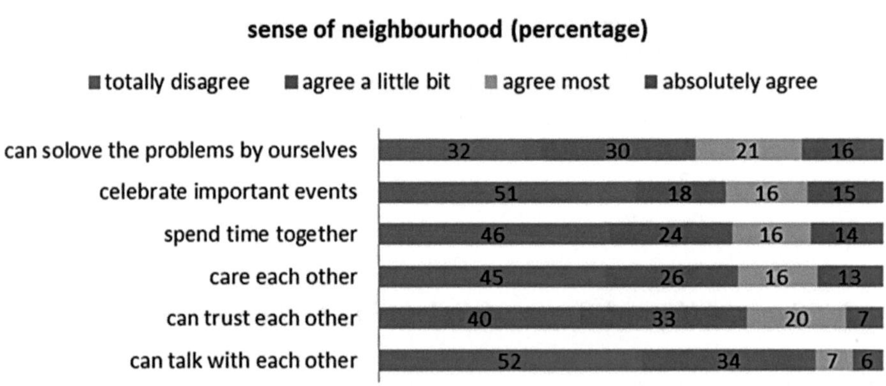

Fig. 3 Residents' sense of community (percentage)

4.4 Demand of Space, Management and Social Activities

In questioning participants about their pressing issues at the household level, over 50 % of households surveyed said they have no specific hardship. Among those who answered yes, the most frequently mentioned issues were low household income (17.8 %), poor housing conditions (12.3 %), lack of ideal employment opportunities (6.8 %) and caring for children (4.1 %).

The most pressing issues at neighbourhood level mentioned were noise (primarily the shuttle bus taking migrant workers to factories), sanitation and the need for a food market (Fig. 4).

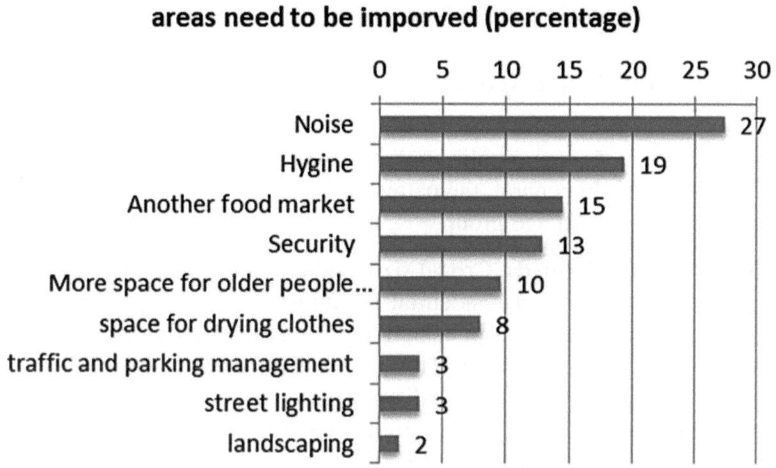

Fig. 4 Scoring of the pressing issues

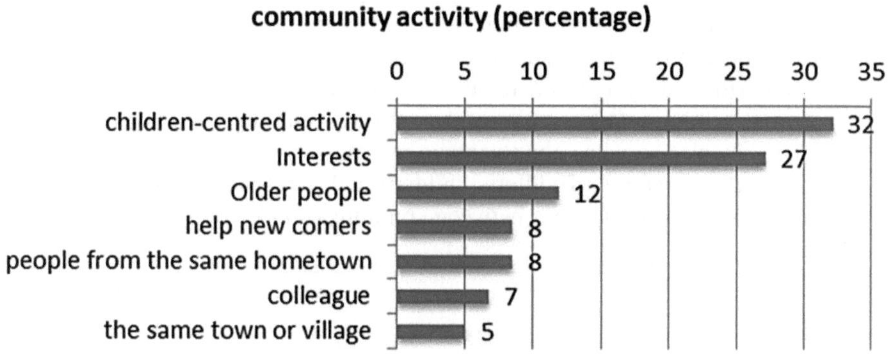

Fig. 5 Preference for different community activities

In focus group meetings, local residents strongly expressed the need for more public spaces, as well as better maintenance and management of these facilities. For example, they wanted a public library and a room to play table tennis. The main obstacles to them using current public space included smoking and inappropriate uses e.g. being used as weddings receptions or funerals.

Figure 5 shows that children-centred activities were the most popular in the neighbourhood. From this perspective, it is clear that children are at the centre of many social activities (49 %) and can be seen to bind the community together. The most popular self-organized interest groups among those surveyed residents included a conversation group (30.4 %), a mother's club (16.1 %), a travel club (14.3 %) and a dating club (8.9 %).

5 Findings from Distance Sampling

5.1 Multi-used Open Spaces

As illustrated in Fig. 6, open public spaces in the neighbourhood were used all day from 8 AM to 8 PM. In general, the patterns of interactions varied based on crowd size, activity purpose and age group.

- **Different Sizes of Crowds**. Crowd size gradually increases when moving from immediate open spaces of residential buildings to central open spaces. At the immediate open spaces in front of buildings individual activities are most likely to occur. Group activities for two to five residents often gather at the corner of buildings with good visibility. Mediam-sized open spaces accommodate group sizes of five to ten for more prolonged and meaningful events. The largest crowds of 20 or more tend to choose only central open spaces to gather.
- **Different Purposes**. More than half of the neighbourhood public open space users participated in social activities. A clear tiered social assembly pattern could also be found on the map of public open space used by social activity participants (Fig. 7). The map indicates that crowd sizes increase from immediate open spaces to central open spaces. In contrast, optional activities (Fig. 8) are mainly performed by individuals alone at immediate open spaces adjacent to residential buildings. Open spaces at the corner of buildings are the most popular because they also serve as good vantage points.
- **Different Age Groups**. Toddlers, children and teenagers mainly gathered at medium-sized open spaces between buildings and central open spaces.

Fig. 6 Map of public open space by all users over time

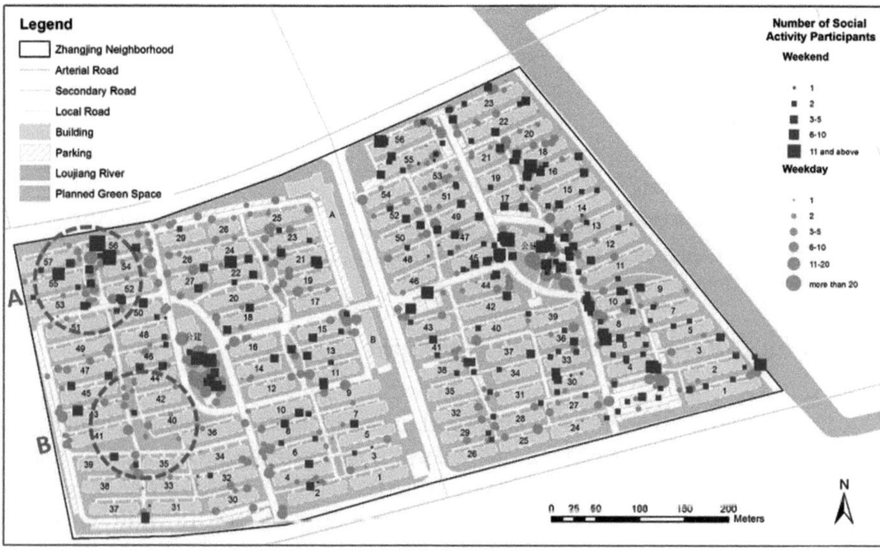

Fig. 7 Map of public open space used by social activity participants

Fig. 8 Map of public open space used by optional activity participants

Conversely, the elderly often spend their time in immediate open spaces in front of buildings, open spaces at the corner of buildings and medium-sized open spaces between buildings. A number of senior residents, especially those with disabilities, did not feel comfortable in the central open spaces.

Table 1 Participants by age groups

	Age groups			
	Toddler (%)	Child and teenager (%)	Adult (%)	Elder (%)
Registered residents[a]	5	13	62	20
Total open space users[b]	12	11	29	48
Social activity participants[b]	19	15	25	41
Optional activity participants[b]	6	9	28	57

[a]Data Source: case study neighbourhood committee's statistics
[b]Data collected from distance sampling of this study

5.2 *The Elderly – Primary Open Space Users*

One in five registered residents with a local *hukou* in the case study neighbourhood were over 60 years old. However, the elderly accounted for half of the population using public open space during sampling periods (Table 1). In this sense, the elderly were the primary users of neighbourhood public open spaces. In addition, senior residents participated in three-fifths of optional activities. Given the fact that optional activities are mostly individual events (Fig. 8), it could be further concluded that the elderly are mainly passive users of public open spaces. In other words, the extent and quality of social activities could be improved if the public open space can be designed in a way to encourage/enable passive users into positive contacts for social purposes.

5.3 *Children Are a Socially Binding Force*

As is listed in Table 1, less than one quarter of open space users were in the toddler to teenager age range. Comparably, the youngest population made up one third of social activity participants. Map patterns also demonstrate similarities between social activity distribution and public space use by toddlers to teenagers. In this regard, the presence of toddlers, children and teenagers in public open spaces facilitates the occurrence of social activities to some extent. It was also observed that adults tend to socialize while children are playing together regardless their backgrounds.

5.4 *Quality of Space Affects Quantity of Participants*

Generally speaking, green spaces are more favoured by residents than are parking lots. For instance, among the eight largest popular gathering points in the study area, six of them are green spaces. By comparing Green Space A and Green Space B in Figs. 9 and 10, it is obvious that Green Space A was preferred by residents.

Fig. 9 Pictures of Green Space A

Fig. 10 Pictures of Green Space B

A closer examination indicates that the design quality of Green Space A outweighs Green Space B in several regards.

- **Accessibility**. Because bushes did not dominate the periphery of Green Space A (Fig. 9), it was more accessible to residents, both physically and psychologically, compared to Green Space B (Fig. 10).
- **Visibility**. Considering that open spaces at the corner of buildings are often popular gathering locations, integrating these into a design can reinforce social and cultural activities in the centrally located open spaces. As a result, Green Space A was well enclosed by a number of residential buildings (Fig. 9). Good visibility allowed activities in corner open spaces at each building to be seen, as well as Green Space A in the middle. This increased the popularity of both sides. On the contrary, poor visibility leaves Green Space B (Fig. 10) on its own with no consistent effect.
- **Social Effects**. An enclosed inner space was formed at Green Space A (Fig. 9) which enabled social interactions to develop, whilst at Green Space B there merely exits paths that cut through the spaces, leaving no space or opportunity for any activities to take place other than simply passing through.

- **Placement and Orientation of Seating**. Although stone benches are not comfortable to sit on for an extended period of time in winter, the benches located at Green Space A were still used from time to time. With corner open spaces integrated around Green Space A, activities taking place at corner open spaces could be easily viewed by residents sitting on benches placed at the edges. On the other hand, benches oriented away from potential activities in Green Space B were constantly unoccupied (Fig. 10). Instead, residents took their own stools from home to sit at corner open spaces that were not originally designed for lingering.

6 Conclusion

Mainly owing to the convenient location and the free shuttle bus to workplaces, the resettlement neighbourhood has changed into a primary location for migrants from other provinces. Nearly half of the migrants were living with their family and the other half shared one apartment with others. The rapid increase in population density, combined with poor management, has resulted in issues concerning noise, hygiene and security. Surprisingly there was a trend for migrants to settle down and stay in the same place for a considerable period. Neighbourhood attachment and sense of community was found to be higher than expected. However, the research also found that the participation (or attendance) level for neighbourhood activities was low, whilst most social activities tended to take place inside the social networks originating in the village or towns or in workplaces. Additionally, children were found to be a socially binding force for the community and children-centred activities were in comparatively high demand. However, from both the questionnaire survey and observation, it was found that the existing open space, owing to the poor design, had failed to meet the demand of local residents. This was particularly the case for the young and the elderly. There was a pressing demand for better management of the existing public rooms and spaces, particularly for places which can accommodate chatting. Moreover, the existing provision and management of the public facilities and space had not adapted to respond to the transformation from a resettlement neighbourhood to a migrant-dominated one. The needs of the residents were overlooked. It is very possible for neighbourhoods of the same type to improve the quality and the usage of public open spaces in a way that enables and encourages more and better quality social activities to take place. A simple measure to do this could include changing the bushes into hard surfaces or grassland that is accessible and suitable for activities of different purposes. Further improvements could be made by positioning reasonable seating places in front of each wing of buildings, at the corner of buildings and at the edge of centrally-located medium and large open spaces with considerations of visibility, activity use and orientation.

Neighbourhood development should also consider the varying demands of different groups. Such groups include people from the same village or town, migrant families from the same provinces and single migrant who work in the same workplace. Owing to the large size of the community, there is no easy way to socially integrate the three different groups together in the short term. However, it is possible to foster community societies or clubs among different groups. Autonomy could be granted for each group, following the establishment of a transparent and collaborative community governance system. There is a great opportunity for community development in Chinese neighbourhoods undergoing change if the equal rights of migrants are recognized, considered, and appreciated by the local authorities.

References

1. Zhang H, Tong X (2006) The social adaptation of landless farmers in the urbanisation process in China. Soc Sci Stud 2006(1):128–134 (in Chinese)
2. Chambers R, Conway G (1992) Sustainable rural livelihoods: practical concepts for the 21st century. IDS Discussion Paper 296. Brighton: IDS
3. UNCHS (Habitat) (1994) Sustainable human settlements in an urbanizing world, including issues related to land policies and mitigation of natural disasters. Resolution 15/11 of the 15th session of the United Nations Commission of Human Settlements, 25th April to 1st May, Narobi
4. Carney D (1998) Sustainable rural livelihoods: what contribution can we make? Department for International Development, London
5. Ashley C, Carney D (1999) Sustainable livelihoods: lessons from early experience. DFID, London
6. Zheng F, Sun J. (2006) From survival to development-the supporting framework for landless farmers in China. Economics 2006(001):54–61 (in Chinese)
7. Zhang G, Gu C (2006) The study on rapid urbanisation and involuntary urbanisation. Urban Plan Forum 30(005):48–54
8. Xu Q (2006) The weak capability and the constraints of the landless farmers. Soc Sc Jiangsu Prov 2006(4):140–144 (in Chinese)
9. Cernea MM (1998) Improverishment or social justice?: a model for planning resettlement. Oxford University Press, Delhi
10. Guo D (2006) The living and life of farmers in rapid urbanisation area- a survey of the transformation of farmers in South Jiangsu Province. J Coll Bejing Constr Eng 22(001):48–51 (in Chinese)
11. Ye J (2007) A survey of the satisfaction of landless farmers in the outskirts of Nanjing. J Jiangsu Broadcase Telev Univ 18(2):70–73 (in Chinese)
12. Ye J (2007) The factors that affect the adaptation ability of landless farmers. Rural Econ 2007(003):114–116 (in Chinese)
13. Ye J (2008) Livelihood and adaptation: a survey of the landless farmers in the outskirts of Nanjing. China Economics Press, Beijing. (in Chinese)
14. Cao H et al (2007) A study of the rural residential development in South Jiangsu Province. Planners 23(2):18–21 (in Chinese)
15. Song Y (2008) Case study of the community development of neighbourhoods of landless farmers. City Issues (009):73–76 (in Chinese)
16. Tan S, Li J (2009) The function of city open space to release stress. Landsc China 2009(6):79–82 (in Chinese)
17. Xi G et al (2011) The relation between city park and mental health of residents. J Northeast For Univ 39(11):3 (in Chinese)

18. Beatley T (2011) Biophilic cities – intergrating nature into urban design and planning. Island Press, Washington, DC, p 191
19. Schilling J, Logan J (2008) Greening the rust belt – a green infrastructure model for right sizing America's shrinking cities. J Am Plann Assoc 74(4):17
20. Banerjee T (2001) The future of public space – beyond invented streets and reinvented places. J Am Plann Assoc 67(1):16
21. Coutts C (2010) Green infrastructure and public health in the Florida Communities Trust public land acquisition program. Plan Pract Res 25(4):11
22. Reese RF, Myers JE (2012) EcoWellness: the missing factor in holistic wellness models. J Couns Dev 90:8
23. Abkar M et al (2010) The role of urban green spaces in mood change. Aust J Basic Appl Sci 4 (10):11
24. Grinde B, Patil GG (2009) Biophilia: does visual contact with nature impact on health and well-being? Int J Environ Res Public Health 6:12
25. Mitchell R, Popham F (2008) Effect of exposure to natural environment on health inequalities: an observational population study. Lancet 372(9650):1655–1660
26. Liao X (2011) Construction the open space for eco and health city-the case study of green road in Zhuhai. Constr China. 2011(06):76–77 (in Chinese)
27. Miao P (2011) Brave new city: three problems in Chinese urban public space since the 1980s. J Urban Des 16(2):29
28. Zhang Y, Dai S (2010) A review of the demand of ourdoor open space for children in China. Survey Anal Landsc 2011(02):82–85 (in Chinese)
29. Largo-Wight E (2011) Cultivating healthy places and communities: evidenced-based nature contact recommendations. Int J Environ Health Res 21(1):21
30. Li D, Fan W (2001) The relation between retirees and city park in Beijing. Urban Environ Ecol 14(3):3 (in Chinese)
31. Sun Y (2003) The measures to the demand of urban green land of elder people. Resour Sci 25 (3):8 (in Chinese)
32. Gehl J (2010) Life between buildings. The Danish Architectural Press, Copenhagen, p 200

The Unprecedented Age Wave and Challenged Fixed-Route and Specialized Transportation Services: The Case of Richmond, Virginia

Xueming Chen

1 Introduction

The public transit and paratransit operations in Richmond, Virginia urgently need to be reformed in order to better meet sustainable transportation and other legally mandated requirements. This work is becoming critically important since the U.S. has entered the population aging society ("the Age Wave") in the early twenty-first century, which demands a more efficient, equitable, and affordable public transit and paratransit system [1].

This paper focuses on the three key issues faced by the Greater Richmond Transit Company (GRTC): poor regional service coverage; costly and inadequate paratransit services; and ineffective governing structures. Based on this empirical study, it will come up with a list of improvement recommendations.

2 GRTC's Facts

2.1 History and Governance

Founded in 1860 and incorporated in 1973, GRTC Transit System is the major transit operator serving the Richmond region. According to the 2011 National Transit Database (NTD), GRTC's service area covers 227 mile^2 and 449,572 residents.

GRTC's overall direction is guided by its Board of Directors, with a management team that conducts its day to day operations. The GRTC directors are

X. Chen (✉)
L. Douglas Wilder School of Government and Public Affairs, Virginia Commonwealth University, Richmond, VA, USA
e-mail: xchen2@vcu.edu

X. Chen, Q. Pan (eds.), *Building Resilient Cities in China: The Nexus between Planning and Science*, GeoJournal Library 113, DOI 10.1007/978-3-319-14145-9_3

appointed annually, in October, by the Council of the City of Richmond and the Board of Supervisors of the Chesterfield County, acting in their capacity as stockholders.

Regarding the Chesterfield County's involvement in GRTC business, a brief history is introduced here. It began on June 6, 1989, as an initiative to create the regional transportation authority. The GRTC Board of Directors approved amendments to the company's by-laws and a purchase agreement for the sale of five shares of stock to the County of Chesterfield at a price of $10,000 a share. Therefore, the Chesterfield County owned 50 % of GRTC. The Henrico County Board of Supervisors decided not to participate in purchasing GRTC in 1989, even though it initially agreed with the purchase agreement in principles [2].

2.2 Budget

In FY 2011, GRTC's total operating revenue was $18,839,778. The agency's most important revenue item was its farebox revenue, especially its fixed-route service (51.45 %), followed by purchased service from the surrounding jurisdictions (29.24 %).

In FY 2011, GRTC had a total operating expense of $43,813,570. The most important expense item was transportation-related expense (47.82 %), followed by administration and general costs (23.26 %).

To cover most of the financial deficits, GRTC received operating contributions from local, state and federal governments. In FY 2011, local government, especially the City of Richmond, provided the most important contribution to support GRTC's transit operations (44.31 %), followed by state (30.50 %) and federal (25.19 %) governments.

2.3 Fixed-Route Services

At present, GRTC's fixed-route bus service consists of a fleet of 186 buses travelling over 36 routes within the City of Richmond, Counties of Henrico, Chesterfield, and City of Petersburg.

Figure 1 shows the GRTC local fixed-route service area. It is very clear that most of the GRTC bus services are provided on the north side of James River. As a general rule-of-thumb, public transit customers are willing to walk a ¼ mile to a bus stop. When placing a ¼ mile buffer zone around local GRTC bus routes, the GRTC fixed-route local bus service area covers approximately 64 $mile^2$ and 191,000 potential customers. Of these customers, about 27,000 are elderly, 47,000 have a disability, and 37,000 are at or below poverty.

Table 1 shows the GRTC's system-wide ridership from FY 2006 to FY 2011. The average annual bus ridership is about 10 million passengers.

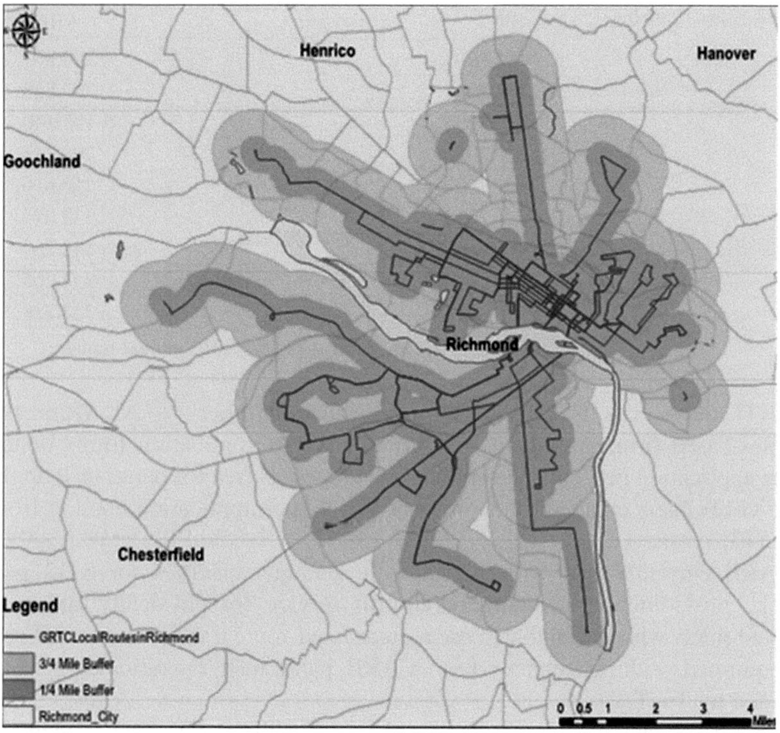

Fig. 1 GRTC local fixed-route service area

Table 1 GRTC systemwide ridership

FY	Systemwide ridership
FY 2006	10,738,378
FY 2007	10,306,201
FY 2008	10,280,212
FY 2009	10,444,498
FY 2010	10,193,867
FY 2011	9,887,800

Sources: GRTC. Transit system 2011 annual report

2.4 *Paratransit Services*

The Americans with Disabilities Act of 1990 (ADA) recognizes that some users of public transit, due to the nature of their disability, will be unable to use fixed-route services even with full accessibility. To ensure equal access for these riders under extraordinary circumstances, public transit operators are required to offer a complementary paratransit service.

In terms of its service area, response time, fares, trip purpose, service hours and days, and capacity, GRTC's Community Assisted Ride Enterprise (CARE) service strictly complies with and even exceeds ADA mandates.

Table 2 Care ridership

FY	CARE ridership
FY 2003	200,887
FY 2004	202,548
FY 2005	197,140
FY 2006	208,783
FY 2007	210,616
FY 2008	232,074
FY 2009	242,560
FY 2010	237,065
FY 2011	258,738

Sources: GRTC. Transit system 2011 annual report

GRTC outsources its ADA eligibility certification process to ADARide, the Los Angeles-based firm. Applicants can either mail in application forms or use the online application process. The application process is free for paratransit applicants while GRTC gets charged an average of $70.00 per application coming from this region [3].

CARE currently has 46 vehicles with seating capacities for 8–12 persons/ vehicle, providing curb-to-curb paratransit service for physically and mentally disabled riders who are unable to use regular fixed-route transit service. All vehicles are equipped with wheelchair lifts. CARE paratransit operations are currently provided by TecTrans.

Table 2 shows the CARE Ridership from FY 2003 to FY 2011. Except for FY 2005, CARE's annual ridership has been stabilizing between 200,000 and 250,000.

Figure 2 shows GRTC's fixed-route access routes spreading across the city. As stated earlier, the ¼ mile buffer zone is the service area for fixed-route transit service. And the ¾ mile buffer zone was added to the layer to assess the overall service coverage provided by complementary CARE paratransit service. Overall, GRTC's fixed-route ¾ mile buffer zone covers most areas that require the transit agency to provide paratransit services, except the western portion of the City.

As illustrated in Fig. 3, the western portion of the City has high densities of disabled seniors, which suggests the necessity of providing CARE services there even though there are currently no fixed-route services provided yet [4].

3 GRTC's Issues

This section examines the three key issues identified in the introduction: poor regional service coverage; costly and inadequate paratransit services; and ineffective governing structures.

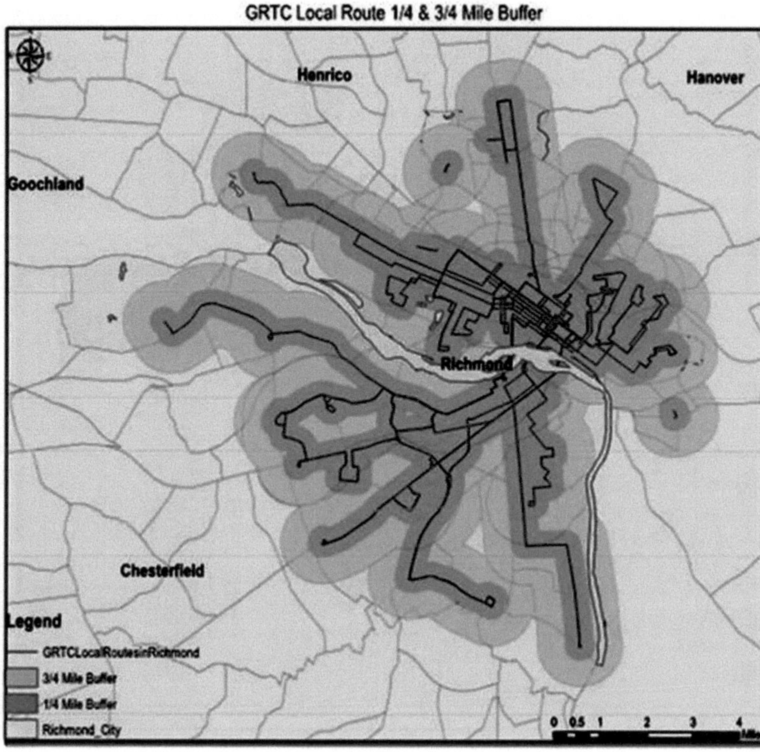

Fig. 2 GRTC – local route ¼ & ¾ mile buffer service area

3.1 Poor Regional Service Coverage

According to the Brookings Institution study, the Richmond area ranks No. 95 out of the 100 metropolitan areas in the U.S. in terms of the proportion of working-age residents who have access to transit, with only 30.8 % of geographic coverage [5].

For fixed-route bus services, GRTC currently does not provide extensive service coverage in the suburban employment centers of the Richmond region, such as Short Pump/Innsbrook and Midlothian areas. Beyond downtown Richmond, only I-64/Broad Street and Hull Street corridors are adequately served [6]. See Fig. 4 for details. Except for a few express bus services, both Henrico and Chesterfield Counties have virtually no transit services at all. Despite the fact that more suburb-to-suburb trips are taking place in the Richmond region, most GRTC bus routes are downtown-bound, which makes it very difficult if not impossible to make suburb-to-suburb travels by taking public transit.

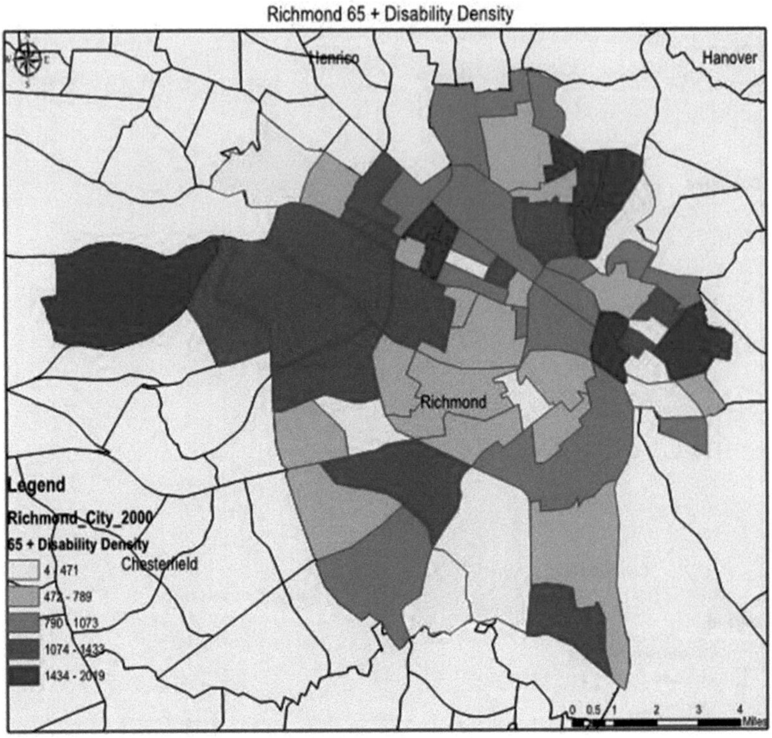

Fig. 3 Richmond 65+ disability density map

3.2 Costly and Inadequate Paratransit Services

The cost per trip shown in Table 3 is for GRTC's fixed-route service and specialized transportation services (Note: most GRTC specialized transportation services are CARE paratransit services). Specialized transportation services were fairly stable from 2003 to 2007 with cost per trip being about $17 and annual ridership being about 200,000. However, its cost per trip dramatically increased twice: in 2008 and 2011. In 2008, the cost per trip jumped from $16.44 to $23.54. The cost per trip further skyrocketed to $28.00 by the end of 2011, and is projected to reach $30.00 by 2013, which is attributable to the expansion of CARE services throughout the region beyond the ¾ mile buffer zone of the fixed-route service. In contrast, fixed route cost per trip has been more or less steady from 2003 to 2012, costing an average of about $4.00 per trip for GRTC's regular bus service.

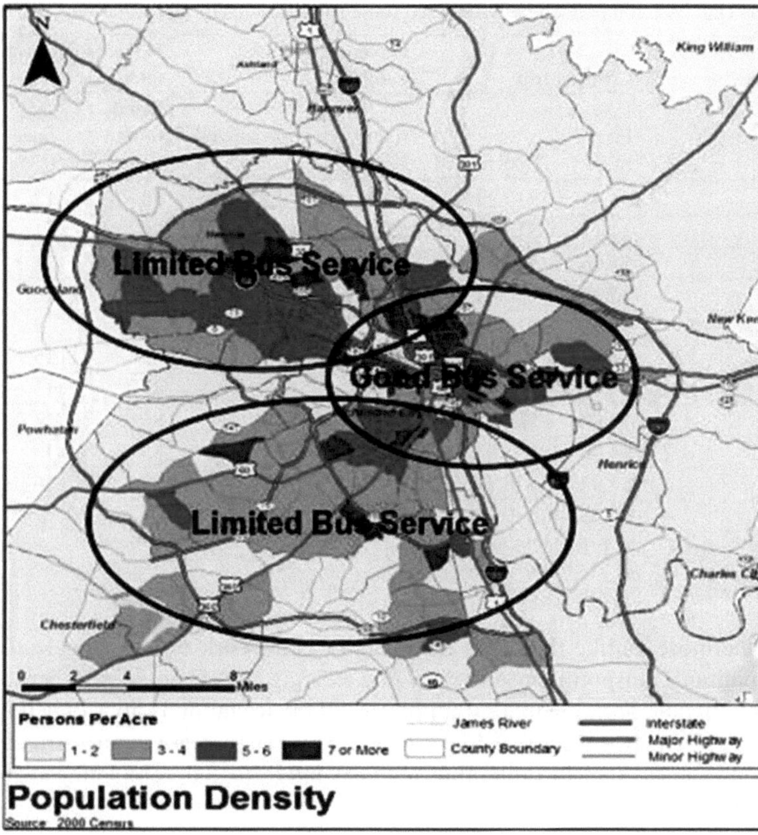

Fig. 4 Bus service and population density

3.3 Ineffective Governing Structures

Even though the idea of forming a regional transportation authority has been floating around this region for the last two decades and beyond, GRTC is not yet a regional transportation authority and therefore cannot levy taxes, issue bonds, or implement any other innovative financing strategies to increase its revenue base.

The commonly perceived greater Richmond metropolitan region includes the following jurisdictions: Town of Ashland, Charles City County, Chesterfield County, Goochland County, Hanover County, Henrico County, New Kent County, Powhatan County, and the City of Richmond. However, GRTC currently primarily only serves City of Richmond, with very limited transit services provided in Henrico and Chesterfield Counties. Ironically, the representatives of Chesterfield County sitting on the GRTC Board of Directors are not promoting transit in that county, but rather preventing GRTC bus lines from being extended into the county.

Table 3 Unit cost comparison between fixed route and specialized transportation

Year	Cost per trip		Percentage of total budget		Percentage of actual total	
	Fixed route	Specialized	Fixed route (%)	Specialized (%)	Fixed route (%)	Specialized (%)
2003	$3.98	$17.22	90	10	88	12
2004	$4.11	$16.83	89	11	89	11
2005	$4.06	$17.22	89	11	89	11
2006	$3.08	$16.33	90	10	90	10
2007	$3.55	$16.44	91	9	91	9
2008	$3.71	$23.54	84	16	85	15
2009	$3.90	$22.66	86	14	87	13
2010	$3.92	$23.10	87	13	86	14
2011	$4.05	$27.03	87	13	85	15
2012	$3.93	$28.31	86	14	84	16
2013	$4.23	$30.12	84	16		
Change % from 2003 to 2013	6.2 %	75 %				

Furthermore, unlike the board members of New York or Los Angeles County Metropolitan Transportation Authority, the existing GRTC board members are not elected officials. They are merely appointees of the Richmond City Council and the Chesterfield County Board of Supervisors. The existing governing structures of GRTC are ineffective in promoting regional interests and completing regionally significant transit projects.

4 Improvement Strategies

To address the above three issues, this section proposes ten improvement strategies for potential implementation.

4.1 Establish the Richmond Regional Transportation Authority (Political Category)

This study recommends that the City of Richmond lobby the Virginia General Assembly to authorize the establishment of the Richmond Regional Transportation Authority. The Authority should be governed by the Board of Directors consisting of the city and county elected officials within the Richmond metropolitan region.

The Commonwealth of Virginia should take the lead in this endeavor. To make sure the region is equitably represented, the seats of Board members should be allocated based on the population of each jurisdiction. The City of Richmond and the Counties of Henrico and Chesterfield should have more seats than other smaller jurisdictions. Some smaller jurisdictions may share one seat.

The Authority should be able to levy taxes, issue bonds, establish benefit assessment districts and implement any other innovative financing measures that are deemed necessary to improve and solidify its revenue basis.

Even though this strategy will initially face a strong opposition from the surrounding counties, it will benefit the entire region in the long run.

4.2 *Coordinate Public Transit and Human Service Transportation (Political Category)*

This study encourages better coordination between GRTC and other human service agencies, such as Adult Care Service, American Red Cross, Bethlehem Community Centers, Inc., Chesterfield Community Service Board, and many others.

In particular, GRTC needs to work with non-profit aging-related service agencies such as Senior Connections, the Capital Area Agency on Aging (SCCAAA). SCCAAA is a private nonprofit organization that has been helping the elderly citizen population of the greater Richmond area to live an improved and healthier lifestyle.

This strategy is feasible for implementation, but it requires lots of coordinating and consensus-building efforts from GRTC and other stakeholders.

4.3 *Reform Fare System for Care Riders (Economic Category)*

The existing fare per CARE trip for the customer is $2.50 for one-way, but the cost per CARE trip for GRTC is $30.00 for one-way. GRTC has completed a peer review of other localities' paratransit service and determined that GRTC is the only system of those reviewed to both provide non-ADA required service and not charge more for it or have it subject to capacity.

Therefore, this study recommends the distance-based fare structure for the service area outside of the mandated ¾ mile buffer zone to allow GRTC to continue serving CARE eligible riders while generating additional revenue to support this service. For those low-income CARE eligible riders, some types of fare discounts should be provided to foster social equity.

Levine (1997) found that in Ann Arbor, Michigan, the ridership of fixed-route transit by ADA eligible riders is very sensitive to price. Elimination of even a very low fare of $0.35 had dramatic effects during the free months [7].

Therefore, this study also recommends providing fixed route service to CARE eligible riders free of charge. Currently, these CARE eligible riders must pay 75¢ for local service. Providing fixed route service to CARE eligible riders free of charge would induce more CARE riders to use fixed-route services rather than CARE paratransit services. This will reduce CARE demand, thus saving CARE operating cost.

4.4 Restructure Existing Bus Routes (Planning Category)

Based on the 2010 population census data, most recent transit on-board data and other data, GRTC should conduct a new bus restructuring study and thoroughly overhaul the existing fixed-route system to make it more closely aligned with transit origin–destination flows, better serving important employment centers and other attractions.

4.5 Public-Private Partnerships (Planning Category)

Public-private partnerships can be very helpful in countering ridership demand for CARE paratransit service. Taxi services that are accessible to disabled and elderly passengers could offset demand for paratransit services. This can either be done through a public entity or encourage private enterprise to meet the needs of the disabled elderly [8]. Since private contractors providing paratransit services will not be under ADA guidelines, GRTC will have to develop minimum standards and regulations for them to ensure quality of service [9].

4.6 Volunteer Driving Programs (Planning Category)

GRTC currently has no volunteer driving program in place. It is very important for GRTC to work with the faiths-based organizations and others to launch volunteer based driving programs that serve the elderly and disabled.

4.7 *Travel Training (Planning Category)*

Currently, GRTC has no travel-training program for the elderly and disabled population living in Richmond City. Travel training provides a promising approach for moving persons from paratransit to fixed-route transportation services. CARE services have seen high increases in paratransit ridership over the last few years and this can be a great tool in managing demand.

4.8 *Subscription-Grouping Rides (Planning Category)*

GRTC has a subscription-based service for CARE passengers who use the service at least four times a week. These types of service are prescheduled and can let the agency plan ahead, allowing other trip requests to be based on confirmed requests. This can also have a major impact on reducing trip requests and calls to the paratransit operator improving general productivity and efficiency.

4.9 *Rely on New Technology (Technological Category)*

GRTC has acquired and implemented its Advanced Communication Project, which includes Computer Aided Dispatch/Automatic Vehicle Locators (CAD/AVL), advanced vehicle monitoring (AVM), Automatic Passenger Counters, (APC), Voice Annunciation, Internet Visual Signs, and stop level bus arrival signs [10]. This will allow GRTC to increase its operational efficiency and effectiveness through improved trip scheduling and routing.

In addition, vehicle replacement programs are a great way to reduce operating costs and save money in the long run. Quicker conversion of diesel based systems for paratransit systems may bring some relief to operating costs as well in the case of GRTC.

4.10 *Continue Improving Fixed-Route Transit (Technological Category)*

GRTC needs to work with local and state agencies to improve access to bus stops by installing low-floored vehicles, or ramps. Bus stops and sidewalks need to be disability friendly with adequate lighting and benches to support elderly disabled passengers. The success of these programs will result in a reduction in the demand for paratransit services.

5 Conclusion

To address Richmond region's severe ADA transit and paratransit issues, this study proposes ten improvement strategies for their potential implementation.

First, GRTC needs to reform its fare structure for ADA paratransit eligible riders. Charging the distance-based CARE fare (adjusted by income level) for the service area outside of the ¾ mile buffer zone of the fixed-route bus service and providing free regular bus service to these riders would bring immediate benefits to GRTC in both revenue generation and cost containment.

Second, a series of planning and technological improvements should be made. In particular, those planning studies and bus restructuring activities should take place first because of their low costs.

Third, it is necessary to establish the regional transportation authority which will provide the ultimate transit solution to the Richmond region. In addition, GRTC should work with human service agencies and other stakeholders to coordinate public transit, paratransit and other human service transportation.

References

1. Richmond Area Metropolitan Planning Organization (2012) 2035 long-range transportation plan. Richmond Regional Planning District Commission, Richmond
2. Greater Richmond Transit Company (GRTC) (1990) Transit development program for the Richmond urbanized area 1991–1995. Greater Richmond Transit Company, Richmond
3. ADARide (2012) FAQ. Retrieved 10 Jan 2012 from ADARide: https://www.adaride.com/FAQ.aspx
4. Pande A (2012) Addressing increased ridership and demand: GRTC – CARE paratransit service sustainability for the city of Richmond. The unpublished master thesis submitted to Virginia Commonwealth University
5. Tomer A, Kneebone E, Puentes R, Berube A (2011) Missed opportunity: transit and jobs in metropolitan America. Metropolitan Policy Program at Brookings, Washington, DC
6. Chen X, Suen I-S (2010) Richmond's journey-to-work transit trip-making analysis. Manag Res Pract 2(3):234–248
7. Levine J (1997) ADA and the demand for paratransit. Transp Q 51(1):29–43
8. Center of Urban Transport & Research (2008) Creative ways to manage paratransit costs. Retrieved 10 Mar 2012 from University of Florida: http://www.nctr.usf.edu/pdf/77606.pdf
9. Project East Action (2012) Innovative practices in paratransit services. Retrieved 10 Mar 2012 from Transit Access Project: http://www.transitaccessproject.org/InternalDocs/Paratransit/Innovative%20Practices%20in%20Paratransit.pdf
10. Connetics Transportation Group (2012) Transit development plan fiscal years 2012–2017. Final report, GRTC Transit System, Richmond

Resilient Planning Frame for Building Resilient Cities

Zhiduan Chen and Baoxing Qiu

1 Introduction

1.1 Global Climate Change and the Energy Crisis

With the increasingly severe impacts of global climate change, cities must not only bear the impacts of such change, they must also establish effective planning systems and responses, understand urban planning theories and compile strategies for tackling climate change.

Due to poorly planned transport and ineffective architectural design, many cities in the world consume large amounts of energy. Currently, however, mineral energy resources cannot meet the demands incurred with rapid city growth. The purpose of pursuing sustainability and coping with disasters are the same. A recovery system should be established for resilient cities to adapt to various changes, for instance, the diversity of traffic and land use systems as well as power generation by many renewable energy sources. In such cases, a city can survive when energy runs short.

Human beings have entered an era full of high risks. Effective urban planning strategies need to be explored to continuously improve the recovery capacity of cities in a bid to cope with future impacts of climate change and energy crises.

Supported by the Shanghai Science and Technology Development Soft Science Doctoral Dissertation Fund (No. 12692192900).

Z. Chen (✉)
College of Architecture and Urban Planning, Tongji University, Shanghai, China
e-mail: zhiduanchen@gmail.com

B. Qiu
Ministry of Housing and Urban-Rural Development of the Republic of China, Beijing, China

X. Chen, Q. Pan (eds.), *Building Resilient Cities in China: The Nexus between Planning and Science*, GeoJournal Library 113, DOI 10.1007/978-3-319-14145-9_4

1.2 Traditional Urban Planning Reform

As complex systems, cities become more fragile as the economy grows more powerful. Long-term gradual changes and unexpected emergencies disrupt city development, such as global climate change, increasingly frequent disasters and energy crises. Cities need to rapidly solve such problems to survive and maintain development vitality under critical and challenging conditions. Qiu (2010) considered cities to be complex adaptive systems (CAS) due to the presentation and reasons of modern urban planning's dilemma, and, on that basis, generalized the methodology of urban planning transformation. Classic resilient urban systems can be classified as complex adaptive systems, which apply resilient thinking and provide a theoretical basis for urban planning transformation.

2 Resilient Theory

2.1 Origin and Development of Resilient Theory

The concept of resilience is derived from early ecological theory and was first proposed by Hollings (1973), leading to related research in many different disciplines and the belief that the essential meaning of resilience was the capacity of a system to resolve external impact and maintain major functions. Research from different disciplines has emphasized numerous study objectives due to diverse academic origins and different traditional research systems [1]. Hollings (1973) emphasized buffering power [2] while others have stressed the speed of post-disaster recovery [3]. Hollings (1973) applied resilient thinking to actions, described world status by resilient thinking and proposed the possibility of global release phase. Hollings (1973) also stated the importance of multi-scale adaptive circle and reverse circle to aspects of space and new opportunities creation [4]. Adger (2005) discussed the advantages of the resilient social-ecological system and how such systems undergo changes and accidental impacts relying on the advantages and learn lessons to improve the resilience – although the impacts may be significant from large-scale natural disasters such as tsunamis and hurricanes [5]. On the basis of complicated system theories, Berkes (2003) studied how human society handled and improved on the ability to adapt to changes in the inter-related social-ecological system [6]. "Panarchy; Understanding transformations in human and natural systems" is complied by many experts, the contents of which forms the foundation of resilient theory framework and describes the thinking, contents and influence of resilient theory, and challenges the theories and experiential viewpoints of dynamic features of the social-ecological system. Jen studied robustness of systems in relation to their ability to maintain their features under adverse conditions. Although the emphasis and contents of resilience and robustness are different, they are inter-related concepts. The book studies their differences and discusses the robustness of complex systems with various forms and scales.

2.2 Sustainable Development and Resilient Thinking

To meet the demands of rapid population growth, humans have adopted resource management to achieve considerable success and continued development. In doing so, however, large areas of forest have disappeared, desertification has increased, shortages in food, energy, water and living resources have become more prevalent, and global ecosystem service functions (such as water quality purification, wind barrier, dam fortifying or waterlogging reduction) have significantly deteriorated. Our current use of the biosphere is obviously unsustainable.

In relation to sustainable development, resilient thinking provides a different approach to understanding the world and controlling natural resources. It explains why efficiency improvement cannot solve the world's resource problems, and provides a constructive option to facilitate the sustainable development of social-ecological system under control.

2.3 Important Points of Resilient Thinking

Over-improving efficiency may impact social-ecological system resilience. Improving efficiency helps a system reach "optimal status" and attain maximum benefits. To achieve this, system administrators reduce redundancy, such as by the "zero inventory" approach (i.e. production management approach of stocking where necessary to reduce the inventory of raw materials and other goods). In these cases, spare parts and supplies will be purchased where necessary. Although it is deemed as an effective and optimal system that can save on costs and inventory expenditure, such a system is very weak and sensitive to accidents. Usually, it will cause extreme shortage of raw materials and personnel and serious chaos in production. Thus, an urban system can be resilient as long as it has proper redundancy.

Human behavior cannot exceed the social-ecological system threshold. Threshold is a core concept of resilient thinking. When controlling the social-ecological system, the behavior of human beings must not exceed the threshold of the system.

Figure 1 is a representation of the ball-basin model, with the ball showing the status of the system. The ball keeps moving in the basin. One basin represents a group of statuses, which have the same function and feedback to balance the movement of the ball. The dotted line refers to the threshold separating the different basins.

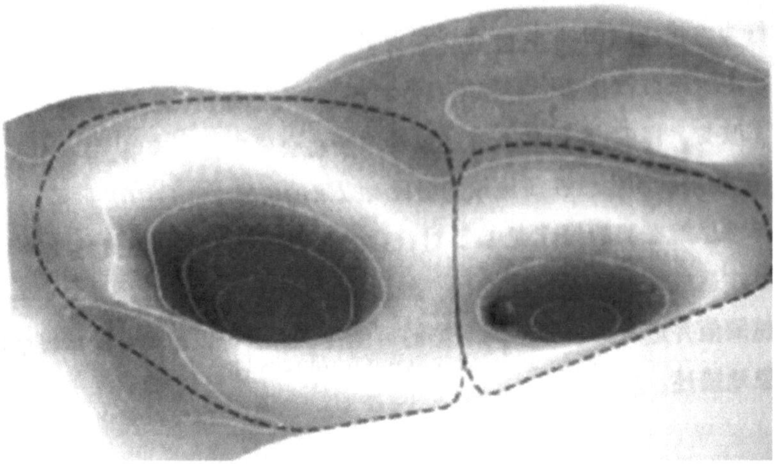

Fig. 1 Ball-basin model of a system

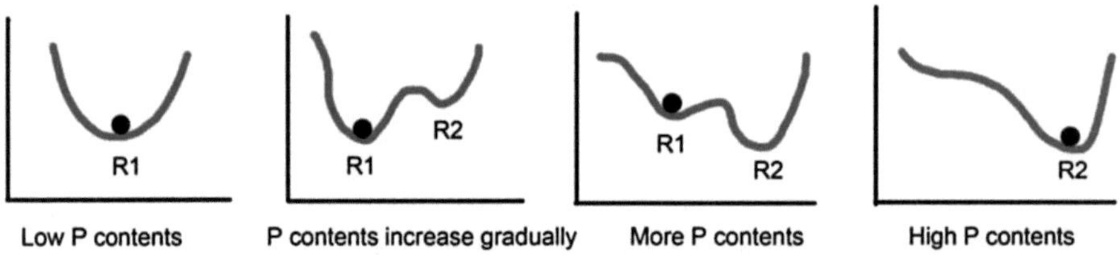

Fig. 2 2D sketch map of ball-basin model

The sustainable input of phosphorus (P) leads to changes in an ecosystem. In Fig. 2, status 1 (R1) (low P content of sediments) has complete resilient restoration ability for the sudden input of P from outside (depth of the ball of the single basin shows the restoring force). With increased accumulation of P, lakes gradually lose resilience. The basin becomes flat and a new basin appears. At that time, lakes are easily disturbed by sudden increases in P content (such as by rainstorms). Finally, the system is easily pushed to a new gravitation region (Status 2, R2) and the lake water becomes muddy (adapted from Folke et al. 2004).

3 Urban Resilience Study

The combination of resilient theory and urban systems broadens the content and vision of urbanology. Alberti (2003) defined urban resilience as "the degree of solving and resolving changes before a series of structure and process change restructuring" [7]. Resilience Alliance defines urban resilience as the ability of a city or urban system to "digest and absorb the external interference and . . . maintain its original features, structures and key processes" [8]. However, urban resilience not only contains self-adjustment of an urban system and the ability to cope with various uncertainties and emergencies, but also contains the capacity to change positive opportunities into capital [6] (Berkes and Folke 2003).

3.1 Urban Ecological Resilience

Urban ecological resilience refers to the degree of an urban ecosystem to solve changes before restructuring and forming new structures and processes (Hollings 1973, 1996). Climate change and urbanization [9] are two of the most important factors influencing urban ecological resilience. The intertwining of ecological processes with social, economic and political processes leads to the fragmentation of natural habitats, homogenization of species and interruption of energy flows and nutrient cycles of cities and urbanized areas, reduces the resilience of cities, and increases the vulnerability of ecosystems. Therefore, urban transformation can be accelerated in response to uncertain and nonlinear external impact, which can improve urban self-organization capacity and coordinative development of human and environmental systems [10]. Urban ecological resilience pays special attention to the influence of urban structure (including urban form, land use distribution and connectedness) on dynamic mechanisms and resilience of ecosystems [11] (Fig. 3). As a result of interactions between ecosystems and human systems, the relationship between urban structure and ecological resilience can be clarified by studying spatial-temporal evolution such as urban form, urban sprawl and land use mode. For example, as an external impact, the construction of large infrastructure generates huge influence on urban structure and the ecosystem.

3.2 Urban Engineering Resilience

Urban engineering resilience mainly emphasizes the ability of an urban infrastructure system, urban population or community to rapidly and effectively recover from natural and man-made disasters, including assessing urban infrastructure resilience and adopting disaster reduction technologies to fortify urban infrastructure and ensure normal operation in the global economy.

In intensively industrialized societies, cities are key nodes connecting crucial infrastructure, and are the most important infrastructure of society. Resilient infrastructure systems, especially the public service departments of water, electricity and medical care, are critical for reducing the social influence of extreme events (earthquake, flood and terrorist attack). Provided that the infrastructure system can bear the impact of extreme events, it can avoid mutual influence between systems and minimize the dangerous domino effect of extreme events to cities [12].

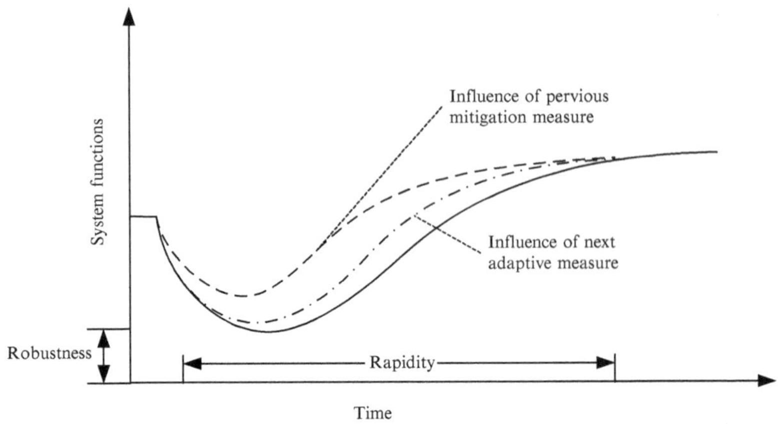

Fig. 3 Sketch map of infrastructure resilience

Infrastructure resilience is mainly related to robustness and rapidity. The Multidisciplinary Center for Earthquake Engineering Research (MCEER) proposed a frame diagram of infrastructure resilience. Robustness represents the capacity of a system to not deteriorate or lose function under certain pressure. Rapidity represents the time required by a system to recover to normal function [13] (MCEER 2005). On that basis, Bruneau (2003) further supplemented "Redundancy" and "Resourcefulness" to the features of infrastructure resilience.

4 Study on Resilient Urban Planning

Resilient urban planning studies can be classified into the "resiliency" studies of urban planning and the "planning" studies for resilient city.

4.1 "Planning" Studies for Resilient City

"Planning" for resilient city refers to people in the face of climate change, energy shortages, population growth challenges, in order to improve the resiliency of the city, through urban planning strategies make the city has a stronger adaptive ability to respond to economic, social and ecological problems well.

Studies on "resilient city" planning are focus on urban disaster reduction planning (Godschalk 2003) and post-disaster reconstruction planning (Vale 2005).

Applegath, the member of founders and host of ResilientCity.org, is the member of Greene Building Council, Canada and the former Chairman of Ontario Architects Association. The website of ResilientCity.org points out that the world is facing significant challenges in relation to climate change, resource shortages and population growth. To improve city resilience, it is necessary to increase city adaptability to better cope with future economic, social and ecological problems through urban planning and architecture design strategies. The website lists 11 principles for urban design and 8 principles for architecture design.

4.2 "Resiliency" Studies of Urban Planning

"Resiliency" of urban planning refers to the use of a more flexible, adaptable approach to deal with dynamic problems in the implementation of urban planning, which requires both the rigidity of the law, such as a city function, urban structure, development direction, main road network, important infrastructure, public service facilities and significant impact block plan, as well as strain, flexibility and inclusiveness to adapt to market.

The concept of "resiliency" was introduced to the field of urban planning in response to market changes. Sheng (2006) stated that economic development in modern cities is occasional. Traditional rigid urban planning cannot adapt to the requirements of urban construction, and resilient working methods will become increasingly important in future urban planning. In resilient urban planning, the urban population scale forecast, urban land layout and urban development policies, as a collection of unit values and value ranges, are flexible and uncertain to maintain overall stable adjustability of urban planning. Deng (2003) opined that planning is static and purely technical in traditional approaches and planners select indexes by subjective judgment in accordance with the principle of "environmental optimum", which lacks flexibility in response to changes in market factors. However, under a market economy, comprehensive coordination of economic and social factors is required for urban development, and consequently deviation during planning implementation can occur. Neither implementing legal planning with the idea of "would rather make a mistake following planning than take responsibility", nor the significant costs in manpower and money to frequently revise planning, can adapt to the rapid and effective development of cities. Therefore, it is important to adopt more flexible and adaptive approaches to solve dynamic problems during implementation of urban planning. Management must comply with the rigidity of laws.

Resilient planning is discussed here in a narrow sense. Market changes are only one aspect that cities need to respond to, with changes in and impacts of other factors increasing continuously. More extensive "resilient planning" is required to construct a "resilient city".

The disadvantage of rigid planning is remedied by resilient planning. Xu (2002) summarizes the theoretical basis, procedures, finite force and infinite repeat features of urban planning resilience, and proposes mastering resilience in different planning phases by three aspects. Xu (2002) states that the resilience of planning is in its flexibility; in other words, planning should have the ability to be expanded and complemented continuously.

5 Conclusions

In applying resilient thinking to urban planning to construct resilient cities and to respond to various threats and disasters, we propose two important viewpoints of resilient thinking and emphasize studying urban ecological resilience and urban engineering resilience. Resilient urban planning studies can be classified into "resiliency" urban planning studies and "planning" resilient city studies. Most studies on resiliency urban planning refer to the flexibility of planning, i.e. resilient urban planning with a stress on planning approaches. However, studies on the planning of resilient cities focus on the objectives of planning.

This paper finally recommends a major resilient cities initiative, including improvements on the concepts and methods of urban planning, expanding urban systematic research and strengthening cooperation among different disciplines involved in resilient urban planning and development in a bid to provide powerful theoretical and practical guarantees for constructing resilient cities.

Acknowledgment The authors would like to express appreciations to my supervisor and my husband for their valuable comments and other helps.

References

1. Zhou H, Wang J, Wan J, Jia H (2010) Resilience to natural hazards. A geographic perspective. Nat Hazards 53(1):21–41
2. Holling CS (1973) Resilience and stability of ecological systems. Annu Rev Ecol Syst 4:1–23
3. Bruneau M et al (2003) A framework to quantitatively assess and enhance the seismic resilience of communities. Earthq Spectra 19(4):733–752
4. Holling CS (2004) From complex regions to complex worlds. Ecol Soc 9(1):11. www.ecologyandsociety.org/vol9/issl/artll/
5. Adger W, Hughes T, Folke C et al (2005) Social-ecological resilience to coastal disasters. Science 309:1036–1039
6. Berkes F, Colding J, Folke C (eds) (2003) Navigating social-ecological systems. Building resilience for complexity and change. Cambridge University Press, Cambridge, UK, p 416
7. Alberti M, Marzluff J, Shulenberger E, Bradley G, Ryan C, Zumbrunnen C (2003) Integrating humans into ecosystems. Opportunities and challenges for urban ecology. BioScience 53 (4):1169–1179
8. Resilience Alliance (2007) Urban resilience research prospectus: a resilience alliance initiative for transitioning urban systems towards sustainable futures. CSIRO, Australia — Arizona State University, USA — Stockholm University, Sweden. https://www.ehs.unu.edu/file/get/3648
9. Ernstson H, van der Leeuw SE et al (2010) Urban transitions. On urban resilience and human-dominated ecosystems. AMBIO 39(4):531–545
10. Folke C (2006) Resilience. The emergence of a perspective for social–ecological systems analyses. Glob Environ Chang 16(3):253–267
11. Alberti M, Marzluff JM (2004) Ecological resilience in urban ecosystems. Linking urban patterns to human and ecological functions. Urban Ecosyst 7:241–265

12. McDaniels T (2008) Fostering resilience to extreme events within infrastructure systems. Characterizing decision contexts for mitigation and adaptation. Glob Environ Chang 18 (7):310–318
13. MCEER (2005) White paper on the SDR grand challenges for disaster reduction. MCEER, Buffalo

The Construction of Neighborhood Unit System of Flood Control and Drainage Based on Landscape Infrastructure

Guan Shaoping and Zhang Xi

1 Introduction

In recent years, the flood and water logging happened in many cities of china have caused huge loss for the nation. Many former researches have analyzed the causes of flood and water logging in cities. Such as climate change led to the frequent occurrence of extreme weather; consequently the frequentness and intensity of downpour have a huge upsurge; the functions of nature have been restricted markedly due to the destruction of the nature from the rapid development of cities; the engineering infrastructures cannot meet with the demand of cities rapid expansion; and the unsuitable location and unreasonable overall arrangement of a city, the sedimentation of waterways, inconsiderate design, the bad emergency management [1]. These researches also discussed the corresponding countermeasures against these causes. Such as controlling the urban heat island effect, protecting the climate environment, increasing water seepage area to reduce the speed of runoff, improving the standard and capability of engineering infrastructure of flood control and drainage, adjusting the city layout to evade flood disaster, dredging river system to improve the ability of flood discharge, perfecting the design to enrich the flood control and drainage facilities, updating the management to improve operating efficiency [1]. But dams and pipelines are still the most important and directive way to control flood and drainage according to the construction pattern and operation mode of cities in china. We can see from the surveys of causes

G. Shaoping (✉)
College of Architecture and Urban Planning, Tongji University (TJU), Shanghai 200092, China

College of Design, South China University of Technology (SCUT), Guangzhou 510641, China
e-mail: guanshaoping2000@163.com

Z. Xi
College of Design, South China University of Technology (SCUT), Guangzhou 510641, China
e-mail: zhangx0303@gmail.com

X. Chen, Q. Pan (eds.), *Building Resilient Cities in China: The Nexus between Planning and Science*, GeoJournal Library 113, DOI 10.1007/978-3-319-14145-9_5

for flood and the response measures of Beijing, Guangzhou, Xian and Kunming in recent years that the main problem is still the drainage pipelines cannot meet the rapid development of cities and the main response measures are still increasing the number of drainage network or improving the design standards of drainage network [2].

The mode that control flood and drainage by means of gray infrastructure has increasingly revealed its limitation even destructiveness in the background of the city expansion and climate change. Some scholars have concluded a few problems about just the drainage mode by means of pipelines [3]: First, the pipelines are sealed, centralized and frail. Urban drainage pipe network is a centralized and closed transport system that were designed according to a standard. So, it cannot respond resiliently to the excessive rainfall because the flow rate and flow has its limits in unit time and the rainwater has to run too long distance in the pipes. In addition, the system will lose efficacy even paralysis when one or some of them were blocked or damaged. So, the pipeline mode has high stiffness and fragility. Second, the construction and maintenance will cost too much. The drainage pipelines in the city are usually up to hundreds or thousands of kilometers, for which the high costs always have to pay. Furthermore, the drainage pipelines frequently been blocked and destroyed, their structure model also led to the difficulty and high cost to manage and maintain them. Third, single function and enormous waste. The drainage pipelines are merely for drainage and function only in case of rainfall; along with unreasonable design and construction, it will cause massive waste inevitably. Again, it goes against water usage. We drain the rainwater out of city by massive pipelines promptly. At the same time we exploit underground water and divert water from outside the city to solve the shortage of urban water use by gigantic water conservancy project. These ways have brought huge resource waste and environment disruption. At last, this drainage model will cause pollution. It will bring massive water pollution because the water flow throughout the city will catch mass pollutants and the pipelines in the moist and dark environment will create massive bacterium. It will led to water pollution seriously and the destruction of the ecological environment. So, it not only cannot control flood and drainage effectively but also led to huge costs of ecology, economy and society just by improving the capability of the traditional gray infrastructures with some ways such as adding quantity, upgrading technology and raising standards.

The limits and drawbacks of drainage pipelines also existing in city's various kinds of hard engineering facility systems in different forms and contents. We can find a problem from the development process of infrastructure in modern city that the infrastructure constantly pursues high standard technical model. The design of these infrastructure systems aimed to ensure that they can achieve a single goal most efficiently by means of engineering. People consider more about the technical requirements of infrastructures and make them more standard. The main goal of most of the roads is to passage, the cutoff and engineered of massive rivers in cities is to meet with flood control standard. This kind of infrastructure has formed its own unique way to control natural environments and urban physical energy cycle, the intimate relationship between them and nature is disappearing. The main function of this kind of infrastructure is usually to exclude other weak functions in a strength position even at the cost of destroy these functions [4, 5, 6]. So, this kind of infrastructure model actually is a way of "from crisis to crisis" [7] to solve problems.

So, it is very urgent to explore new approaches to get rid of the single mode that manage flood mainly by means of dams and drainage pipelines. This paper proposes the neighborhood unit mode of flood control and drainage aiming to take advantage of nature and artificial nature to deal with urban flooding and keep away disturbance and destruction of ecological system from the engineering infrastructures. Meanwhile, this mode will enhance the city resiliency, create economic, social and cultural benefits to promote the urban sustainable development.

2 Methods

This paper explains the feasibility of neighborhood unit mode of flood control and drainage by Combining qualitative and quantitative methods in three steps. Firstly, aim at the limitations and problems of traditional model which control flood by dams and drainage by pipelines, we propose the conception of neighborhood unit model of flood control and drainage based on landscape infrastructure which combines the idea of "landscape infrastructure" and "neighborhood unit". Secondly, the four key elements of neighborhood unit mode of flood control and drainage including a certain amount of rainwater storage and retention area, the balanced distribution of rainwater storage and retention areas, the connection of rainwater storage and retention areas and the construction of landscape infrastructures of neighborhood unit system of flood control and drainage will be used to analysis four cases qualitatively by dividing them into positive and negative group and then illustrate the feasibility of this mode in practice. At last, build the functional relationship model between the rainfall and water drainage by pipelines base on 3 h rainfall in 1998 in Wuhan as an indicator, and then take Yanxi Lake and East Lake in Wuhan as example to get the relation curves between the rainfall and water drainage by pipelines under different water storage and retention conditions, then validate the effectiveness of neighborhood unit model of flood control and drainage through the comparison of these curves.

3 Idea Formation

The circulation of rainfall, evaporation, runoff and infiltration could be regarded as the natural hydrological cycling. Certain percentage of evaporation and transpiration is for the formation of rainfall in the future. Certain percentage of infiltration is for the replenishing groundwater. Certain percentage of runoff may also be used as a supplemental source of water for rivers continuity. They are the "natural pathway" of rainfall. The rapid development of cities had replaced the original water bodies in cities and led to a large amount of nature ground covered by cement, as a result, large areas of impermeable land surface and "concrete forest" has interrupted this natural cycling and most of the rainfall has no way out but to form runoff and converged to flood [8]. Accordingly, on the surface, the forming of flood in cities is due to inadequate and poor management of the engineering infrastructures such as dams and drainage pipelines, but neglecting and even ignoring the "natural pathway" of rainfall is the internal reason [8].

The expanding of city size always stays ahead of the updating of the flood control and drainage ability of engineering infrastructure such as dams and pipelines, as a result, the overburdened drainage pipelines can only operate overloaded, along with the intrinsic limitations of dams and drainage pipelines, over-dependent on them cannot solve the problem of the flood disaster in cities fundamentally. To solve the problem effectively, we should go back to the root of the issue and find solution by recovering the "natural pathway" of rainfall. But city as human activities gathered area is an artificial subsystem which exerts huge destruction to the existing eco-systems of natural water system. So, it is difficult to repair the "natural pathway" of rainfall [8]. But cities have not completely away from nature, this article seeks to build a flood control and drainage model which according with natural hydrological cycling and create multiple benefits by using nature and even artificial nature resources and combing the engineering infrastructure of urban stormwater management.

This research would benefit from the urban planning philosophy of "neighborhood unit" and "landscape infrastructure".

Neighborhood Unit It was proposed by the American planner Perry in early twentieth Century to adapt the planning structure changes in modern city because of the development of automobile traffic, which has changed the conventions that residential structure subservient to roads and square themselves which result in traffic congestion. It unified the planning of the residential area in a bigger realm, making the neighborhood unit as a cell to live in. The scale of neighborhood unit refers to the service radius of a primary school, and the complement public utilities in the neighborhood unit are built for the daily life of the dwellers, so the daily needs of dwellers in neighborhood unit can be satisfied internally, then the most of the traffic can also be digested internally to avoid the great pressure from massive traffic leading to accidents [9]. Draw lesson from this thought we can look the city as a huge region for flood control and drainage and dividing this region into several "neighborhood units" of flood control and drainage based on its own basic conditions, each "neighborhood unit" is a "cell" of the huge region. The "neighborhood unit" here refers to the infrastructure unit which centered on a couple of flood storage and retention areas such as lakes, reservoirs, wetlands and artificial rain carrier and the city region they served. The rainfall inner each neighborhood unit should being digested in the unit to avoid too much rainwater drains into drainage pipelines within a short time led to flood.

Landscape Infrastructure It is a new term emerged in America academia recently. It was been first proposed by Garry Strang in 1996 [10]. As a new term, it does not mean to a new type of infrastructure but an idea or a strategy of planning. Combine with the ideology of landscape urbanism [11], we can interpret the idea that landscape refers to the complexes constituted by land and the space and matter on the land. It is the marks on the land engraved by complex nature process and human activity; all ecological processes necessarily pass through it [12]. So it is the carrier of all nature process and human activity. In that sense, landscape is not only the scenery and symbol to interpret the interaction between human and nature but also the space and environment human lived in and an ecological system. Therefore, it has multi- attribute and functions of ecology, material space, language [13] and aesthetics and so on. Whether the animate green infrastructures based on natural conditions or the abiotic gray infrastructures based on artificial technology and materials are all the elements of landscape or landscape itself. So these infrastructures should possess the multi- attribute and functions of landscape and to support and participate in the ecological process and results of the nature, society, economy and culture of city. We can conclude based on the analysis above that landscape infrastructure is the organic whole combined landscape with infrastructure which can be built by using landscape design methods to improve the comprehensive functions of urban infrastructure and led it to become both infrastructure and landscape with multiple values of environment, economy, society and culture and so on [6].

Based on the functions and features of landscape infrastructure and followed by the planning theory of neighborhood unit, we connect the neighborhood units of flood control drainage which are mainly consist of lakes, rivers, and green system around them, to become a huge system like the connective tissue (Figs. 1 and 3). Aiming at this huge system of infrastructure, using landscape design methods to give it more comprehensive functions beside flood control and drainage. Similar to the connective tissue that widely distributed in the human body, the huge system widely distributed in the city and plays an important role in connection, supporting, nutrition, protection and other functions, thus it will becomes a system combined landscape with infrastructure of flood control and drainage and creates the multiple values for environment, economy, society and culture. This is the neighborhood unit system of flood control and drainage based on landscape infrastructure.

4 Construction

Considering the description of the basic concept of the model in Sect. 3. The construction of the model would be described in four steps including the principles, elements, structure and performance.

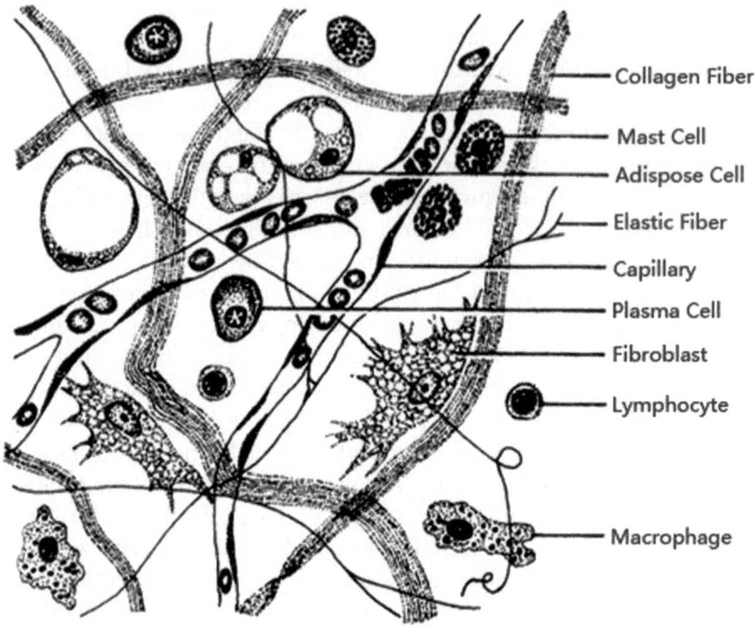

Fig. 1 A diagram of connective tissue (Source: http://baike.baidu.com/view/178915.htm)

4.1 Principles

- The neighborhood unit model of flood control and drainage mainly use nature, artificial nature and part of engineering infrastructures as supplementary to manage rainwater.
- The neighborhood unit model of flood control and drainage recover the "natural pathway" of rainfall according to natural hydrological cycling.
- The neighborhood units should been evenly distributed in the whole city.
- It makes sure the interconnection between neighborhood units and the interconnection among neighborhood units, the river system and drainage pipelines in the city.
- It combines with landscape design to build resilient and sustainable infrastructure which has multi-benefits of ecology, society, economy and culture.

4.2 Elements

There are three main constituent elements of neighborhood unit system of flood control and drainage. They are neighborhood unit, interconnection infrastructure and the operating system.

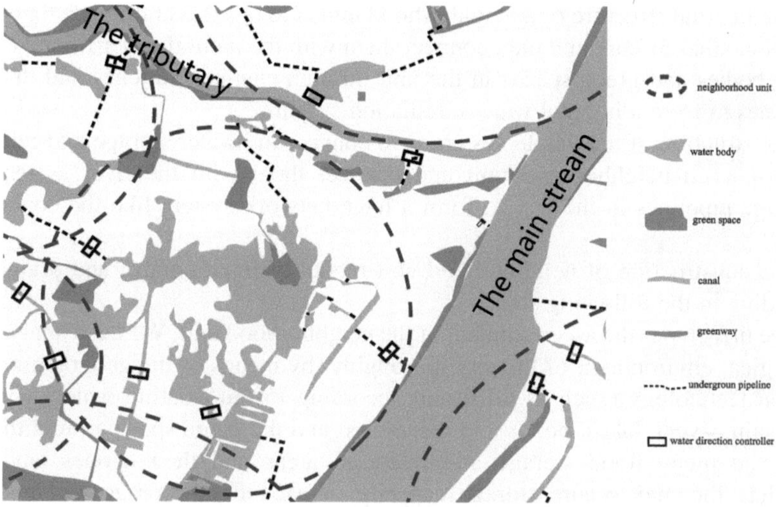

Fig. 2 A conceptual diagram of neighborhood unit of flood control and drainage (Created by myself)

Neighborhood unit. It is the basic unit to construct the neighborhood unit system of flood control and drainage (Fig. 2). It consists of the infrastructure unit which centered on a couple of flood storage and retention areas such as lakes, reservoirs, wetlands and artificial rain carrier and the city region they served. The infrastructure unit of flood control and drainage consists of varies natural and artificial natural water bodies, green space and part of the engineering infrastructure which have the function to regulate rainwater. The scale of each "neighborhood unit" is determined by the regulation ability of the infrastructure unit of flood control and drainage and turban geography.

Interconnection infrastructure. It refers to channels, ditches, pipelines and greenways which have the function of connection (Fig. 2). The interconnection infrastructure should manage to adopt a natural friendly form, and minimize the compositive impact on environment through landscape design if it bound to have engineering process.

Operating system contains the infrastructures and equipments to monitor and regulate the distribution and flow direction of water (Fig. 2).

4.3 *Structure*

The basic structure of neighborhood unit system of flood control and drainage include the internal structure of a neighborhood unit and the structure of the whole system.

The internal structure of a neighborhood unit is to take several flood storage and retention areas as core and then connect them with the natural and artificial natural water bodies and green spaces in the unit through channels, ditches and drainage pipelines to form a internal water circulation system.

The structure of the whole system is to connect the water storage and retention areas in each neighborhood unit and connect them with the river system and drainage pipelines in the city to form a huge network system like the connective tissue.

The construction of neighborhood unit model of flood control and drainage is according to the following steps.

The first step is the ascertainment of the neighborhood unit. We need analysis the geological environment of the city thoroughly, by means of the geographic information technology (such as GIS) and according to the existing waters (such as mountain, rivers, lakes, ponds and reservoirs) and the green space around them to designate many flood storage and retention areas and their scopes, and then calculate the max secure storage capacity of each flood storage and retention area. Then we suppose the seepage and storage capacity of the hard paving areas is 0. According to the hard paving area scope that each flood storage and retention area served in condition of a certain max rainfall (such as the 3 h rainfall once in a century), we can designate many neighborhood units of flood control and drainage in the city.

We suppose the drainage facilities in each area of the city are the same, so the districts without covering by neighborhood units in the city are just those lacks of flood storage and retention areas. They are all vulnerable to the hazard of flood and water logging. So we need combine with the requirements of urban economy and culture etc. to build up the artificial nature flood storage and retention areas in the future urban renewal planning of these districts to satisfy their demand for storage and retention flood, and then, forming many neighborhood units of flood control and drainage in these districts. The flood storage and retention areas in these districts should be equally distributed to avoid constructing giant flood storage and retention areas based on design flood storage capacity of these districts. So we need evenly distribute the design flood storage capacity of these districts. We can choose some fairly suitable areas to build flood storage and retention areas based on the geographical and construction condition.

The factors which influence the storage capacity of a flood storage and retention area in the detailed flood storage area design are: the water area, the water depth, the area and structure of green around the water bodies. The variation of these factors will influence the storage capacity of the flood storage and retention area and then the ascertainment of neighborhood unit, so we should regulate these factors based on the conditions of geography, construction, ecology and culture.

The second step is the connection of the neighborhood units. The flood storage and retention areas inside neighborhood unit or among the neighborhood units connected with each other through waterways, the flood storage areas which cannot connected with each other through waterways due to the geological environment could be connected with each other by means of underground pipelines to form the

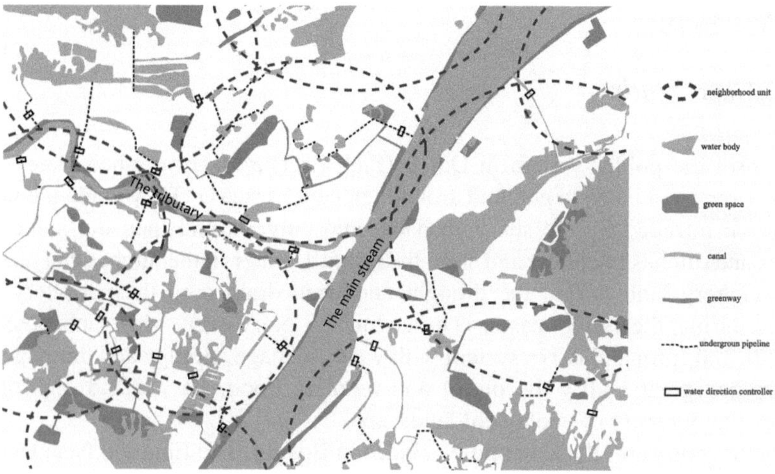

Fig. 3 A conceptual diagram of neighborhood unit system of flood control and drainage (Created by myself)

water network system and ensure the dynamic balance of water in the city. Furthermore, the flood storage and retention areas are connected with each other through greenways to ensure the integrality of the park system, so the park system with the function to detain and seep water would form the landscape infrastructure throughout the city (Fig. 3).

The last step is the regulation and usage of the system. We regulate and use the rainfall through a series of effective measures. There are control facilities such as flood gates and pump stations between each two flood storage and retention areas and among the flood storage and retention areas, river system and drainage pipelines to regulate the water flow direction, and then it builds hydrological monitoring network to act on these water direction controllers based on the system. These control facilities and hydrological monitoring network form a intelligent control system to control and monitor the flood in the whole system.

In addition, the construction of the system from beginning to the end should combine with landscape planning and design, to build the connective and supportive frameworks for the system to the urban organism from the aspects of ecology, society, economy and culture at the same time of meeting the function requirement of flood control and drainage.

So, the whole city is a huge neighborhood unit system of flood control and drainage based on landscape infrastructure.

5 Assessing

5.1 Case Study

First case: the polder system in Dutch (Fig. 4). The Dutch drained water out of shallow seas and lakes to get land around ten centuries AD. They made the land no longer be intruded by the sea with dams and culverts and then dig some water bodies and ditches to collect and store the rainfall, water in the ditches can replenish the periphery land during the drought and been drained to the periphery water bodies during the rainy season. These ditches interwoven with each other to a network and form a massive system with water seepage, storage and drainage. This system has great ability to control and regulate flood. Its wetland environment creates the diversity of ecological landscape.

The Second case: the Emerald Necklace in Boston. The Emerald Necklace park system in Boston connects the original Parks, marshlands and drainage ditches. The green system extending 16 km chained 9 park systems [14] (Fig. 5). With decades of construction, the river corridors have penetrated in the whole city and became the nature flood storage areas. Its long linear system have reduced flood in storm period, frostless period and flood period effectively. It bettered the water quality of river and the ecological quality of periphery environment, provide rich outdoor activity spaces for citizen, too.

Third case: Guangzhou is in the Pearl River network area and crisscrossed by rivers and waterways, but many waterways have disappeared or be severely deposited and then the phenomenon of dirty, smelly and unrest was very striking in the last 20 or 30 years because of rapid development of urban area. The water logging in the city was serious whenever the rain storms. Guangzhou has been involved in

Fig. 4 The diagram of polder in Dutch (Created by myself)

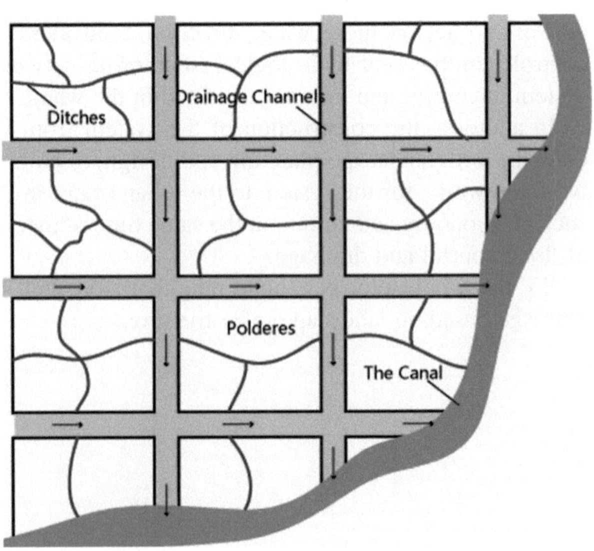

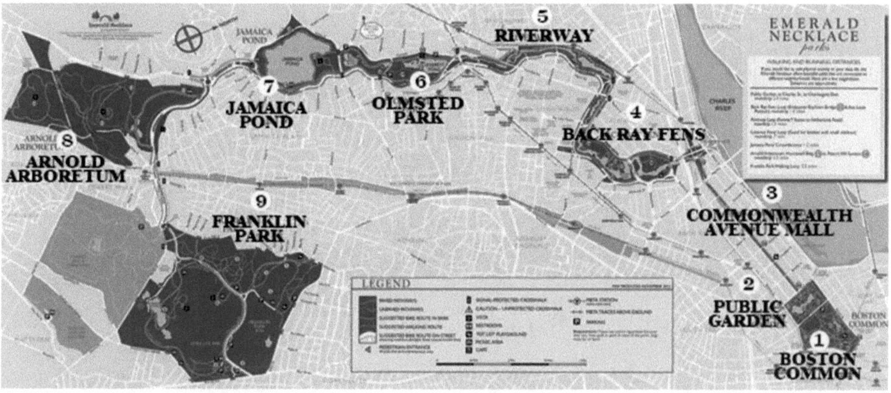

Fig. 5 The Emerald Necklace in Boston (Source: http://www.jx216.com)

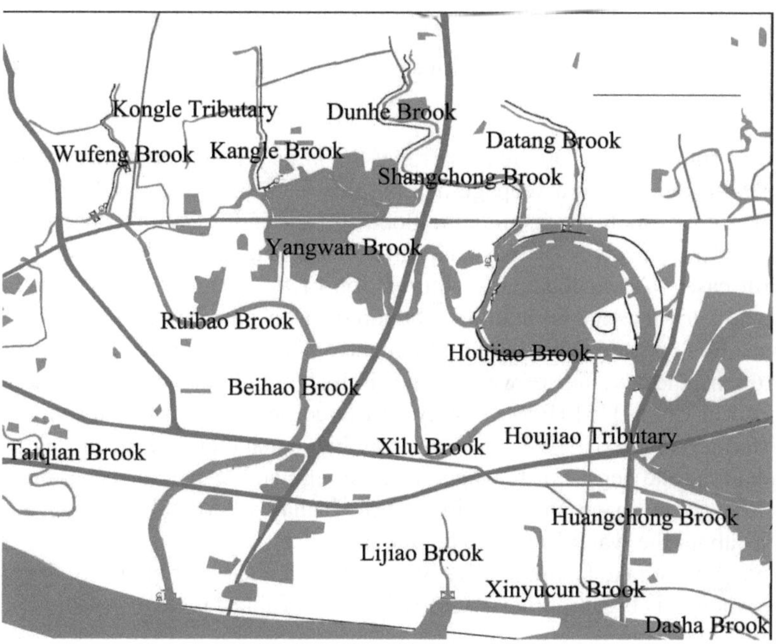

Fig. 6 An analytical diagram of the system of water and green in Guangzhou Haizhu (Redrawing of the map of Haizhu)

the harness of waterways, now the condition has significantly changed through high standard comprehensive harnesses that pull down the buildings along banks, clean up, dredge and connects of the rivers. Figure 6 is the situation of the harness of waterways in Haizhu district. We can know from this figure that the water network has been dredged and connected, the water quality has improved significantly after harness, but the problem that water logging whenever rain storms still haven't

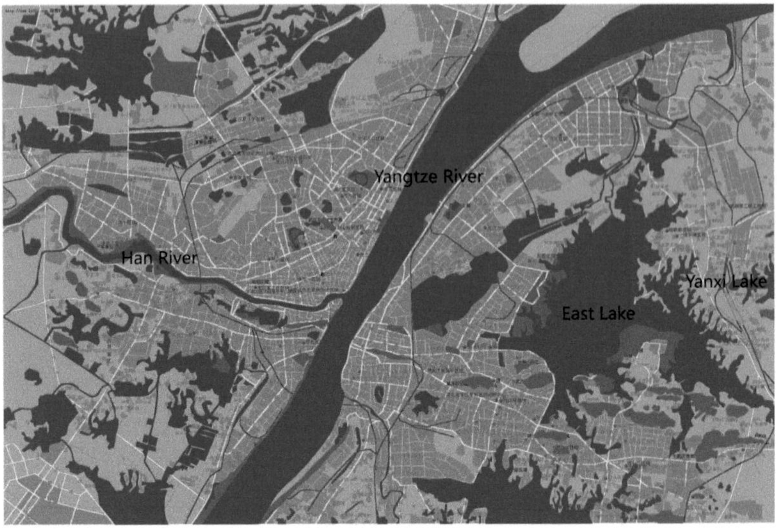

Fig. 7 An analytical diagram of the system of water and green in Wuhan (Redrawing of the map of Wuhan)

improved significantly. One reason is that the large amount of hard ground in the city prevent rainfall from seepage nearby, this lead to huge water logging on the ground surface, another reason is the flood storage and retention capacity is weak because of the narrow waterways and little green space along the banks.

Forth case: Wuhan located in intersection of the Yangtze River and the Han River and known as the reputation of hundreds lakes city. Nevertheless it has been damaged by flood and water logging frequently. The water surface area in Wuhan is 2,117.6 km^2, the coverage of water bodies is 26.1 % and the water surface area ratio is topped in China [15] (Fig. 7). Why still flood here with such enormous nature flood storage areas exist? We can know from Fig. 4 that there are much water bodies in Wuhan but the degree of breakage is quite high and uneven distribution, what's more, the green space shortage in central Wuhan is critical. Here is a simple analysis about the water breakage degree in Wuhan:

The index we take is: the density of splashes(C), the water breakage degree will higher when C is bigger, otherwise it will lower (Table 1).

$$C_i = n_i / A_i$$

Where, C_i = the breakage degree of I, n_i = the patch number of I, A_i = the area of I, in km^2.

We can know from Fig. 7 and Table 1 that the lakes in Wuhan are innumerable but uneven distributed; there is little green in the city even along shores; lakes lack of connectivity and the breakage degree of water bodies is high. These have caused the problem that massive water bodies cannot form a connected huge water system to equalize the digest capacity of flood in the city's different areas just like polder

Table 1 The breakage degree of lakes in Wuhan

i	n_i	$A_i(km^2)$	C_i
Lakes	166	803.17	0.2

system in Dutch. What's more, the green coverage ratio in Wuhan is very low, so the districts with little water bodies will vulnerable to flood in case of rain storms. Therefore, Wuhan will have great flood hazard even the nature storage condition here is good and municipal facilities here are complete.

We can draw inspiration from the first case and second case: firstly, we should take full advantage of the nature service function of the ecological infrastructure (such as nature water body and green), this could play an important role in flood control and drainage, and what's more, it would produce ecological, economical and cultural values. Secondly, the breakage and solitary of flood storage areas will seriously weaken the function of flood control. Strengthening the connection of them to form bigger scale network system and disperse the partial flood pressure that attribute to the disproportion of rainfall through system balance function, its flood control and drainage capacity will be enhanced largely. Lastly, the areas of seepage and storage that are distributed in balance would promote the digest capacity of flood nearby and alleviate the flood disaster caused by the flood control and the disproportion of drainage capacity.

5.2 *Performance Analysis*

5.2.1 Model Deduction

After the rainfall, as a result of the surface coverage difference, some of the water seep into earth, some evaporate into air, some are detained in the low ground and the rest drained through the drainage pipelines. The factor of evaporation and the amount of water seepage in hard construction areas are been neglected in this article, so we can get the functional relationship between the total volume of rainfall and the water storage and the water drainage as follows:

$$V = Q_{drainage} + Q_{retention} \tag{1}$$

Reference [16] provides that the rain flow formula is:

$$Qy = \Psi qF \tag{2}$$

Where, Qy = rainwater design discharge, in L/s; Ψ = runoff coefficient, assign a value of 0.9; q = design rainwater intensity, in L/hm^2; F = drainage catchment area, in hm^2. From Formula 2 we can get that:

$$Q_{drainage} = Qy \times T(60/1,000) = 0.9qF \times 60T/1,000 = 54qFT/1,000 \quad (3)$$

$$Q_{retention} = D_{river} + D_{plant} \quad (4)$$

Where, $Q_{drainage}$ = the amount of water from drainage pipelines, in t; T = duration of rainfall, in min; $Q_{retention}$ = the amount of water be retained, in t; D_{river} = the amount of water be stored by rivers, in t; D_{plant} = the amount of water be retained by plants, in t.

The study supposes that the designed rainstorm intensity of drainage pipelines in Wuhan is according to 10 years 3 h downpour. It discusses the functional relationship between 3 h' V and $Q_{drainage}$ of in 1998 in the case of the $Q_{retention}$ vary in Yanxi Lake zone and two lakes zone which consist of Yanxi Lake zone and East Lake zone.

Reference [16] provides that the designed rainstorm intensity formula of Wuhan is $q = 983(1 + 0.65lgP)/(T + 4)^{0.56}$, where, P = the design recurrence interval; t = the duration of rainfall, in min. So the design rainfall intensity q = 87 ($L/hm^2 \cdot s$). If the recurrence interval is 10 years, the duration of rainfall is 3 h. In the real circumstance, the drainage pipeline would not meet with the requirement of rainstorm intensity because of blocking, deformation and other influential factors. So we set an adjustment coefficient $\alpha(0 \leq \alpha \leq 1)$. Meanwhile, it need time that the rain water flow into the rivers and lakes, so we also set an adjustment coefficient for the water storage, $\beta(0 \leq \beta \leq 1)$. We suppose β equal to 0.3, as in [17], the capability of water storage capacity of woodland is 5,461.22 t/hm^2, shrub land is 5,092 t/hm^2, and the grassland is 5,203.23 t/hm^2. So we can conclude Eqs. (1), (3) and (4) into:

$$V = Q_{drainage} + Q_{retention} = [\alpha 54qFT/1,000] + 0.3(D_{river} + 5,464.22A_1 + 5,092.63A_2 + 5,203.23A_3) \quad (5)$$

Where, V = volume of rainfall, in m^3; D = the storage capacity of lake, in m^3; A_1, A_2 and A_3 respectively refers to the area of woodland, the complex of woodland, shrub land and grassland, grassland, in hm^2.

We had known the design rainfall intensity (q) and the design rainfall duration (T), so we can conclude Eq. (5) into:

$$V = 845.64\alpha F + 0.3(D_{river} + 5,464.22A_1 + 5,092.63A_2 + 5,203.23A_3) \quad (6)$$

5.2.2 Empirical Research

Equation (6) is the mathematical model of this research, and then we apply this model to Yanxi Lake and East Lake in Wuhan province. Yanxi Lake located the acreage of Yanxi Lake is 1,333.3 hm^2, the catchment area is 6,830 hm^2; the acreage of East Lake is 3,300 hm^2, the catchment area is 12,170 hm^2 [18]. The storage capacity of Yanxi lake (D_y) is 29,897,500 m^3, the storage capacity of East lake (D_d) is 68,709,200 m^3 [19].

We should definite some key concepts before we begin with the empirical research. Yanxi Lake zone (Z_y) refers to the lake area of Yanxi Lake and the land area in the catchment area of Yanxi Lake. East Lake zone (Z_d) refers to the lake area of East Lake and the land area in the catchment area of East Lake. Two lakes zone refers to the zone consist of Yanxi Lake zone and East Lake zone. The area of Z_y (A_y) refers to the sum of the area of Yanxi Lake and the area of land area in the catchment area of Yanxi Lake. The area of Z_d (A_d) refers to the sum of the area of East Lake and the area of land area in the catchment area of East Lake. The catchment area of Yanxi Lake (C_y) refers to the precipitation area which rain water flowing into the same valley including Yanxi Lake. The catchment area of East Lake (C_d) refers to the precipitation area which rain water flowing into the same valley including East Lake. The catchment area of design drainage pipeline in Yanxi Lake (F_y) equals to the catchment area of Yanxi Lake. The catchment area of design drainage pipeline in East Lake (F_d) equals to the catchment area of East Lake. The green space of Yanxi Lake Park (G_y) refers to the green space of park if we build a park centered on Yanxi Lake. The green space of East Lake Park (G_d) refers to the green space of park if we build a park centered on East Lake. Yanxi Lake Park (P_y) refers to the area includes the green space of Yanxi Lake Park and Yanxi Lake. East Lake Park (P_d) refers to the area includes the green space of East Lake Park and East Lake.

The catchment area of a lake is the area surrounded by watershed borderline and cross-sections. So, $F_y = G_y = A_y; F_d = G_d = A_d.$

There were four steps in this research. First, we suppose the rainstorm in Yanxi Lake zone can be digested only through drainage pipelines, we regard the 3 h maximum rainfall in 1998 in Wuhan as the 3 h rainfall in 1998 in Yanxi Lake zone which is 159 mm [20]. So we can conclude the following result based on Eq. (6):

$$V = 10859700 > 5775721.2\alpha$$

The relationship between the volume of rainfall and the amount of rainwater drained by pipelines shown in Fig. 8 curve I.

Second, we suppose Yanxi Lake could store water in Yanxi Lake zone. So we can conclude the following result based on Eq. (6):

$$V = 10859700 \geq 8969250 + 5775721.2\alpha$$

The relationship between the volume of rainfall and the amount of rainwater drained by pipelines shown in Fig. 8 curve II.

Again, we suppose to build a park centered on Yanxi Lake, the area of the green space of the park equal to the area of the lake, so the area of Yanxi Lake Park is the double of the area of Yanxi Lake. The green structure in the park is according to the structure of green space of parks in Shanghai [21]. The area proportion of woodland landscape is 59.09 %, the area proportion of the complex of woodland landscape, shrub land landscape and grassland landscape is 28.79 %, and the area proportion of the grassland landscape is 12.12 %.

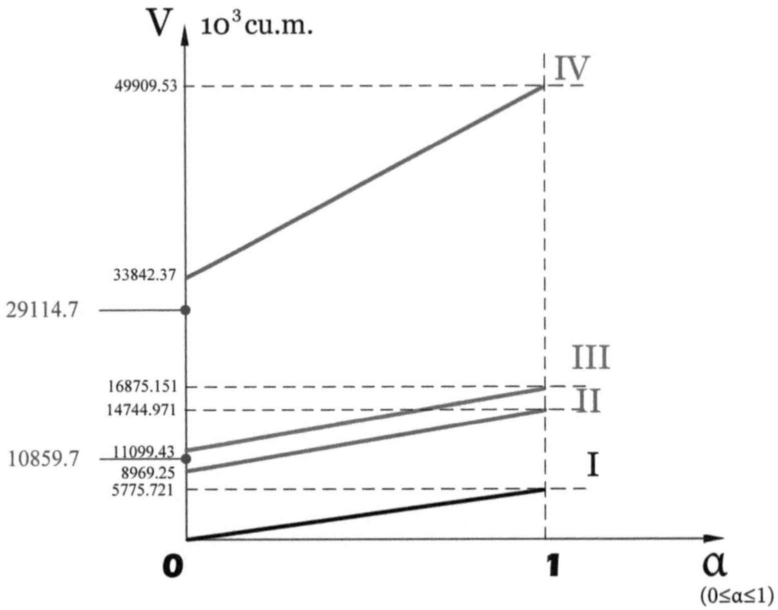

Fig. 8 The relationship between the volume of rainfall and the amount of rainwater drained by pipelines

So we can conclude the following result based on Eq. (6):

$$V = 10859700 < 11099430 \ + \ 5775721.2\alpha$$

The relationship between the volume of rainfall and the amount of rainwater drained by pipelines shown in Fig. 8 curve III.

The above analysis is based on the premise that Yanxi Lake is isolated and without connection with other lakes (Fig. 8). At last, we assume that Yanxi Lake connected with East Lake, there were a lake park in East lake zone, the area of the green space of East Lake Park is 1,333.3 hm^2. The rainfall intensity is uneven in space and time, so we suppose the 3 h maximum rainfall in East Lake zone is 150 mm, the 3 h maximum rainfall in Yanxi Lake zone is still 159 mm, we can conclude the following result based on Eq. (6):

$$V = 10859700 \ + 18255000 = 29114700 < 33842370 \ + 16067160\alpha$$

The relationship between the volume of rainfall and the amount of rainwater drained by pipelines shown in Fig. 8 curve IV.

5.2.3 Results Analysis

We could know from curve I in Fig. 8 that its highest point hasn't reached to the volume of 3 h rainfall in 1998 in Yanxi Lake zone which is $10{,}859.7 \times 10^3\ m^3$. We could know from curve II that its highest point has surpassed the volume of 3 h rainfall in 1998 in Yanxi Lake zone. We could know from curve III that its lowest point is surpassed the volume of 3 h rainfall in 1998 in Yanxi Lake zone. We could know from curve IV that its lowest point has far surpassed the volume of 3 h rainfall in 1998 in two lakes zone which is $29{,}114.7 \times 10^3\ m^3$. Therefore, Yanxi Lake zone cannot drain all rainwater of 3 h in 1998 in Yanxi Lake zone through drainage pipelines only. It can digest the rainwater of 3 h in 1998 in Yanxi Lake zone through the storage of Yanxi Lake and the drainage of pipelines when Yanxi Lake could store water. It can digest the rainwater of 3 h in 1998 in Yanxi Lake zone through the storage and retention of Yanxi Lake Park without drainage pipelines when Yanxi Lake Park store water as landscape infrastructure. When connected it with East Lake, the water will achieve to a dynamic balance in the whole district, but the range and intensity of precipitation are not changed, so the ability to resist the flood in the district will be reinforced because the amount of rainfall is distributed equally. Thus the two lakes zone can digest the rainwater of 3 h in 1998 through the storage and retention of Yanxi Lake Park and East Lake Park and the balancing action when connection them with each other without drainage pipelines. The analysis above clearly presents a certain amount of flood storage and retention areas which consist of water bodies and green spaces, the balanced distribution of flood storage and retention areas and the connection between them. They are precisely the key points of neighborhood unit model of flood control and drainage and show the effectiveness of this model.

6 Conclusion

The neighborhood unit model of flood control and drainage based on landscape infrastructure play a positive role in flood control and drainage of city, the construction of resilient city and the healthy development of ecological city. Firstly, it follows the rainwater natural circulation and use natural condition combine with part artificial infrastructure to decentralized and centralized manage rainwater through the dividing and connection of neighborhood units. It could reduce the flood in cities and improves water protection and utilization efficiency. Secondly, it combines landscape with infrastructure. On the one hand, it emphasis the coexistence and cooperation of human and nature and the construction of flood control and drainage infrastructure using natural conditions to reduce the construction of dams and pipelines. Therefore, it reduces the interference and destruction to natural environment and the wasting of resources. On the other hand, this model integrates the water network, green network and engineering infrastructure based on the neighborhood unit system of flood control and drainage. Thus, it increases the diversity of urban area, and promotes the capacity of urban area in coping with uncertainty and nonlinear external impacts, with these, it could increase the ability of self-organization and actualize the coordinated development of people and environment [22]. Lastly, it exerts landscape design idea on the infrastructure system of neighborhood unit model of flood control and drainage to build multifunctional and multiform urban open spaces and cultural and living facilities. It embodies multi-benefits such as promote social justice and people's physical and mental health level, cope with crisis and post-disaster reconstruction. In a word, whenever at normal or emergency situation, neighborhood unit system of flood control and drainage could make city more resilient and promote the sustainable development of the city.

References

1. DAI Shen-zhi, CAO Kai (2012) Countermeasure of urban flood and drainage in China. Mod Urban Res (1):21–28 (in Chinese)
2. http://www.weather.com.cn/life/2013/07/gdt/1927154.shtml
3. Li Jing, Xu Xi (2013) Urban distributed stormwater landscape infrastructure. In: Proceedings of the annual conference of Chinese Society of Landscape Architecture, Nanjing, Nov 2013, pp 1053–1058
4. Zhai Jun (2009) The city as landscape based on landscape infrastructure – the landscape urbanism approaches. Landsc Archit Front (7):46–51 (in Chinese)
5. Ying-Yu Hung (2009) Landscape infrastructure: in plain view. Landsc Archit (3): 48–53 (in Chinese)
6. Li Jing (2011) Research on the design and practice of modern urban landscape infrastructure, (in Chinese). Ph.D. dissertation, Department of Urban Planning BJFU University, Beijing
7. http://m.ammoth.us/blog/2010/04/reading-the-infrastructural-city-chapter-one-index/
8. Dai Yulong, Sun Minqing, Lu Debao. Simple opinions on urban stormwater management Problem. http://www.doc88.com/p-7098075546206.html
9. Zhiqiang W, Dehua L (eds) (2010) Fundamental principles of urban planning, 4th edn. China Building Industry Press, Beijing, pp 33–34
10. Bélanger P (2010) Redefining infrastructure. In: Mostafavi M, Doherty G (eds) Ecological urbanism [M]. Lars Muller Publishers, New York, pp 238–239
11. Weller M (2004) Landscape urbanism: polemics toward an art of instrumentality [A]. In: Blood J, Raxworthy J (eds) The Mesh book: landscape/infrastructure [C]. RMIT University Press, Melbourne, pp 66–75
12. Hensel M (2003) Ocean north – surface ecologies. In: Mostafavi M, Najle C (eds) Landscape urbanism: a manual for the machinic landscape. Architectural Association, London, p 111
13. Spirn AW (1998) The language of landscape. Yale University Press, New Haven\London, p 15
14. http://baike.baidu.com/view/4096625.htm
15. http://news.sina.com.cn/c/sd/2013-07-24/114527758054.shtml
16. http://wenku.baidu.com/view/f77c31d1240c844769eaee1e.html
17. Long Wu, Yang Yun-hua, Wang Ke-qin, Li Jian-zeng, Li Bao-rong, Li Yun-jiao (2007) Moisture-holding capacity of different plant cover types in the Jianshan River drainage area. J Zhejiang For Coll 24(2):135–139 (in Chinese)

18. Lu Shangong, Qiu Hanming, Tao Shouxiang, Chen Songshan (2003) A preliminary study on lake-river connectivity in Wuhan. In: Proceedings of the national water conference, Shanghai, Sept 2003, pp 197–204
19. Wang Xionghui, Li Min (2010) Drainage capability analysis of pumping station in the "Big East Lake" ecology river-network project in Wuchang. Urban Water Supply (5):65–67 (in Chinese)
20. http://221.232.78.243/rain/overYearPeriodRain.asp
21. Che Shengquan, Song Yongchang (2002) Landscape pattern analysis of Urban Park Green Land in Shanghai. J Shanghai Jiaotong Univ (Agric Sci) 20(4):322–327 (in Chinese)
22. Folke C (2006) Resilience: the emergence of a perspective for social – ecological systems analyses. Glob Environ Chang 16(3):253–267

Community Life Regression Under the Background of Rapid Urbanization

Liang Guo and Min Li

1 The Emergence of Community as a Social Integration Mode

Before the founding of our country, the traditional urban community was built on the bases of blood and geographic relationship as an ascription integration mode. After the founding of PRC and during the planned economic period, the work unit became the central nervous system to regulate and control the social operation. This function of regulation and control reflected not only in the economic operation, but also in the management of urban society and molding city residential space. The unit engulfed the community; the community was internalized in the unit. This is the community type that we call WORK UNIT COMMUNITY. But in face, it is not a kind of urban community on firm sense. During this period, the state integrated the social members through the UNIT. This integrative management of politics, economy and society, and the unit compound residential morphology had objectively maintained and developed various social functions of neighborhood. The unit as an extensive government agency dealt with the resource allocation according to the established allocation rules on behalf of the government. It formed a situation that the unit was dependent on the government meanwhile the social members relied on the unit, from which the country's political integration was fully reflected [1].

However in 1980s, the advantages of UNIT COMMUNITY had gradually disappeared and increasing exposed many shortcomings. On the one hand, the

L. Guo (✉)
College of Architecture and Urban Planning, Huazhong University of Science and Technology (HUST), Wuhan 430074, China
e-mail: 438963805@qq.com

M. Li
Shenzhen New Land Tool Planning and Architectural Design Co., Ltd., Shenzhen 518100, China
e-mail: 13623862305@126.com

X. Chen, Q. Pan (eds.), *Building Resilient Cities in China: The Nexus between Planning and Science*, GeoJournal Library 113, DOI 10.1007/978-3-319-14145-9_6

unit carried too many social functions so that it was in the plight of production and management. On the other hand, the UNIT COMMUNITY lost many chances to grow up due to the restrictions of unit system, such as the high degree of source control, over-bound of individual members, and the rent-seeking phenomenon caused by close dependent relationship. Along with the reform and opening up of China, the development of market economy, and the change of urban social structure, the WORK UNIT BUILT society had gradually disintegrated and the primary groups like neighborhood began to decline. The community has emerged as a new social integration mode.

2 The Present Types of Community Life in China

Urban-village, unit-system community and commercial residential area are the main types of present community life. The conclusion of them can give a relative authentic reflection of the current urban community status and characteristics, and then provide some ideas to seek the solutions.

2.1 Urban-Village

Urban-village is what used to be a village but has been gradually incorporated into the city system in the process of urbanization. Though its village committee has been changed into residents' committee, it still retains the collective allocation of the collective land rental income. It is a typical community which has been least urbanized but most effected by institutional factors.

2.1.1 Main Characteristics of Urban-Village

(a) Community resource. On the one hand, the social security system here is low and imperfect. Due to the abuse of land expropriation and improper resettlement of landless peasants, they lost their social security for the possession of land but in the meantime can not enjoy the same social security with the urban residents. On the other hand, the economy of urban-village usually bases on renting, which has no competitive force. And the lagging community plan, disorderly and unsystematic house layout, poor sanitation, insufficient infrastructure facilities, narrow roads, blocked fire and ambulance channel, and acute social order are the main problems there.
(b) Community governance. Urban-village bears both characteristics of village and city. Though it has been in the city, it still implements the management of village. People retain their ideas, lifestyle, housing construction and management mode. The primary party organization is lax while the awareness of villager's self-government is weak.

2.2 *Unit-System Community*

It is a typical community medium affected by urbanization and institutional factors. Although along with the system reform, the unit community is no longer mainstream mode of urban community, but actually some of them haven't disappeared yet. They remain as the basic unit of social space in the micro urban life after the structural change. Therefore, unit-system community shouldn't be simply excluded from the planning of community construction. It should be reconstructed according to its realistic development and the residents' living demands.

2.2.1 Main Characteristics of Unit-System Community

(a) Residential constitution. One part consists of the original unit employees without opportunities and expectations to improve the housing including the retired workers, the laid-off workers in the process of enterprise reconstruction, those who still work but are unstable and with low income. The other part includes the migration and floating population who difficultly survive in the city. Viewing from the current residential constitution, it has become an obscure area in the urban residential space showing the demographic characteristics like aging, marginalization and underclass increased.

(b) Community resource. In current urban space planning, the unit-system community has relatively fallen behind in the city development. The community environment is dirty, disorderly and bad. There are no sufficient modern welfare facilities to satisfy the residents' living demand, no convenient transportation network and no decent public activity areas and facilities, not to say those artificial natural landscapes symbolized modern city.

(c) Community governance. In addition to the disadvantaged residential constitution and the lack of community resource, viewing from the later development and management of the unit community, the unit has already early withdrawn from the housing management and community maintenance in the context of institutional transformation. However as to various historical issues, there is no new social institution willing to take over the property management of those communities, which brought a vacuum of the unit community management and led to the accelerated deterioration of the living environment and conditions.

2.3 *Commercial Residential Area*

It is a typical community with high degree of urbanization and population mobility but less affected by the institutional factors. It refers to those residential districts built by the developers and circulated through the real estate market. It completely follows the operation of marketing principle that the economic means replaced the administrative means. Now the unit people obtained the right to choose living environment freely, which is a huge social reform. The personal independent choice began to affect the urban structure and the residential space has been rearranged in accordance with the differentiation of classes.

2.3.1 Main Characteristics of Commercial Residential Area

(a) Community resource. Commercial residential communities are generally equipped with complete and perfect facilities. The high-income ones have superior facilities and some are with intelligent management and supervising system. They have sound social service system and high degree of specialization, industrialization and socialization. But the relationship between the members is loose and the neighborhood activities are the least.
(b) Community governance. Commercial residential community implements neighborhood committee management or owners' committee management, or both of them. In the high-income communities, the owners' committee has increasingly played the leading role. Their spatial layout usually absorbs the latest community planning philosophy and some of them are really distinctive.

3 Problems Caused by High-Density Living Environment in the Context of Rapid Urbanization

China has a large population but relatively less land. With this limitation, high-density development becomes a tradition of most cities. In the context of rapid urbanization, a large amount of rural population rushed into cities. Many of them, especially the central ones are constantly growing taller and denser. The urban space becomes three-dimensional and densely developed, which made great negative impact on the urban life quality. Under the background of overwhelming urbanization, we are bound to prepare for the crowded culture it spawned. These problems are performing at:

3.1 Spatial Differentiation

Through the filtering of housing price and sorting mechanism of social economic disparities, the residents with similar professions, income and value orientation tend to live together. The whole city formed a living differentiation even isolation from each other. Otherwise, the split of state residential policy aims to different groups, the various develop modes and design standards of developers for different grades of residence, to some extent exacerbated the differentiation and isolation [2].

3.2 Community Protest

The current land leasing policy is imported from Hong Kong, which has accumulated funds for large-scale city construction in the meantime resulted in protests for compulsory relocation.

3.3 Weak Community Awareness

At this stage, strong sense of community is still built on the original ties such as occupation and relatives. The residents especially the new ones are lack of communication. As the emigration of people and the disintegration of previous social relationships, the community awareness might be weakened rapidly.

3.4 Lack of Community Vigor

The rapid urbanization has improved people's living conditions, whereas the cold high mansions estranged people from each other. The former neighborhood which occupied an important place in the interpersonal interaction has been increasingly weakened. The mutual concern and help become absent from people's daily life [3].

Thus we can see that with the further development of industrialization, urbanization, marketization and internationalization, our country is experiencing a profound change of social structure, interest pattern and ideas. Community has gradually become a intersection of various interest, a focus of different social conflicts, a force point of social construction and the support of party's primary governance. However, confronting the standardized, mechanized and specialized communities which replaced previous ones that people closely linked to each other, the increasing crimes and environmental pollution, it has a great significance to explore how to solve the disadvantages caused by high-density urban development through new community planning [4].

4 The Root of Issues Caused by High-Density Living Environment

4.1 Institutions

1. Urban-rural duality. Due to the existence of urban-rural duality, the migrant workers are separated from the urban residents, so that their status in politics, education, employment and social security are weakened.

2. Real estate's lack of livelihood value orientation. Firstly there is a deviation of real estate's development policies orientation. Secondly the great benefits of real estate promote local government's intended or unaware guidance that the alliance of rights and capital makes it gradually breaking away from its nature of supplying residence.
3. The vacuum and incompleteness of community management. There are some communities which have the vacuum of management or authority, and some are even without any management administration. Meanwhile the management of some new residential community is not complete that the property administration took the place of management agency which makes those new communities are separated from the management of urban government and overall society to form new social problems.

4.2 Planning

1. Land development mode. Our country use to be compounds of office and army, but now compounds of real estates and commercial land projects. This kind of development mode cut the city into pieces, for example the spring up of self-closed and large-scale gated communities. They fragmented the urban space that affected by the long-term function-classified thinking, the residential lands are developed as isolated function space.
2. Code of conduct. From the exploration of modern city planning such as neighborhood unit, neighbor and community, Chinese settlements are restricted by those doctrines of city and architecture. For example, the neighborhood unit theory of planning design guidance is different from the neighbor scale range in daily life. Neighborhood unit theory considered that the best neighborhood unit scale is 5,000–6,000 people, whereas the anthropologist proposed that once the neighbors exceed 300 households (1,000 people), there would be no neighbor interaction at all. The rigid planning and architecture code seriously impeded the development of architecture style and its system of streets and alleys.
3. Supporting facilities. In the connotation of urban community construction, community resource is a requisite part, which affects people to choose the residential place. But some of the communities are lack of supporting facilities like insufficient medical treatment, schools, shopping, and employment conditions. The new residential construction especially in the suburbs exceeds the real demand so that it becomes the bubble economy of city and regional development.
4. Spatial environment. Nowadays, some residential areas over-designed the outside space in order to pass the approval of relative planning department smoothly. The developers provide "perfect" space one by one as fixed place with fixed function without permission of any change, at the meantime the most vital job of the administrator is to keep outside spatial environment unchanged. This ultimate and immobilized construction mode rejected the possibilities of reflect and adjustment toward residents' demand in daily changing interaction. It is not humanized or sustainable.

4.3 *Social Factors*

1. Residential commerce and increase rent business make the community members complex. With different backgrounds, religions and values, diversity of behavior and psychological barrier formed between neighbor residents which became the obstacle of their interaction. At the same time, the individual thought and consumerism culture erosion are the most important reasons for public space shrinking
2. As the long-term household function served such as clothing, food, shelter and transportation, or to raise children and support elders undertook by the specialized social organizations, the need for mutual help on production and living between the neighborhoods decreased sharply without the motives for interaction. Otherwise, the workplace and residential area are separated, which makes the people who go back early and come back at dusk has no spare time to take activities in the neighborhood.

5 Community Construction in the Context of Rapid Urbanization

The sharp social construction change brought communities the demand of integration. At present, cities in our country are experiencing the differentiation of unit functions and the increasing expectations towards communities. After the multi-functional unit organizations gradually transformed into a pure functional one, the individuals will face to the fierce market competition directly [5]. According to this transformation, people expect that the community will become an independent social system away from the enterprises and institutions where not only people could get psychological relax and comfort, but also the disadvantaged social groups in economic field could find a public place to express self-will. As a result, the community construction has its specific social significance in China at this stage.

5.1 *Rigid Planning Level*

5.1.1 Community Planning Theory

The planning design theory of residential areas in our country mainly focuses on the function of "living" and "residence". One of the theoretical foundations is the thinking of function partition. The large-scale residential area in the suburbs resulted in the pendulum city traffic which makes the residents struggling forth and back from work [6]. Therefore, the community planning theory should emphasize the mixed and balanced relationship between residence and work, the diversity and openness of community and the cultivation of community sense.

5.1.2 Community Development Mode

Affected by early "neighborhood unit" thought, the residential areas in our country pay attention to the integrity of its own system regardless of size. It stressed passable internal roads to block external crossing traffic. This approach to some extent ensures the tranquility and safety of the community while actually destroyed its function too. To develop street network in small groups can keep the cities safe and combined with each other. This requires us to finding an appropriate scale of community in the high-density environment. In the United States, there is a new theory called "walkable" city or 20 min community, where people can get to the destination in 20 min by walk.

5.1.3 The Construction of Low/Multi-layer Buildings and Street System in High-Density Urban Area

High-density is a basic element of urban environment. With the development of industrial cities, the multi-layer internal urban architecture type which conducted service to traditional feudal families has gradually simplified into single courtyard or row house. The former internal traffic had been replaced by urban street and allays of different levels. However since 1950, it is proved that whether in western countries or in China, the low/multi-layer high-density architecture type and street system are most effective in the modern urban construction.

5.1.4 Spatial Environmental Design for the Promotion of Neighborhood Interaction

The construction of community spatial environment should create a sense of belonging and security, which can form a multi-leveled progress as "surrounded but not closed, transparent but not sparse". Considering the observation and conclusion of the residential interaction in the community, and the specific requests of spatial environment for different activities, the author believes that the residential environment conductive to neighborhood communication should as follows: be in accordance with the population structure; establish a sense of territory; provide different levels of privacy but at the same make the neighborhood communication a necessity while the residents traveling through the public domain to privacy space [7].

5.2 Elastic Management Level

5.2.1 Community Management

In the decline process of unit community, new issues emerged in Chinese cities. Besides, from the international experience, social action oriented by the state is adopted by majority of the countries in the world. To form the new social construction and management setup like "government decides, society acts" to a large extent depends on the changing of the government action and whether the government can explore and apply a series of new governing instruments to adopt the social organization into their framework. In the community system practice, the street and community system reform of "Jiang Han mode" followed the unity and diversity of the levels, which avoided the "hierarchy paradox" and built a organic cohesion and fine interaction between the public administration and the residential self-governance [8]. Though there are still many problems and difficulties of community self-governance, we should make a long-term adherence. It not only benefits to our primary community administration, but also can add the sense of belonging and pride of the residents.

5.2.2 Community Integration

Community integration is a very important issue in the community construction. On the one hand, with the problems emerged from the declination of unit community and high-density urban living, the solutions need to depend on the construction of new community. But due to the psychological dependence formed under the unit system, withdrawal of government can not bring a bottom-up effective connection of the community residents. So in this case, in order to keep the social stability, the government should get involved into the community to ensure the participation through its mobilization and to promote new construction of social community [9]. In a different way with the government control under the unit system, the social construction guided by government would proceed from the residents' requirement to form the organization cultivated by government which could be further developed into organized system, then finally to form a community of social integration [10].

On the other hand, it includes the integration of new community urban residents. Rural surplus labor is the main source of new urban members that is what we usually call migrant workers. For their integration, in addition to their own limitations of economy, psychology and social life we should break the current system and social barriers. This requires us not only to establish the social policy and legal system from the system perspectives, but also to support from the work of neighborhood committee to supply education and help to the new residents. What's more, we could promote the social integration from the community participation, including the work-market integration, social security integration and urban community integration.

5.2.3 Community Culture

Community culture is the soul of the community construction. An important issue of current community activities is that it can not integrate into residents' life. It emphasized on society more than life. The more profound meaning of community cultural construction is to jointly cope with the common problems in our lives and to establish an internal living order to strengthen the sense of community identity. Therefore, we should fully extend the function of community culture, cultivate community identity awareness, encourage community participation, improve the social supporting system based on social capital, complete organization system, and realize social value of public culture to make community culture participation identified with society.

5.2.4 Community Public Service

In the development of community governance, it demonstrated its internal limitation of bad balance between demand and supply. The key point to solve the problem is to form an integration of government, community and social organizations to make a socialization of public service. The social transformation gives an opportunity to it. The development of community makes a preparation for taking over the government function. It is an important way to meet various demands of residents. Now Xu Hui District of Shanghai and "care system" of Beijing communities became very good example for practice [11, 12].

6 Conclusion

Community construction has a 20 years history in our country. During this period, community has changed from a scholar notion into practice of government. It has gradually become a living mode of people and a basic organization of society. Though community construction has become an important social strategy and a variety of community organizations and social service facilities has been established, the community awareness and participation are still weak under the rapid urbanization. Community members do not help, trust or cooperate with each other, but become more and more indifferent and isolated because the people pay more attention on the search of private interest [13]. In this circumstance, it has a great significance to do research on construction and return of community. This essay concludes the root of living issues in high-density community and proposes the solutions in two aspects. The author believes that there are still lots of things to explore.

References

1. Zhang Hongyan (2002) On contemporary status and trend of Chinese urban community differentiation and change [J]. Planner 18(8):5–8
2. Zhang Sen (2009) Analysis on evolution of urban residential space in social transformation [M]. China Social Sciences Press, Beijing

3. Zhang Runjun (2010) On the incitation of Chinese urban community vigor [J]. Northwest Norm Univ J 47(6):125–130
4. Li Qing (2010) New concepts of community planning: a case study of Playa Vista in USA [J]. City Plan (9):42–48
5. Zhang Jie, Lu Jie (2003) Frome the large-scale urban design to "daily living space" [J]. Urban Plan 27(9):40–45
6. Wang Meiqin (2011) New explanation of traditional unit community reconstruction in late unit times – based on the survey of one state-owned unit community [J]. Acad Forum 34 (2):134–138
7. Wang Hua (2004) The analysis of "neighborhood interaction" and "spatial environment" in residential community [J]. Archit Knowl (3):46–48
8. Chen Weidong et al (2010) Chinese Harmony Community – Jiang Han mode [M]. China Society Press, Beijing
9. Peng Nu (2010) Discussion on "Community Building" from the perspective of "Citizen Participation" – resume of the second session of Chinese architecture thought [J]. Mod Archit (1):129
10. Zheng Qi (2010) On the citizen community – analysis on generation of community and impact of government nurturing [M]. China Society Press, Beijing
11. Li Guiqing (2007) Community types and characteristics of neighborhood – a case on Beijing [J]. Jiangsu Adm Sch J (2):59–65
12. He Haibing (2010) Harmony community – report on harmony community construction in Shanghai [M]. Academia Press, Shanghai
13. Yu Yanyan (2011) Blue book of Beijing – report on Chinese community development [M]. Social Science Academic Press, Beijing

Addressing Social Revitalization in Conservation of Historic Quarters in China

A Social Capital Initiated Urban Conservation Approach

Jie Han, Liangliang Wang, and Chye Kiang Heng

1 Introduction

Nowadays, conservation of urban historic neighborhoods has become progressively significant to sustain the urban social ecology. Particularly in contemporary Chinese cities, where urban development is extremely rapid, mainstream urban conservation practices favor the "demolition" approach through a top-down process. Through land transactions, parcels of inner city land, which usually accommodate traditional settlements, are sold to developers by local governments. Urban development, understandably, is a complicated process. Many social issues are raised through profit-driven redevelopment of historic areas when higher socioeconomic groups, who are served by high-end development, replace existing lower-income social groups and their traditional mixed-use neighborhoods. On one hand, the spontaneous process of socializing is interrupted by one-time mass demolition and relocation of original residents to urban fringe areas where employment opportunity and civil infrastructures are inadequate. The result is unbalanced economic growth and deteriorated social spatial justice, which reduces the affordability of a city and increases burdens to cities in the long run. On the other hand, social replacements diminish the social values attached to the physical fabric of neighborhoods and damage the openness and inclusion in a society.

J. Han (✉) • L. Wang
Department of Architecture, School of Architecture and Civil Engineering, Xiamen University, Xiamen 361005, China

Department of Architecture, School of Design and Environment, National University of Singapore (NUS), 117566 Singapore, Singapore
e-mail: aki.hanjie@gmail.com/hj@xmu.edu.cn; leonwang@xmu.edu.cn

C.K. Heng
Department of Architecture, School of Design and Environment, National University of Singapore (NUS), 117566 Singapore, Singapore
e-mail: akihck@nus.edu.sg

X. Chen, Q. Pan (eds.), *Building Resilient Cities in China: The Nexus between Planning and Science*, GeoJournal Library 113, DOI 10.1007/978-3-319-14145-9_7

Based on the discussions above, this paper asserts the necessity for urban conservation to address the social revitalization in urban China. It further attempts to challenge the methods and ethics of conservation planning in contemporary social-economic context. It will discuss the challenges of urban conservation in terms of whether there should be more conscientiousness attached to the process of managing change in historic places so that a decayed historic place could be revitalized and transformed in a way that would be resilient, economically, culturally, and socially.

Based on the study of social capital theory and its adaptations in other academic fields, this paper proposes an alternative conservation approach, namely a social capital-initiated conservation approach. Through a pilot study of the conservation of Xi'an Hui Fang which is a historic district of the Ethnic minority-Hui people, the roles of different forms of social capital are explored. The dynamic combinations of different forms of social capital are examined through the use of social network analysis, as well as how the social capital-initiated urban conservation could improve social cohesion and promote a sustainable transition under a decision-making structure of path dependence.

2 Literature

2.1 Approaches in Urban Conservation

Scholars have discussed several approaches to contemporary urban conservation. In terms of "what to conserve," an area-based approach was adopted to enlarge the scale of conservation objects from single historic monuments to groups of buildings, markets, and towns, etc. [1–4]. Adapted by many researchers, the townscape has become a powerful instrument to manage and instruct urban conservation and renewal [5–8]. With the growing research interest of conserving the identity of urban areas, urban morphology has been adopted to understand the hidden structure and relationship of various elements within a historic area, and in the identification of layers of historical development [9, 10].

Economic approaches explore the topics of economic restructuring, obsolescence, revitalization, reuse, and demolition concerning historic areas. According to Tiesdell [2], the economic growth of historic areas can be gained through three basic methods: functional restructuring, functional regeneration, and functional diversification. There is an abundance of studies on the functional reuse of historic areas [2, 3, 11]. Tourism-led, housing-led, and culture-led revitalizations are three popular forms of functional initiatives [2]. However, these approaches, in general, lack the concerns of social aspects, which have led conservation into an unbalanced situation.

As urban conservation is increasingly influenced by other urban categories, scholars argue that conservation should be an integral part of urban planning [2, 3, 8, 12]. Conservation planning and control guideline, while easy to adopt, lacks an historical and contextual view and is obliged to political interest [1, 3, 11, 13, 14]. Approaches, such as policy-led conservation planning, public intervention and participation, and place management, are characterized as relatively "soft" approaches, which promote participation and understanding of sense of place, compromise the top-down approach, and balance the conflicts among different interest groups [1, 2, 15, 16].

By integrating community needs and resources into conservation planning, the community development approach addresses the issue of equity and aims to provide multiple development choices with fewer impacts on community identity and daily life [17]. As a bottom-up approach, community development typically lacks the connections to government agencies and other institutions for proper guidance, and, as a result, the efficiency of plan implementation is affected.

Gentrification is often seen as one result of capital and real estate led revitalization of historic areas, and is usually accompanied with social replacement [18–21] and spatial restructuring [19]. This approach of gentrification provides perspective in understanding the economic, social, and political context of urban development, and examines the limitations of urban conservation as a tool to manage change under the dominant concern of economic gains.

Sustainable development, as an approach to conservation planning, requires a new perspective to understand urban conservation; rather than keeping a place physically untouched or solely economically revitalized, urban conservation is about place management in view of the process. The process consists of managing change and the development path of an historic place towards sustainability. It has been verified in many cases that without a solid social network and social justice, urban conservation could not achieve sustainability, and the economic and environmental dimensions would decline progressively [22–24]. However, literature shows that social sustainability concerns are largely absent in urban conservation, and there is a certain lack of feasible approaches to promote social sustainability.

3 Theoretical Underpinnings

3.1 Understanding Social Sustainability in Urban Planning Via a Process Perspective

Urban social sustainability refers to "the city as a backdrop for lasting and meaningful social relations that meet the social needs of present and future generations" [25]. Social justice and equity, as well as social structure and organizations within the community, are highlighted as dimensions to assess urban social sustainability; and particularly social justice and equity are largely discussed as a reflection of urban planning outcomes [26–29].

The political economy approach explains the phenomena of the polarization of society and examines the inequality in planning as a result of income redistribution and political decisions [26, 28]. Under the market economy, it was assumed that the market could define justice and rationality automatically, and, guided by this economy, modern planning would lead to socially just outcomes [27].

During the social economic transition in the 1990s, a postmodernist approach re-conceptualized social justice and inequity [27, 29–31]. The postmodernist approach supported a socially inclusive process rather than solely focusing on "substance and material outcomes" [32]. To quote Young [30], "social justice must consider not only distributive patterns, but also the processes and relationships that produce and reproduce those patterns."

3.2 Social Capital and Its Adaptations

In an effort to strengthen the social dimension in sustainable development, Putnam [33] recommends that public policy makers pay more attention to the formation of social capital. Cuthill [34] also argues that "social capital provides a theoretical starting point for social sustainability." Accordingly, understanding social capital theory and the adaptations are essential in strengthening social sustainability in urban conservation.

Social capital is understood as resources and assets of relations [35, 36], network structures [36, 37], and trust and norms [38, 39]. Social capital as a public good will benefit individuals, group members, and the community, and lead to collective action and cooperation [36, 40]. Rather than a one-time feat, social capital is often produced in a process of repetition where the norms of reciprocity and trust emerge in light of obligation and expectation [38]. As individual trust is transformed into social trust, which resolves the dilemma of collective action under the logic of rational choice, transaction costs are reduced, and the efficiency of governance is also increased [38, 40].

In addition, social capital would have the potential to be converted into economic, cultural [35], and human capital [36]. Putnam et al. [38] identified two ideal forms of the social network structure: horizontal and vertical. The three forms of social capital are categorized as bonding, bridging, and linking social capital [33]. Bonding social capital consists of "the strong ties connecting family members, neighbors, close friends, and business associates" [41]. Bridging social capital can "bring together people across diverse social divisions" [42] and "generate broader identities and reciprocity" [33]. Linking social capital is formed by "vertical connections that connect individuals and groups with institutions" [43].

3.3 The Adaptation of Social Capital Theory

The research scope of social capital has extended from sociology and politics to many other fields, such as "attacking poverty" and inequality [41, 44], solving economic development dilemmas [44], community governance [38], institutions and organizations and policy making [45, 46], and so on.

In policy making, the dynamic combination and interaction between three social capital forms have been highlighted [39, 46]. Social capital highlights the process rather than the outcome [46], and its adaptation of policy helps achieve collective action and reciprocity, promoting dynamic relationships between networks. There exist potentials in adapting a social capital approach to urban conservation under a decision-making framework of path dependence. An observation of the different roles and combinations of social capital might reveal how decision-making was influenced and retargeted in the urban conservation process.

3.4 A Social Capital Approach to Urban Conservation

To conduct a social capital approach, several methods are employed. Firstly, a qualitative social capital survey is conducted in a Meso level. Groups, institutions, associations, and organizations are identified as nodes of communication and cooperation, which could produce different forms of social capital.

Secondly, social network analysis [47] in a Meso level will examine the complex interactions and the dynamic composition patterns of the identified social capitals in the different stages of conservation practices. Two parameters will be measured: overall density and degree of centrality. Density is "the percentage of all possible ties that are actually present in a network graph" [48, 49]. Based on social network theory, a high density of relationship ties positively indicates the implementation of cooperation and innovation [50, 51]. Degree of centrality means "the number of ties that every node has, or the number of organizations with which each stakeholder exchanged information or collaborated" [48, 49]. Degree of centrality can indicate the level of a node's communication and cooperation with other nodes.

Thirdly, the transformation process is revealed through a path dependency method [52]. According to Hoyem [53], the path dependency method enhances understanding in historic district planning. Through an inspection of the lock-in effect of the mainstream conservation approach, which is profit-led mass demolition and mass relocation, attention is paid to verify whether actions initiated through social capital in the transitional stage function as the breaking point or a transitional point of a stable path and reshape a new path of decision-making.

4 Case Study

4.1 Background of Urban Conservation in China

In China, studies on conservation of urban historic areas can be categorized in the areas of conservation theory [54], conservation policy and planning principle [55–57], urban renewal and urban design approach [58–60], and adaptive reuse of historic buildings [61–63]. In terms of conservation practice, the following approaches are usually adopted: the "spot-line-area" classified protection approach, the adaptive reuse approach, and the approach of strict preservation of selected buildings with mass redevelopment to the surrounding area. Regarding historic districts, most Chinese cities have adopted a top-down approach consisting of mass demolition and the relocation of historic neighborhoods with protected buildings. Nevertheless, it is noticeable that some cities have witnessed the emergence of alternative approaches within the last two decades. The following case study in the Xi'an Hui Fang is examined to understand the social capital-initiated approach in urban conservation.

4.2 *The Conservation Process of Xi'an Hui Fang-a Historic Quarter of Ethnic Hui*

Xi'an is a famous historic city located in northwestern China. The current townscape as a historic city is originated from Xi'an Fu during the Ming Dynasty in fourteenth century AD. The townscape consists of fourteenth century circumvallation with a moat, the Drum tower, the Bell tower, and several avenues.

Located in the city center, Xi'an Ethnic Hui Historic District covers 1.3 km^2 and accommodates over 60,000 residents, among which approximately 35,000 are Ethnic Hui people who are followers of Islamic religion. The history of this area as a district of Ethnic Hui People concentration can be traced back to Song Dynasty (960–1279) [64, 65]. Within this district, there are 11 mosques situated among 13 neighborhoods of traditional courtyard housing. A unique social-spatial unit "SI-FANG" (mosque-neighborhood) was formed gradually as Ethnic Hui people settled around these mosques, running some small-scale business alongside neighborhood market streets.

Since 1949, this district experienced self-construction, the courtyard subdivision, and small-scale replacements of work-units, which lead to serious physical degradation. In 1980, Beiyuanmen historic street within Hui Fang Historic District was designated as a protected area [66]. The conservation activities within this traditional Hui Fang historic District and its development path can be examined in three stages: the initial stage, the transitional stage, and the target stage.

4.2.1 The Initial Stage (1980–2000)

The initial stage witnessed the ending of the Cultural Revolution, the recovery of social and economic activities, and the beginning of reform and opening-up. It is under such socioeconomic background that urban conservation is first mentioned in the Master Plan 1980–2000 to enhance the historical features and townscape of Xi'an. In the Master Plan, Beiyuanmen Historic Street, a prominent historic street within Hui Fang, together with another traditional inner city street, were designated as the first two historic neighborhood streets. The Master Plan changed from the previous single physical spatial planning approach [66]. Due to the lock-in effect of previous policies under a planned economy, during the initial stage, conservation practices continued to adopt short-term single physical revitalization strategy with a top-down process. Aside from the awareness of tourism-led conservation, the initial stage also saw collaboration with overseas institutions and foreign funding as emerging trends. The initial stage of conservation is observed in the Xi'an Hui Fang Historic District Protection Project.

In 1997, the "Sino-Norwegian Cooperative Plan for the Protection of Xi'an Hui Fang Historical District" was issued by the central government of China. Subsequently, an official sub-district office, the Xi'an Hui Fang Historic District Protection Project Office, was established to execute the implementation of the XHHHDPP. During the implementation of this project, concentration was placed primarily on physical upgrading, the improvement of infrastructure and the environment, and the restoration of three traditional courtyard houses.

4.2.2 The Transitional Stage (2000–2008)

In 2002, the Xi'an Historic City Conservation Regulation (XHCCR) under Master Plan 1995–2010 was issued for implementation as law, in which the boundaries of the Beiyuanmen Historic District (BHD) were designated (Fig. 1).

The transitional stage saw deep social and economic transition. With the process of transitioning from a planned economy to a market economy, institutional change favored the market-led economy and overlooked the function of the planned economy. Along with the decentralization of power from central to the municipal government, real estate has been given priority in urban development. Although many historic quarters are located in central urban areas, redevelopment with mass demolition and relocation was widely adopted. Accordingly, this transitional stage witnessed the weakening of conservation regulation as local governments adjusted the scope of the pre-existing conservation plan to favor real estate. The redevelopment project on Sajinqiao Street within the Hui Fang historic district offers a reflection of this.

In 2004, Lianhu District (LD) initiated the Sajinqiao Redevelopment Plan (SRP). At the same time, the Xi'an City Planning Bureau licensed a land use permit (No. 2004-289) of 98 acres for a road-widening project on Damaishi Street and Sajinqiao Street, which would only require 20 acres. Seventy acres of spare land were planned for the SRP, where 1,900 houses, of which 90 % were ethnic Hui residences, would be demolished. The compensation price offered by LD was rather low compared to the market price. Both private developers and the Development Center of LD got involved.

According to the proposed plan, many ethnic Hui residents were to be relocated to suburban areas, separated from their traditional environment, religious space, and economic practices. Most of the residents who were refusing to move filed a petition to the Provincial Government. As a result, the first round of demolition was ceased. In April 2005 in the newly proposed Xi'an Master Plan 2004–2020 [67], the scope of the Beiyuanmen Historic District was adjusted by the Xi'an City Planning Bureau. Including both Sajinqiao Street and Damaishi Street, eight historic streets, which were previously designated in the BHD in XHCCR, were excluded (Fig. 2). In the wake of this adjustment and fearing the re-emergence of mass demolition acts, residents of Sajinqiao Street, along with religious leaders,

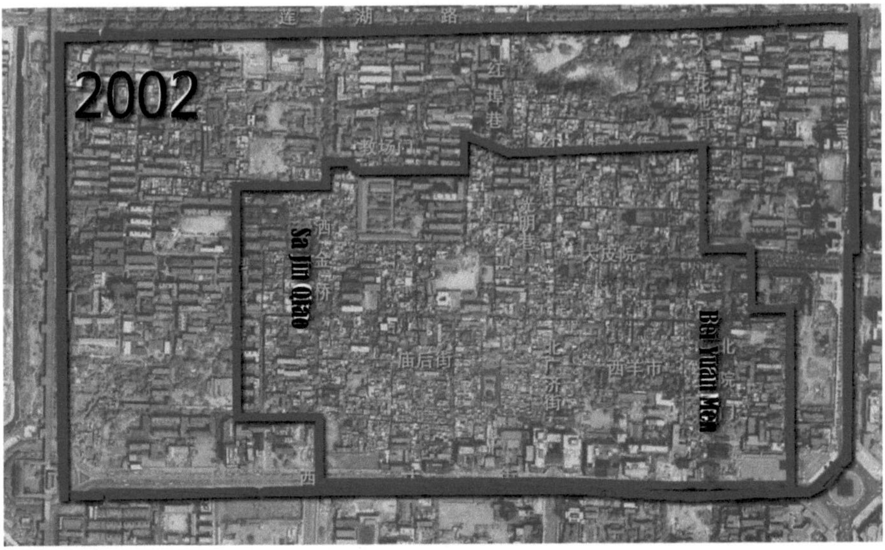

Fig. 1 Scope of BHD in Xi'an Master Plan (1980–2000) (Source: Xi'an Master Plan 1980–2000)

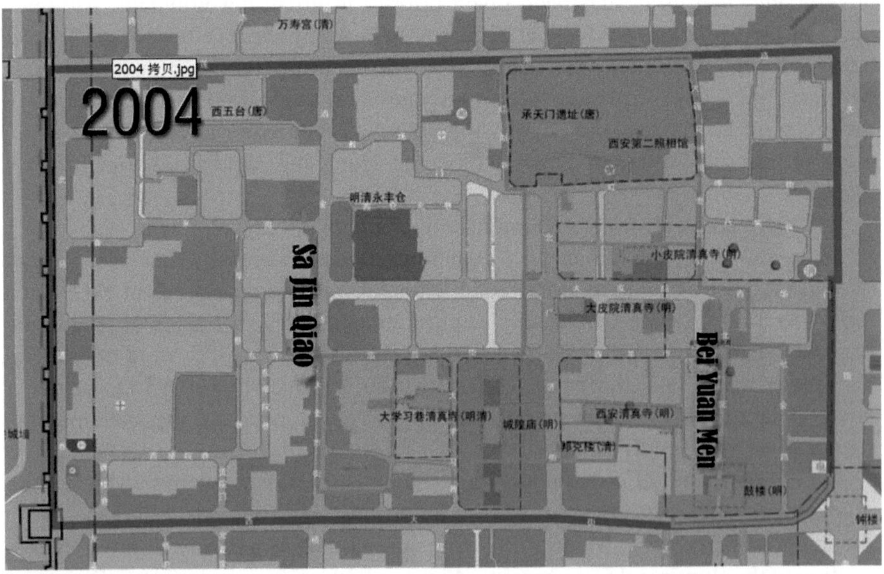

Fig. 2 Scope of BHD in proposed Xi'an Master Plan (2004–2020) (Source: Xi'an Master Plan (2004–2020))

gathered at the West Mosque where resident representatives were elected and 11 Mosques were united in preventing demolition.

The central government intervened and a second round of demolition was stopped in July 2006. The feedback from the Xi'an Planning Bureau in June 2006 asserted that before the Master Plan 2004–2020 was finally approved by the central government, all redevelopment regarding the scope and methods of the Hui Fang Historic District should strictly refer to XHCCR 2002. In this transitional stage, local residents and Mosques actively participated in the conservation of a place, strengthening the weakening conservation regulation, and enhancing their social, economic, and culture needs. This successfully challenged the profit-led redevelopment pattern and largely contributed to shaping new conservation planning for the following stage.

4.2.3 The Target Stage (2000–2008)

In June 2008, the new Xi'an Mater Plan 2008–2020 was approved by the central government. The municipal government assured that the social, economic, and physical needs of ethnic Hui residents would be given priority in future redevelopment plans. In the future, through an inclusive conservation planning approach, entire Hui Fang will adopt broader methods to manage changes, and, at the same time, ensure that the existing community will be socially, economically, and physically resilient in the long run.

5 Discussion

5.1 Main Findings of Social Network Analysis[1]

5.1.1 Identifying Three Forms of Social Capital

Social capital is identified in both the Xi'an Hui Fang Historic District Protection Project (1997–2002), and the Sajinqiao Redevelopment Project (2005–2007). Major organizations and groups, which were involved in the decision-making process of the above projects, are identified as relationship nodes. These nodes communicate or cooperate within or among certain scales, and are categorized by the different relationship ties they could keep within a community.

Table 1 identifies 19 organizations/groups as relationship nodes. The relationships ties among three specific nodes, namely Hua Jue Xiang Mosque, Residents of Renovated Courtyard Houses, and residents in general, are identified as bonding social capital ties within a community. Two nodes are identified as generating bridging social capital. Norwegian University of Science and Technology and Xi'an University of Architecture and Technology, as research institutes and professional consultancies, brought a horizontal network of resources to the community, as well as relationships from the outside of the community. The Hua Jue Xiang

[1] Social network analyses and visualization is achieved by UCINET Social Network Software.

Table 1 Nodes in XHHHDPP

Social capital	Nodes of communication and cooperation producing social capital to community	
Linking	1	Ministry of Science and Technology of China
	2	Department of Science and Technology Shaanxi Province
	3	Xi'an Municipal government
	4	Xi'an Municipal Construction Committee
	5	Xi'an Bureau of Housing and Land Administration 2nd Branch
	6	Xi'an Planning and Design Research Institute
	7	Xi'an Hui Fang Historic District Protection Project Office
	8	Norwegian Ministry of Foreign Affairs
	9	NORAD
	12	Beiyuanmen Street Office
	15	Xi'an Islamic Association
	16	Xi'an Ethnic Affairs Commission
	17	Construction Section
Bridging	10	Norwegian University of Science and Technology
	11	Xi'an University of Architecture and Technology
	13	Hua Jue Xiang Neighborhood Committee
Bonding	14	Hua Jue Xiang Mosque
	18	Residents of Renovated Courtyard Houses
	19	Neighborhood Residents

Neighborhood Committee also functioned in connecting the community with different relationship resources, both official and non-official. The other 13 nodes are mostly governmental organizations and official associations, which could vertically link the community to authorities and powers and be identified as linking social capital. Among these nodes, the Xi'an Hui Fang Historic District Protection Project Office is active in keeping a two-way relationship with other government organizations, such as Xi'an Municipal Construction Committee and the Xi'an Planning and Design Research Institute, through communication and collaboration.

Referring to Table 2, 21 organizations/groups are identified as nodes of relationship ties. The close relationship among three nodes, namely the Resident Representatives, Neighborhood Residents Volunteers, and Neighborhood Residents, are identified as bonding social capital since they are closely connected regarding the demolition issues within the Sajinqiao neighborhood. Five nodes, namely the West Mosque, the Old Mosque, Venerable Elders among the neighborhood, Residents in surrounding neighborhoods, and the 11 Mosques Union, are identified as bridging social capital by helping Sajinqiao Neighborhood connect with outside horizontal resources and relationships. The 11 Mosques Union and the West Mosque are especially active nodes in bridging the Sajinqiao Neighborhood with outside resources. Similar to Table 1, the other 13 nodes are mostly governmental organizations and official associations that could provide the Sajinqiao Neighborhood with vertical links to authorities and powers, and may also be

Table 2 Nodes in SRP

Social capital	Nodes of communication and cooperation producing social capital to community	
Linking	1	Central Government of China
	2	Shaanxi Provincial Government
	3	Shaanxi Provincial Ethnic Affairs Commission
	4	Xi'an Municipal Ethnic Affairs Commission
	5	Xi'an Municipal Islamic Association
	6	Xi'an City Planning Bureau
	7	Lianhu District Bureau of Construction
	8	Lianhu District Center of Development
	9	Lianhu District People's Procuratorate
	10	Lianhu District Public Security Bureau
	11	Lianhu District Bureau of Ethnic and Religious Affairs
	12	Miaohou Street Police Station
	13	Sajinqiao Neighborhood Committee
Bridging	14	The West Mosque
	15	The Old Mosque
	17	Venerable Elders among neighborhood
	20	Residents in surrounding neighborhoods
	21	The 11 Mosques Union
Bonding	16	Resident Representatives
	18	Neighborhood Residents Volunteers
	19	Neighborhood Residents

identified as generating linking social capital. In this case, the linking social capital actually found in the Sajinqiao Neighborhood range from the following four levels: the district level, the municipal level, the provincial level, and the central government level.

5.1.2 Network Analysis

Relationship ties of communication and cooperation regarding this project are recorded by the one-way or two-way patterns that are illustrated in the social network of XHHHDPP 1997–2002 (Fig. 3) and SRD 2005–2007 (Fig. 4). For example, Node A would like to work together or communicate with Node B through the exchange of information. However, Node B seldom responds or cooperates. This relationship would be identified as a one-way pattern. In contrast, if the communication between the two nodes is active and there is information shared between them, the relationship would then be identified as a two-way pattern.

Data shows that the overall density of the entire relationship network of XHHHDPP 1997–2002 (20.7 %) is lower than in the entire network of SRP

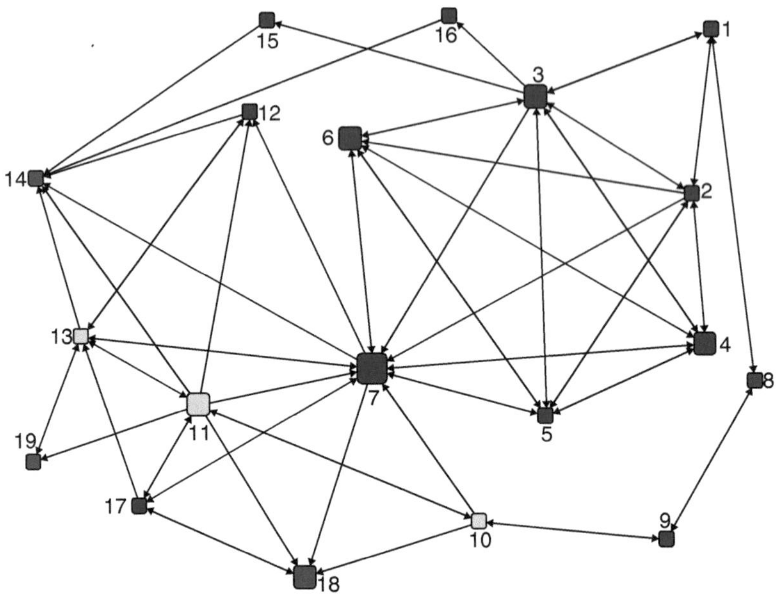

Fig. 3 Network analysis of XHHHDPP 1997–2002

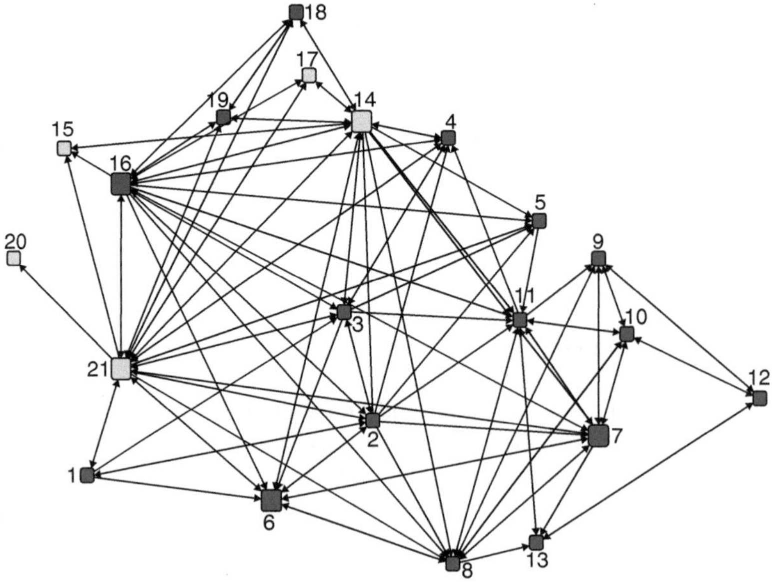

Fig. 4 Network analysis of SRP 2005–2007

Table 3 Density overall

Projects	Density overall	
	Density	Number of ties
XHHHDPP 1997–2002	0.2074	70
SRP 2005–2007	0.2786	117

Table 4 Degree centrality

Nodes	Degree	Nrm degree
	XHHHDPP 1997–2002	
Mean	4.737	26.316
Std Dev	2.53	14.059
Xi'an Hui Fang Historic District Protection Project Office	12	66
Xi'an University of Architecture and Technology	8	44
Neighborhood residents	2	11
	SRP 2005–2007	
Mean	7.143	35.714
Std Dev	4.062	20.312
The 11 Mosques Union	15	75
The West Mosque	14	70
Resident representatives	14	70

2005–2007 (27.9 %), and its ties of communication or cooperation (70) are fewer than that in SRP 2005–2007 (117) (Table 3). These results indicate that, compared to SRP 2005–2007, there is a lack of intensive communication and cooperation among the interest groups in XHHHDPP 1997–2002. Given that the density of communication and cooperation among linking social capital are considerably denser and some major linking social capital ranks top in degree centrality (12) (Table 4), it is assumed that a top-down pattern is dominant.

Compared to SRP 2005–2007, the weak degree of centrality of bonding social capital (2) indicates that local residents are not well informed and do not reveal a strong desire to participate. In SRP 2005–2007, a high ranking degree of centrality of both bonding (14) and linking social capital (15) indicates a better participation pattern, which is highlighted as a positive function of social capital to promote collective actions.

5.2 *Discussions*

The network analysis of the initial stage indicates a clearly top-down conservation pattern. It illustrates a dense connection within linking social capitals, such as the project office and other authorities. Information and decision-making regarding the protection project were shared and made within these groups, which has a negative effect on social inclusion. Linking social capital is found to have played an important role in keeping the project implemented through a top-down approach.

There is no direct connection between linking and bonding social capital, though the Huajuexiang Neighborhood committee functions as a bridge between them regarding the infrastructure improvement issues within the neighborhood. As the community has a low degree of centrality, they may not have the paths required to be well informed and make their needs known. The nodes between them are powerful, controlling information and other resources, which may lead to a potential rent seeking and trust crisis.

A one-way connection pattern between bridging and linking social capital may assert that suggestions from the two collaborative universities to the project office were largely ignored without feedback. Informal links between bridging and community bonding social capital have been identified. Xi'an University of Architecture and Technology, as the consultant university, volunteered to communicate with the foreign university, construction section, resident locals, and the neighborhood committee based on academic and professional interest. However, local residents were reluctant to communicate, as suggested by the one-way connection pattern.

The transitional stage represents a dynamic relationship network that is inclusive, accessible, open, and balanced, as shown by the degree of centrality. When community bonding social capital is strong and well connected with bridging social capital, as is the case with the West Mosque and the Mosque Union, it will function with active participation and negotiation. It is also assumed that the return to the top-down conservation approach will affect local people, who will in turn work as strong social forces to trigger a breaking point.

The implications highlight that the direct link and communication between nodes of linking, bonding, and bridging social capital should be created to mitigate the strong top-down conservation process. Public intervention could be achieved through the strengthening of direct connections between community bonding and higher-level linking social capital. Feedback from both the professional and local levels should aim to reformulate conservation methods or principles that were previously causing many negative effects. Another implication emphasizes that in conservation planning, attention should be paid to establishing a lively connection between bridging and community bonding social capital in order to generate cooperation.

6 Conclusion

This paper has explored the body of knowledge of existing conservation approaches and urges the academic interest towards a socially sustainable concern of urban conservation. Based on this research, a social capital-initiated urban conservation approach is proposed under the frame of path dependency, using a social network analysis to examines the roles of social capital in the decision-making process.

Through a pilot study of the Xi'an Hui Fang Historic District, this paper has revealed the stages of the dynamic combination of social capital and the interaction between different forms of social capital, highlighting the cooperation between the strong bonding and active bridging social capital. The analysis of the transitional stage has indicated that rather than continuing to adopt a mainstream conservation approach, the conservation principle has evolved to become more socially inclusive and feasible.

This paper concludes that, by promoting social cohesion and collective actions in the decision-making process, the social capital-initiated conservation approach is a feasible method to ensure the conservation of historic areas. In future studies, the cognitive nature of social capital, such as trust, respect, collective norms, and so on and so forth, together with the formation of social capital, will be examined, along with additional case studies, in order to provide another perspective to the future of urban conservation planning.

Acknowledgment Jie Han would like to acknowledge the support from Mr. Ma Yunhai.

References

1. Larkham PJ (1996) Conservation and the city. Routledge, London
2. Tiesdell S, Oc T, Heath T (1996) Revitalizing historic urban quarters. Architectural Press, Oxford
3. Hobson E (2004) Conservation and planning: changing values in policy and practice. Spon Press, London/New York
4. Rodwell D (2007) Conservation and sustainability in historic cities. Wiley-Blackwell, Oxford
5. Slater TR (1984) Preservation, conservation and planning in historic towns. Geogr J 150:322–334
6. Cozen MRG (1975) Geography and townscape conservation. In: Uhlig H, Lienau C (eds) Anglo-German symposium in applied geography: Giessen-Wurzburg-Munchen. Lenz, Giessen, pp 95–102
7. Whitehand J (1990) Townscape management: ideal and reality. In: Slater TR (ed) The built form of Western cities. Leicester University Press, Leicester, pp 370–393
8. Cohen N (1999) Urban conservation. MIT Press, Cambridge, MA, p 359
9. Whitehand JWR (1992) Recent advances in urban morphology. Urban Stud 29(3–4):619
10. Whitehand J et al (2011) Urban morphology and conservation in China. Cities 28:171–185
11. Pickard R (2001) Management of historic centres, vol 2. Taylor & Francis, London and New York
12. Ashworth GJ, Tunbridge JE (1990) The tourist-historic city. Belhaven Press, London
13. Barrett H (1993) Investigating townscape change and management in urban conservation areas: the importance of detailed monitoring of planned alterations. Town Plan Rev 64(4):435
14. Mageean A (1999) Assessing the impact of urban conservation policy and practice: the Chester experience 1955–96. Plan Perspect 14(1):69–97
15. Strange I (1997) Planning for change, conserving the past: towards sustainable development policy in historic cities? Cities 14(4):227–233
16. Creighton JL (2005) The public participation handbook: making better decisions through citizen involvement. Jossey-Bass, San Francisco

17. Salazar MP (2006) Promises and challenges of urban community-oriented conservation: the case of greenways in Detroit, Michigan State University, Michigan
18. Appleyard D (1979) The conservation of European cities. MIT Press, Cambridge, MA
19. Smith N (1979) Toward a theory of gentrification a back to the city movement by capital, not people. J Am Plann Assoc 45(4):538–548
20. Gottdiener M, Budd L (2005) Key concepts in urban studies. Sage Publications, London
21. Atkinson R (2000) The hidden costs of gentrification: displacement in central London. J Housing Built Environ 15(4):307–326
22. Mo T, Lu D (2000) Regeneration of Urban Form of Shanghai Lilong-Conservative Development of Xintiandi [J]. Time Archit 3:40–42
23. Shao Y, Zhang L, Dun M (2004) Conservation and social development of Ancient Town of Lijiang [Shijie wenhua yichan lijiang gucheng de baohu he shehui fazhan]. Ideal Sp 6:52–55
24. Ruan YS, Yuan F (2008) Conservation and development of the historic water towns in the south of Yangtze river [J]. Urban Plann Forum 5:52–59
25. Yiftachel O, Hedgcock D (1993) Urban social sustainability: the planning of an Australian city. Cities 10(2):139–157
26. Harvey D (1973) Social justice and the city. Hodder Arnold, London
27. Harvey D (1992) Social justice, postmodernism and the city. Int J Urban Reg Res 16(4):588–601
28. Fainstein S (1997) Justice, politics and the creation of urban space. In: Merrifield A, Swyngedouw E (eds) The urbanization of injustice. New York University Press, New York, pp 18–44
29. Fainstein S (2005) Planning theory and the city. J Plan Educ Res 25(2):121–130
30. Young IM (1990) Justice and the politics of difference. Princeton University Press, Princeton
31. Sandercock L (1998) Framing insurgent historiographies for planning. In: Making the invisible visible: a multicultural planning history. University of California Press, Berkeley, pp 1–33
32. Cardoso R, Breda I (2007) Vazquez, social justice as a guide to planning theory and practice: analyzing the Portuguese planning system. Int J Urban Reg Res 31(2):384–400
33. Putnam R (2000) Bowling alone: the collapse and revival of American community. Simon Schuster, Nova Iorque
34. Cuthill M (2010) Strengthening the 'social' in sustainable development: developing a conceptual framework for social sustainability in a rapid urban growth region in Australia. Sustain Dev 18(6):362–373
35. Bourdieu P (1986) The forms of capital. In: Handbook of theory and research for the sociology ofeducation. Greenwood Press, Westport, CT, pp 242–258
36. Coleman JS (1994) Foundations of social theory. Belknap Press, Cambridge, MA
37. Lin N (2002) Social capital: a theory of social structure and action, vol 19. Cambridge University Press, Cambridge, UK
38. Putnam RD, Leonardi R, Nanetti R (1993) Making democracy work: civic traditions in modern Italy. Princeton University Press, Princeton, pp xv, 258
39. Fukuyama F, I. Institute (2000) Social capital and civil society. International Monetary Fund, IMF Institute, Washington, DC
40. Pena MVJ, Lindo-Fuentes H, W.B.C.A.C.M. Unit (1998) Community organization, values and social capital in Panama. Central America Country Management Unit, Latin America and Caribbean Region, the World Bank
41. World Bank (2000) World development report, 2000/2001: attacking poverty. Oxford University Press, New York
42. Scott J (1991) Social network analysis: a handbook. Sage, London, p 210
43. Babaei H, Ahmad N, Gill SS (2012) Bonding, bridging, and linking social capital and psychological empowerment among squatter settlements in Tehran, Iran. World Appl Sci J 17:119–126
44. Woolcock M (1998) Social capital and economic development: toward a theoretical synthesis and policy framework. Theory Soc 27(2):151–208

45. Cohen D, Prusak L (2001) In good company: how social capital makes organizations work. Harvard Business Press, Boston
46. Woolcock M (2001) The place of social capital in understanding social and economic outcomes. Can J Policy Res 2(1):11–17
47. Franke S (2005) Measurement of social capital: reference document for public policy research, development, and evaluation. Policy Research Initiative, Ottawa
48. Kazmierczak A (2012) Working together? Inter-organisational cooperation on climate change adaptation. Working report for EcoCities project, The University of Manchester, Manchester, UK
49. Borgatti SP, Everett MG, Freeman LC (2002) Ucinet for windows: software for social network analysis. Analytic Technologies, Harvard
50. Jütte W (2007) Co-operation, networks and learning regions. In: Social capital, lifelong learning and the management of place: an international perspective, Routledge, New York, p 95
51. Scott J (2000) Social network analysis: a handbook, 2nd edn. Sage, London, p 208
52. Magnusson L, Ottosson J (2009) The evolution of path dependence. Edward Elgar Publishing, Cheltenham
53. Hoyem H (2004) Path dependency – a useful method for planning and design of historical district? Community 2:75–81
54. Ruan Y, Lin L (2003) Authenticity in relation to the conservation of cultural heritage [J]. J Tongji Univ Soc Sci Sect 2:1–5
55. Su X (2010) Urban conservation in Lijiang, China: power structure and funding systems. Cities 27(3):164–171
56. Zhang S (1999) The objective and method of conservation of historic towns. Urban Plan 7:50–53
57. Ruan Y, Meng S (2001) The study on some issues related to the conservation and planning for the historic streets and areas in China [J]. City Plan Rev 10:25–32
58. Lu J (1997) Beijing's old and dilapidated housing renewal. Cities 14(2):59–69
59. Wu L (1991) Rehabilitation in Beijing. Habitat Int 15(3):51–66
60. Shin HB (2010) Urban conservation and revalorisation of dilapidated historic quarters: the case of Nanluoguxiang in Beijing. Cities 27:S43–S54
61. Wang J (2009) 'Art in capital': shaping distinctiveness in a culture-led urban regeneration project in Red Town, Shanghai. Cities 26(6):318–330
62. Ruan Y, Zhang S (2004) The promotion of industry heritage conservation on cultural business. Urban Planning Forum
63. Chang Q (2003) A conservation strategy of architectural heritage. Tongji University Press, Shanghai
64. Bi J (2004) Reuse of the courtyard homes protected as the heritage in Drum Tower district of Xi'an. Xian University of Architecture and Technology, Xi'an
65. Chen S (2005) Study on conservation of the dwelling culture and living environment of Xi'an Muslim concentrated area. Xian University of Architecture and Technology, Xi'an
66. XCPB (1980) Xi'an Master Plan 1980–2000. Xi'an Municipal Government, Xi'an
67. XCPB (2004) Xi'an Master Plan 2004–2020. Xi'an Municipal Government, Xi'an

The Strategies of Open Space Planning Under the Goal of Building Resilient City

Hu Yuwei and Huang Jianyun

1 Introduction

In the Worldwide, urban population is currently increasing by 2 %. In 2008, half of the world population lives in cities the first time in human history. By 2050, 70 % of population will be urban dwellers, and the future urban population growth will occur in developing countries. China, as the world most populous country, is in a stage of rapid growth of urbanization. We can foresee it will face tremendous pressure and a variety of issues. Suffered 2008 snowstorm, Wenchuang earthquake and 2012 Beijing floods, China shows the vulnerability facing external shocks. As a new concept of development, the own internal mechanism of Resilient city can adapt to changes, especially in response to climate change. Open space, as the main carrier for improving the quality of urban ecology and disaster prevention, the ecological and regulatory functions have important significance in building resilient city.

2 The Concept and Design Principles of Resilient City

2.1 The Concept of Resilient City

The definition of "resilient" in "CiHai" is the character that the object can completely restore after removal of external forces.

H. Yuwei (✉) • H. Jianyun
The Department of Architecture of School of Naval Architecture, Ocean & Civil Engineering of SJTU, Shanghai 200240, China
e-mail: City20136@163.com; huangjianyun2004@163.com

X. Chen, Q. Pan (eds.), *Building Resilient Cities in China: The Nexus between Planning and Science*, GeoJournal Library 113, DOI 10.1007/978-3-319-14145-9_8

In science, "resilient" is derived from ecology. It means that the system can effectively resist outside interferences, and restore the balance, including the ability to withstand functional changes, organize, learn and adapt [1].

Accordingly, this article defines resilient City: the adaptive capacity to external shocks. When natural or man-made disasters happen, cities can make quick response, saving the basic functions of each system in the maximal degree, and restore in a short period.

2.2 *The Design Principles of Resilient City*

Resilient city was originally proposed in response to oil shortages and climate change. With social and economic development as well as interdisciplinary collaboration, the contents of resilient city have expanded. In general, main design principles as follows:

- The Reduction of carbon emissions: It should reduce CO_2 and other greenhouse gas emissions .use fossil fuel less and improve energy efficiency and use renewable energy sources.
- Biological and social diversity: the greater degree of diversity of the city, the higher ability to resist external forces and recover quickly.
- Moderate system redundancy [2]: When natural disasters or terrorist attacks destroy one of city system, the standby system can work. These systems which have influenced people to survive include electricity, water, communications, food supplies and transportation systems.
- Local self-sufficiency [2]: cities and communities need to create sustainable products and service. When disasters happen, specific areas can ensure food, fuel, water and other daily necessities and services.
- Adaptive design: The Planning must take into account of local circumstances, and make targeted designs that can adapt abnormal states of the city in a certain extent.

3 The Definition and Role of Open Space

3.1 *The Definition of Open Space*

According to the "open space law" of London, United Kingdom in 1877, open space is any enclosed or non-enclosed site with no buildings or less than 1/10 of the land with buildings, the remaining land for parks or entertainment or not being utilized areas [3].

In our country for open space study, Dexi Shen, Guoping Xiong (1996), Qi Yu (1998) and Fazeng Wang (2004) [4, 5] all believe that urban space is outside the building entity which contains the garden vegetation, rivers and lakes, squares, roads and other socio-economic functions with some artificial ground in the built-up areas, also including suburban forest land, woodland, lakes and other waters.

Integrated the definition of open space at home and abroad, open space mainly is public space existed outside the building entity in the urban built-up area from the perspective of this article. It includes green space (parks, open green space, producing green space, farmland, etc.), water (rivers and lakes), roads and squares. Those that can actually improve urban ecological environment and provide shelters for human would be the important contents under the goal of building resilient city.

3.2 *The Roles of Open Space*

As the "gray and negative space" corresponding to the physical buildings, open space is especially important in the modern city of high density. In general, the effects as follows:

1. Improve the Urban Ecological Environment and Reduce the Frequency of Disasters

Since the industrial revolution, with human ability to transform nature greatly enhanced, the contradiction between human and natural environment has become increasingly prominent. Through building the systematic open space, let external natural environment in different ways and forms protrude into the city, establish the contact between natural systems and urban man-made systems, which can improve urban ecological environment and greatly reduce the frequency of disasters.

2. Provide Disaster Shelters and Enhance Disaster Prevention

Our country is one of widest range of natural disasters in the world including weather disasters, floods and geological disasters, earthquakes and forest fires. Open space not only can provide emergency shelters, but also prevent and reduce secondary disasters to a certain extent, such as fire plants that can delay the spread of fire.

3. Maintenance of Natural Ecological Balance and Prevent Urban Sprawl

Open space can protect ecological boundaries between the city and the nature, and control the growth helping to prevent urban sprawl and excessive erosion of the natural environment, therefore, preserve habitats inside and outside the city.

4. Create Good Urban Landscape and Show Local Characteristic Culture

Open space provides outdoor public places, meet people demand of leisure and entertainment, and enhance the quality of life. At the same time, it is the place displaying the city image, creating a characteristic urban landscape through the integration of local culture.

4 Strategies of Open Space Under the Goal of Building Resilient City

4.1 Coordination Other Planning with Open Space Planning

Open space planning mainly is green space system planning in our country, which has not legal status. As the composition of urban design, it also has no legal effect. Urban disaster prevention and mitigation planning is the special planning of city master planning, but open space planning has not combined very well with disaster prevention and mitigation planning. First, the categories of open space and disaster sites do not match well. Second, the number and size requirements of refuge open space are not clear and operated difficultly. At the same time, both disaster location and open space should be associated closely with specific urban population size and population density in different regions, only in this way can fully play their roles.

4.2 Build a Network of Open Space System

Open space planning mainly is green space system planning in our country, but as a full sense of the open space system should include at least three kinds of lands: green space, square and road land and water. Meanwhile according to different scale, scope and objects, open space system should divide into four levels: city level, urban level, community level and neighborhood level. For building resilient city, first, the planning should clear each level contents, then connect different levels of open space together through the corridor space (green corridors, river corridors, walking trails), finally combined with anti-disaster mitigation requirements, the relevant contents will be extracted from the three lands, and build up the open space network system.

4.3 Maintain the Integrity and Continuity of Mountains and Water Pattern

In ancient China, our ancestors made the idea of "TianRen He Yi" which "Tian" means natural and "Ren" is human beings. This idea emphasizes that human beings are simply one part of the universe. While the city is as a plaque of regional landscape matrix, we must safeguard the integrity and continuity of mountains and water pattern, and build corridors and habitats with the ecological environment through the "habitat islands" in order to maintain species diversity and stable fauna and flora, thereby enhancing ecological functions and maintaining ecological balance.

4.3.1 Prevent the Privatization of Mountains

Because of urban sprawl and the construction of large hilltop houses driven by economic benefits, the mountain is arbitrarily cut and the wetland system under the mountain has disappeared, leading to more natural disasters. So preventing the privatization of mountains is the basic measure to maintain the integrity and continuity of mountains and water pattern.

4.3.2 Strengthen the Continuity of Water Pattern

Due to large-scale construction and industrial pollution, a large number of rivers are buried, and wetlands have disappeared, resulting to the fracture of river corridor in many cities. To defend against floods, human beings builds dams arbitrarily, further disturbing the ecological balance, thereby causing more floods. In order to rebuild the continuity of water pattern, it should restore the river natural form and the buried rivers, exert the river corridor function connecting wetlands, reservoirs, lakes together, and restore the ecological water cycle.

4.4 Build a Diversified, Multi-Level Green Space System

4.4.1 Contents Expansion

Let farmland into the city: large farmland outside the city is an important regional ecosystem matrix, which not only is an important food source of the city, and also provides a variety of wildlife habitat areas and living spaces. Farmland biodiversity and agriculture production activities are conducive to maintaining the diversity of the urban system, which is in keeping with design principles of resilient city. In order to improve the city resilience, farmland should not be confined the distribution outside the city, and it can extend into cities. First of all, farmland can be the Isolated green between city clusters. Second, an urban farm or urban agricultural areas can be set up in the city-level parks. And a number of fields can also be arranged in the large residential areas for planting corn or local edible crops.

Monuments and heritage sites as far as possible included: Most of China cultural relics protection units are wood, which have a good seismic performance and have covered a large of land with low density, so they are good choices for shelters. So let these protection units be included in the park planning as far as possible. It will protect the cultural relics and surroundings. In certain times, these units can be emergency evacuation places.

Open dedicated green spaces, such as individual units or executive green spaces, and form a diversified multi-level green space system.

Table 1 Cities that urban green coverage is more than 40 % in China

City name	Green area (Ha)	Park green area (Ha)	Per capita green area (square meters)	Green coverage area of built-up area (ha)	Green coverage of built-up area (%)
Beijing	62,672	19,020	53	65,348	55.10
Shanghai	120,148	16,053	89	38,105	44.00
Guangzhou	124,420	9,971	187	38,224	40.15
Shenzhen	96,368	16,987	371	37,384	45.04
Nanjing	77,087	6,773	141	27,456	44.36
Dalian	18,153	3,510	60	17,618	45.17
Nanning	37,125	2,149	137	8,687	40.04

Source: 2010 statistical yearbook of China

4.4.2 Upgrade Indicators and Optimize the Structure

For a long time, the traditional indicators of measuring the quality of urban green space system in our country are: green rate, green coverage, per capita green area and per capita park green area people. Although these indicators can reflect the overall condition of urban green space, they are two-dimensional indicators, so cannot reflect actual results comprehensively and objectively (Table 1).

As we can see, even the city green coverage and per capita green area index are high, we cannot say that the city has a good environment quality, and Beijing is a typical example. According to the domestic effects of urban green space analysis, only when the green rate is higher than 40 %, the city may have good environment quality, and when the green coverage rate is less than 40–60 %, the internal structure and spatial distribution of green space show its importance. First, the spatial distribution should be balanced, and consistent with the region population density. Meanwhile green space system should emphasis on primary biochemical composition and localization, and actively nurture and plant native trees rather than rare or expensive trees blindly. The proportion of green space system that can really provide emergency shelters in the disasters also serves as an important indicator.

4.5 Build the Pedestrian System of Combined "Pedestrian and Green"

As one of the basic human activities, walking should be given more importance under the goal of building resilient city. On the one hand, walking can effectively reduce the dependence on cars and car congestion, thereby reducing greenhouse gas emissions. On the other hand, when the city suffered natural or man-made disasters, residents generally take a walk way to shelters. It may face greater risks by car especially the city roads suffer varying degrees of damage. Meanwhile small

Table 2 List of common disaster plants

Disaster prevention	Plant features	Examples of plant
Fireproof	High shelter high-lighted	Ginkgo, coral tree, privet Pittosporum, Schima,
Shockproof	Dense canopy, climbing	Ginkgo, Pistaciasuperba, ivy
Windproof	Pyramid or curvedcrown, fast growth	Ginkgo biloba, coral tree, acacia, poplar,
Flood control, Mudslides	Dense canopy, interception of rainfall	Metasequoia, oleander, spruce, willow
Prevent the spread of germs	Secrete substances kill bacteria	Osmanthus, birch, Pistacia, rose

Source: www.daidu.com

shelters generally do not have a car parking, more cars will reduce the effective area, and are more likely to bring some security risks to evacuating people.

"Pedestrian-oriented" means that we should fully consider the functions scale and related facilities needs of walking, and establish a continuous pedestrian system. "Green-oriented" is the emphasis of "Green" on the basis of a good combination with parks, city squares, waterfront open space and other city open space by planting disaster prevention trees, and create an attractive and safe environment for walking. The pedestrian system will bring main living areas important offices as well as neighboring large shelters together through the "Green channel" and then connect urban roads system seamlessly through urban public transport system and important pedestrian nodes (Table 2).

4.6 *Improve Public Participation in Open Space Planning*

As public good for all people to share, urban open space system is with distinct social characteristics, not just the engineering problems of urban planning and construction, but a democratic decision-making and management issues. Public participation in urban open space system planning will greatly expand the public awareness of open space system, thus improve people awareness of ecological protection and disaster emergency shelters. In order to improve public participation in open space planning, information openness and public participation in legal proceedings must be implemented. If people can effectively deal with external interference, it will reduce losses and improve city recovery, which is the fundamental goal of building resilient city.

5 Urban Green Space System Planning and Water Planning Practices of Guigang

5.1 City Introduction

Located in the southeast of Guangxi Autonomous Region, middle of Xi River and Xun Yu plain, Guigang has the largest port in South River, another name for He city. East of Wuzhou, south of Yulin and Qinzhou, west of Nanning, north of Laibing, it has a good location and long history. The administrative area of Guigang is 10,600 km^2, which the urban area is 153 km^2. In 2010 the total population is 5.03 million, which the urban population is 400,000.

Generally speaking, the ecological conditions of Guigang is very favorable, while the city forest coverage rate reach 44.69 %. Due to lacking of systematic planning in the past, it has not yet formed a complete system of urban open space. To enhance the city resilience, urban open space needs to be further optimized.

5.2 Disaster Prevention and Mitigation Planning

The disaster prevention and mitigation planning of Guigang include disaster planning, flood control planning, fire planning and civil defense planning. The urban construction of seismic fortification intensity is 6°, and planned city parks, plazas, parking and school playgrounds is the major earthquake evacuation sites. Urban flood control standard has recently reached 50 years, forward 100 years. Drainage standard has recently reached in 20 years, forward 50 years.

In the disaster prevention and mitigation planning, it is stipulated to the flood control gates, pumping stations, fire stations, major evacuation routes and evacuation sites, in which the main evacuation routes for the north and south: Yingbin Road, Xianyi Road, Zhongshan Road, Jiefang Road, Tongji Road; Chengbei Road, Bushan Road, Golden Port Road, Jiangnan Road, Chengnan Road, major evacuation sites: north central Park, national parks, Liyu river wetland park, Nanshan scenic park, Duchon river park, etc. But it did not quantify these infrastructures, nor build a hierarchical disaster prevention system (Fig. 1).

5.3 Green Space System

5.3.1 The Mountains and Water Pattern

Urban green space system should be from a wider range of landscape pattern to consider. Strengthen and safeguard the integrity and continuity of regional mountains and water pattern, and build the structure of the overall green space (Fig. 2).

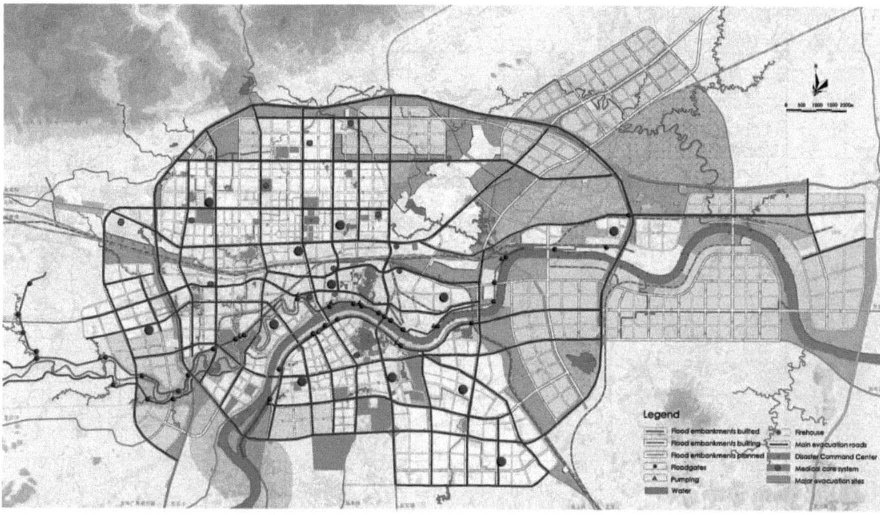

Fig. 1 The comprehensive disaster planning of Guigang center city (Source: Master planning of Guifang (2008–2030))

In this planning, we first studied the regional mountains and water pattern. Pingtian Mountain, Dong Mountain, Gui Mountain and a series of mountains are in the northern of Guigang. Yu River with east–west and Qian River with north–south have a intersection in Xi Mountain of Guiping. Yu River traverses the city, and becomes the most important water of the area. Then we analyzed the relationship between prevailing wind direction and peripheral urban oxygen base through enlarging the wind rose for several times and placing on Guigang maps. Combined annual maximum air northward direction, the city has obtained the best position of the oxygen base. By setting continual pergola as air duct, the good air will be into the city, which is also the biological migration channel (Fig. 3).

Finally, according to master planning and the green status of Guigang, we have obtained the structure of green space system for "eight green wedges, a green ring, five horizontal and vertical green roads, five central parks"(Table 3).

5.3.2 System Components

In accordance with "urban green space classification standard" (CJJ/T85-2002), the central city green space will be divided into five categories: G1 park green space, G2 production green space, G3 protection green space, G4 attached green space, G5 other green space. In this planning, green rate will reach 38 %, green coverage rate of 45.98 %, and per capita park green area of 11.74 square meters / person in 2030 (Fig. 4).

Green nurseries in the production green space are mainly for protecting and nurturing native trees and a variety of plants with disaster prevention (Table 4).

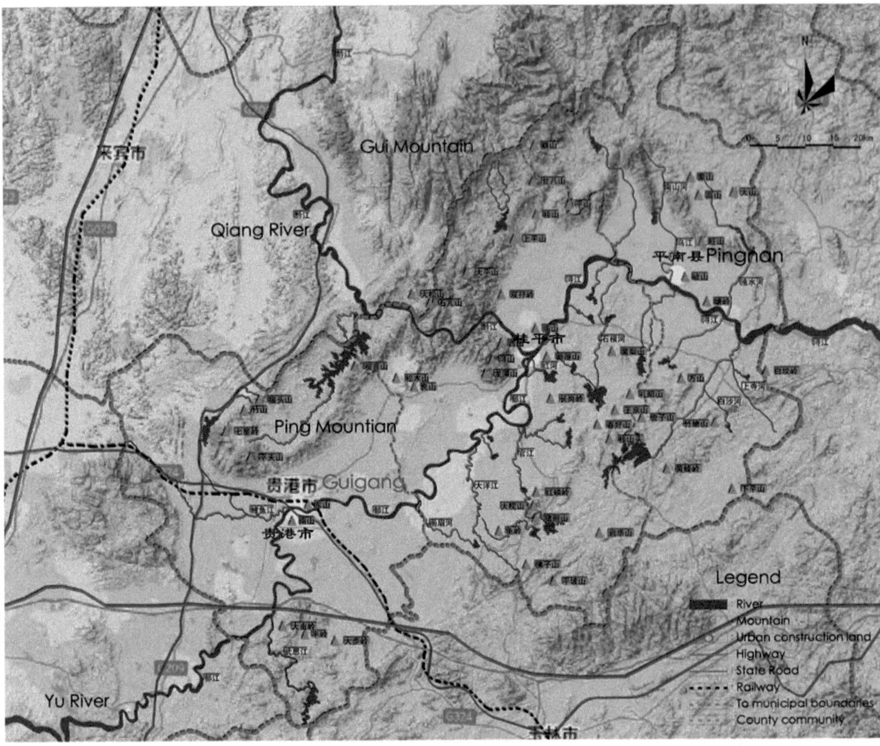

Fig. 2 The mountains and water pattern (Source: Green space system planning of Guigang (2011–2030))

In view of a number of cultural relics protection units of Guigang, let these units be included in planning park. Set a riverside belt around the South River Terminal ancient ruins, which can share part of the recreational function. For Nan mountain Temple Cliff, set the city scenic areas.

5.3.3 Biodiversity Conservation

There are three measures on biodiversity protection by protecting rare and endangered plants, the typical vegetation and enriching plant landscape. Strengthen the protection of the Nan mountain scenic area, Dong mountain Scenic area and MAOER mountain country area. While relying on rare plants and abundant natural plant resources of Liyu River and Niupi River, establish Liyu river breeding center for the foundation for biodiversity conservation.

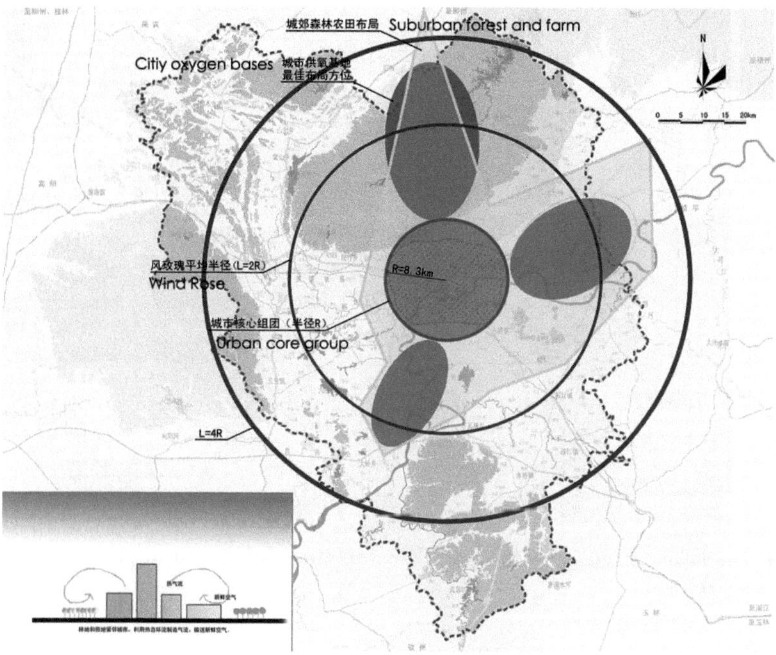

Fig. 3 The relationship between wind direction and urban oxygen base (Source: Green space system planning of Guigang (2011–2030))

Table 3 The constitutes of green space system

Category	Name	Contents
Park green space	Comprehensive park	East lake park, LIyu river park, National park
	Topic park	Nan mountain park, Jiangnan gardens
	Linear park	Riverside parks
	Roadside greenery	Land along the river, lake, street, road
Production green space	Green nurseries	Liyu river nursery, DuChongjiang nursery
Protection green space	Health isolation belt	Industrial land belts
	Roads protective green	Protective green belt on road, rail and sides
Attached green space	Dedicated green	Each unit or administration green
Other	Scenic areas and woodlands	East lake, Nan mountain scenic area
	Source protection zones	Pinglong reservoir, Shiniu reservoir
	Country park	Three country parks

Source: Green space system planning of Guigang (2011–2030)

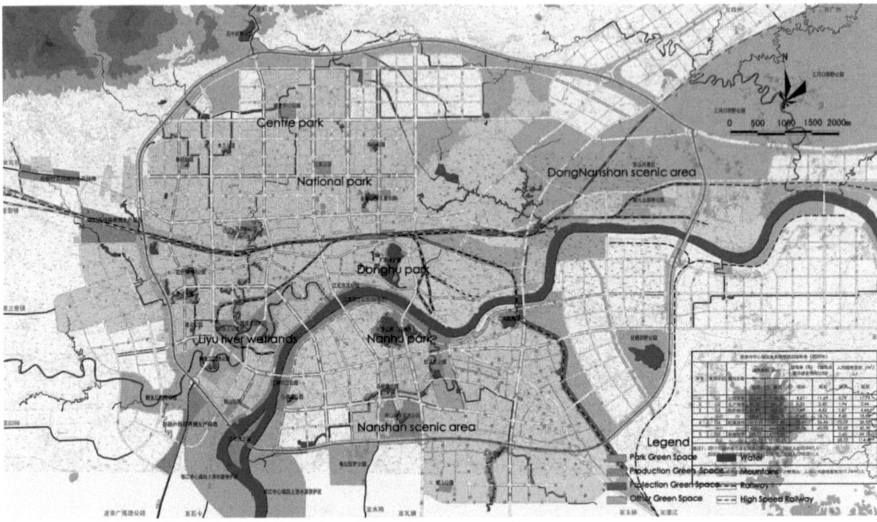

Fig. 4 The layout of green space system (Source: Green space system planning of Guigang (2011–2030))

Table 4 List of native species of Guigang

Category	Name
Palm branch	Dong brown, Palmetto, fish Kwai, Chrysalidocarpus
Arbor	Almond, water stone Rong, eucalyptus, Fatsia flowers Ficus, cycads, banana, flame flower, Bauhinia, magnolia, magnolia,
Shrubs	Excoecaria, Crinum, wood duck feet, big red, gold leaves, Ixora, Fujian tea, red wood, excelsa, mosaic galangal, Greek beauty,
Disaster plants	Fatsia, magnolia, cycads, Araucaria, lotus, coral tree

Source: Green space system planning of Guigang (2011–2030)

5.4 *Water System*

Combined with planned roads and urban water backbone, it has formed the skeleton of "five lakes five line" "Five lakes" means Yangshan Lake, East Lake, South Lake, West Lake and North Lake, while "five lines" means Liyu River, Sha River, Macao River, Du Chong River and Yu River. The water system has connected the wetlands, reservoirs and lakes together to forming a "big connection" (Figs. 5 and 6).

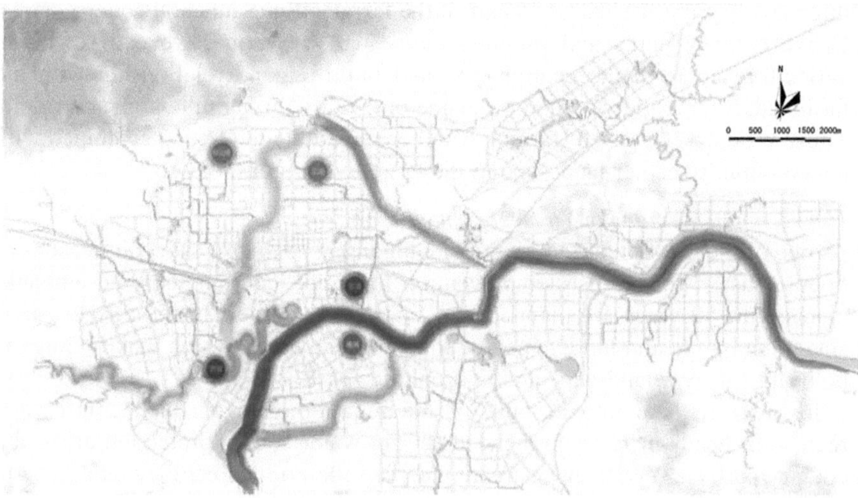

Fig. 5 The structure of water system planning of Guigang

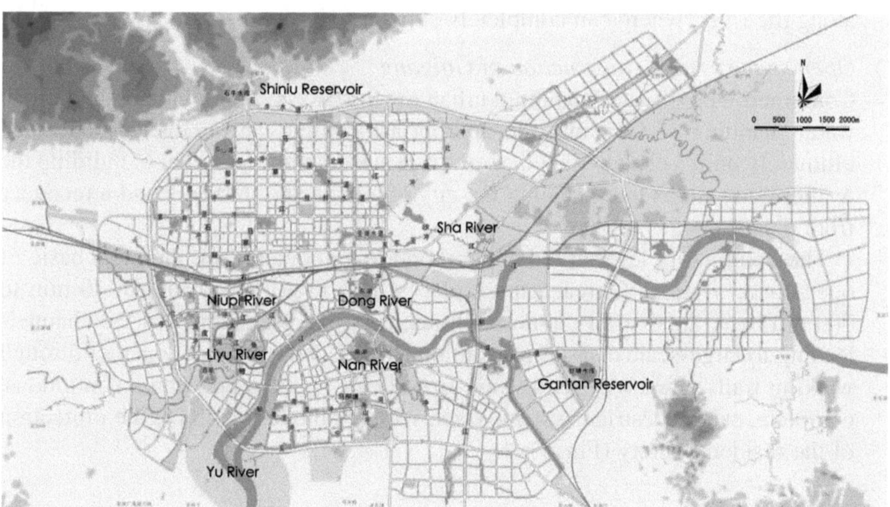

Fig. 6 Water planning of Guigang (Source: Water system planning of Guigang (2011–2030))

5.5 Open Space System Construction of Guigang Under the Goal of Building Resilient City

1. *The recommendations of green space system.*
   Giving full roles on the protection of farmland biodiversity and urban food sources, let the farm extends into the city. Specifically, it can be divided to

three levels. First, we can set a part of the farmland in Sanhekou country parks, Ganye country parks and Shiniu country parks. Second, we can set urban agriculture areas in the central city, and build a good relationship with the farmland, forests and woodlands outside of the city through the green corridors. Third, we can set a part of farmland and create edible landscape combined with the community center landscape in the large living areas.

2. *The recommendations of Yu river remediation.*
   As one of the most important rivers of Guigang, Yu River is the main reasons of flood disaster. In the disaster prevention planning, continuous flood embankments with 50 years are planned two sides of Yu River. It not only causes a single landscape and hinders people touching the water, which makes Guigang no beautiful river landscape and hydrophilic space.
   In order to take into account of the requirements for flood control and recreation needs, it is recommended setting water source protection areas and country parks in Yu River upstream, keeping the original ecology landscape to develop urban tourism. Combined the Liyu River Wetland Park, strengthen the protection and use of the river shoal by setting flood embankments partly. We can plan the local open space for activities, such as courts, waterfront plazas along the river, where can completely submerge in flood season.
3. *Open space system construction of Guigang*
   Combined with disaster planning, urban green space system planning and urban water planning of Guigang, the appropriate contents are extracted, and have ultimately built a closely linked open space system under the goal of building the resilient city, so as to enhancing the city ability facing disasters and a recovery from disasters.
   The community level and neighborhood level contents are the most basic of open space system. To reach these places, it is usually by walk. In 10 min to 700 m range, the district public facilities, emergency shelters and the channels leading to large-scale disaster prevention bases have connected together through walking trails, greenways and water-front belts, so as to building a continuous, complete, safe pedestrian system which will be able to maximize the protection of the residents safety (Fig. 7).

6 Concluding Remarks

Traditional open space planning is mainly green space system planning, by building a variety of green space in order to improve the urban ecological quality and enhance the urban landscape. But the open space system planning should change from reducing outside interferences to the response to external shocks, and focus that how to deal with the city damages and quickly recover from the shocks when the city suffer natural or man-made disasters. And then the open space planning must strengthen the coordination with other planning, especially with the master

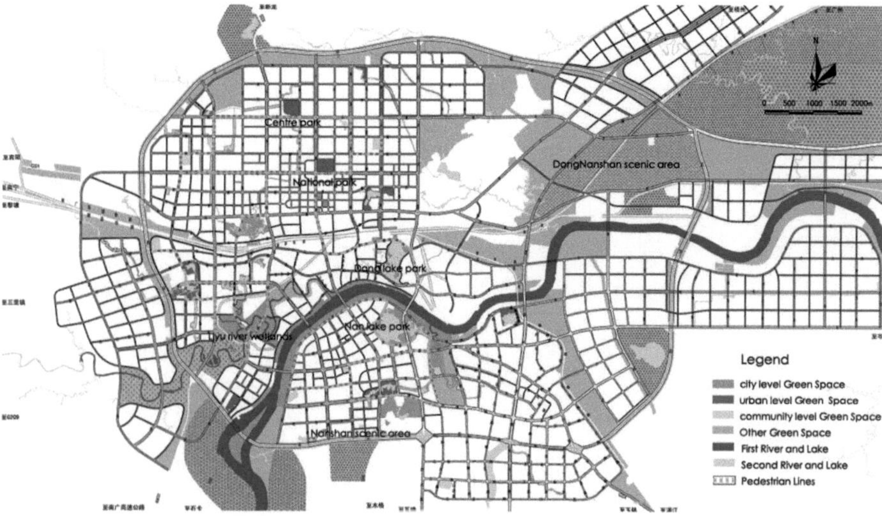

Fig. 7 Open space planning of Guigang (Source: Author made)

urban planning and disaster mitigation planning, so as to guiding urban development and construction. In addition, it must also improve public participation in open space planning system to allowing the public become the city largest think tank, not only can improve the science of the planning, but also provide a strong base for the implementation of relevant policies and measures.

References

1. Carpenter S, Walker B (2001) From metaphor to measurement: resilience of what to what [J]. Ecosystems 4:765–781
2. Hua G, Ren G (2012) Research on the multifunction of urban agriculture under the goal of building resilient city [J]. J Eng Stud 4(1):49–56
3. Bo Xu (2007) Urban disaster prevention planning study [D]. Tongji University master degree thesis, Shanghai
4. Qi Yu (1998) Urban open space system construction [J]. Urban Plann 6:50–55
5. Fazeng Wang (2005) Urban open space system optimization [J]. Geography 2:1–8

A People's Atlas of Muncie: Citizen Representations of Urban Space

Junfeng Jiao, Steven M. Radil, Jenna Harbin, and Yuan Li

1 Introduction

Geographic Information Systems (GIS) are an essential tool for analyzing and representing spatial information. An emerging frontier in GIS deals with integrating qualitative data with GIS and other related geospatial technologies [1]. Situated within a growing body of mixed methods techniques, this emerging field of Qualitative GIS (QGIS) examines how GIS can be used to create new types of representations that incorporate multiple meanings and how these representations can be used to interrogate discussions of social action, community change, and power [1, 2].

Mental maps are a well-established type of qualitative methodology in planning, geography, and similar fields where individuals sketch aspects of a city or neighborhood, producing representations of urban space that blend physical reality with personal experience [3–5]. Akin to a cartographic survey instrument, mental maps have been used to advocate for social justice and as a springboard for citizen-based political interventions [6]. Mental maps also have potential to generate location specific information that can be utilized by a GIS. As such, mental maps may be seen as a potentially important method of data collection for QGIS. What makes mental maps important for QGIS research is that the opinions and knowledge

J. Jiao (✉)
School of Architecture, University of Texas at Austin, Austin, TX 78759, USA
e-mail: jjiao@austin.utexas.edu

S.M. Radil
Department of Geography, University of Idaho, Moscow, ID 83844, USA

J. Harbin
Indiana Housing and Community Development Authority, Indianapolis, IN 46204, USA

Y. Li
Department of Urban Planning, Xiamen University, Xiamen, Fujian 361005, China

X. Chen, Q. Pan (eds.), *Building Resilient Cities in China: The Nexus between Planning and Science*, GeoJournal Library 113, DOI 10.1007/978-3-319-14145-9_9

contained in a mental map is from a person with direct lived experience with a particular urban space or setting [1]. The use of mental mapping can help identify what residents understand as the most important features or aspects of their material and social environment and whether these are understood as positives or negatives [5].

This study used mental maps to explore how QGIS may be used to create meaningful information that could be used by neighborhood-scale organizations in Muncie, Indiana, USA (Fig. 1). Muncie, once seen a symbol of prosperous middle-class American society, is a small city whose manufacturing-based economic fortunes have rapidly shifted in the last two decades as factories have been closed and the population has steadily declined [6]. As a result of persistent economic challenges, Muncie's residents have experienced high unemployment, a steadily deteriorating tax base, lower property values, and little new economic investment. The infrastructure and physical quality of many of Muncie's

Fig. 1 Muncie is roughly 60 miles northeast of Indianapolis

neighborhoods have suffered in the face of these broad patterns of capital mobility and declining investment [7].

Recently, Muncie has also experienced an attempt to revitalize the city as a citizen-based organization was formed to identify challenges to the city and develop plans to improve the quality of life for its residents. In 2010, this organization, which called itself the Muncie Action Plan, surveyed hundreds of Muncie residents about the strengths and weaknesses of the city as a whole. The results of this survey led to the development of several initiatives to attract new investment and improve aspects of daily life for residents [7, 8].

Our project drew upon the city-wide surveys [8] but sought to rescale knowledge and planning efforts to the level of individual neighborhoods. The project partnered an Urban Planning and a Geography course at Ball State University with the Muncie Action Plan to survey residents of seven different neighborhoods in Muncie about the perceived strengths and weaknesses of each neighborhood. The broad goal was to identify issues and strengths particular to those neighborhoods to complement the larger city results from the previous surveys. Students were divided into several teams and assigned to different neighborhoods. The students were then asked to develop a survey that involved the techniques of mental mapping as a means of collecting data from neighborhood residents about how they understood their own neighborhoods. Information generated from the survey process was mapped using GIS and interpreted in partnership with the participants. After feedback from multiple neighborhood meetings, a supplementary online survey was also created to reach more community members. Demographic data was also collected for each neighborhood through Census data.

2 Objective

The goal of this research was to explore incorporating qualitative information into a GIS and to examine what residents felt what was beneficial and what was missing from their own neighborhoods using mental mapping and surveys. This involved deriving mappable data from those answers that could be used to drive neighborhood-specific change and focus the efforts and activities of the neighborhood organizations into the locations or areas where residents' felt it was most important [1]. By combining such small scale and targeted neighborhood data with the larger city-wide data already gathered by the Muncie Action Plan [8], a better idea of the types of efforts needed at multiple scales could be developed. Critically, the project did not aim to develop action plans for the neighborhoods. Instead it strove to provide a spatial framework to residents' own understandings to enable neighborhood groups themselves to develop their own agendas and priorities.

3 Methods

In partnership with the Muncie Action Plan, seven neighborhoods were selected for the project: Anthony, Forest Park, Industry, Morningside, Old West End, South Central, and Westbrier. Under the supervision of the faculty members that led the project, students first developed base layers in GIS for each neighborhood. This included building locations, transportation layers (streets, sidewalks, railroads, walking trails, etc.), zoning, and open space (parks). Next, demographic summaries were prepared in GIS by using layers of neighborhood boundaries provided by the Muncie Action Plan to isolate the census blocks from the 2010 US Census that composed each neighborhood. Both the base map and the demographic summaries provided the students with a level of familiarity with the neighborhoods early in the project.

The next step involved developing surveys that incorporated aspects of mental mapping. A written survey form was designed that asked basic questions about the respondent before asking them to provide either a written list (with locations) of positive or negative features of their neighborhood. On the written survey, residents could also answer this question graphically, by sketching a hand drawn map and providing a legend (see Fig. 2). These surveys were distributed to residents by students that attended a series of neighborhood meetings. These were regularly scheduled meetings held by neighborhood organizations and students were connected with the community leaders in each neighborhood to open the lines of communication and attend meetings where possible.

After attending several meetings in different neighborhoods, it was clear that another method of distributing surveys was needed. While some neighborhoods meetings were well attended, others were poorly attended by residents which limited the number of respondents. In consultation with several neighborhood leaders, students developed an online version of the survey. The online version (see Fig. 3) asked the same questions as the paper version but did not offer an opportunity to provide hand drawn sketch maps. Location information had to be provided in written form in this format.

The paper survey responses ($n = 16$) and online survey responses ($n = 78$) combined to provide a wealth of neighborhood-specific, mappable data. However, the number of responses varied widely between the seven neighborhoods as some had little to no participation, whereas other neighborhoods returned dozens of responses. Two neighborhoods (Morningside and Industry) had no participation at all as the neighborhood leadership did not hold meetings during the time frame of the project. Three other neighborhoods (Anthony, Forest Park, and South Central) had less than 10 respondents each which was largely a function of low attendance of regular meetings. Two neighborhoods yielded over 30 responses each (Old West End and Westbrier). Overall the interest and ease of response increased with the online version of the survey as compared to the paper format. In the remainder of this paper, we focus on results from Old West End, one of the two neighborhoods with the highest levels of participation.

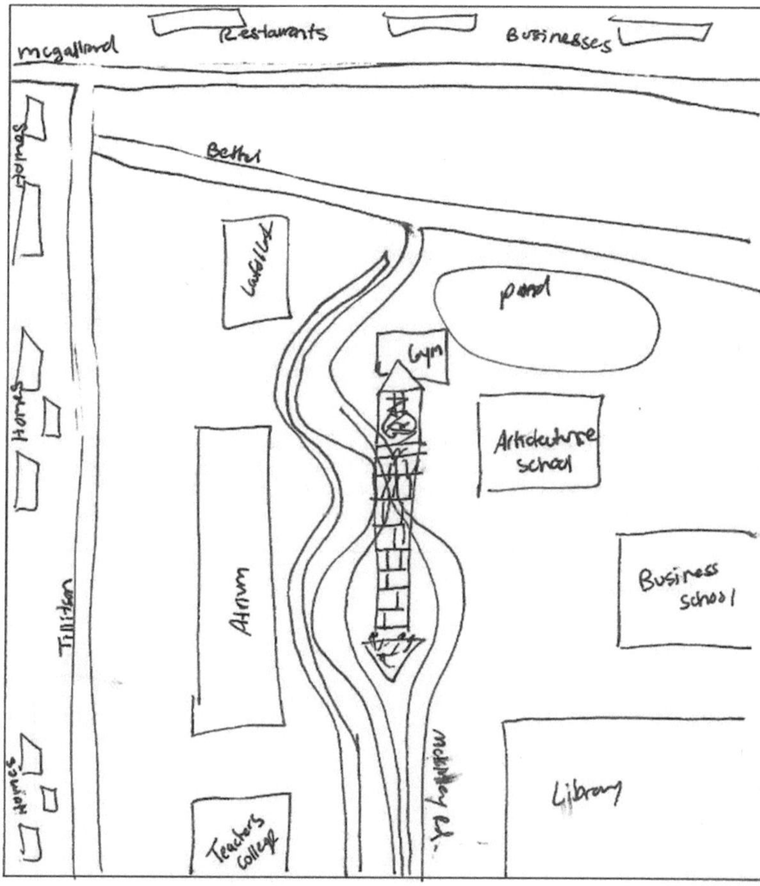

Fig. 2 An example of mental map survey response

Survey results were then synthesized into charts and maps for a report to be given to the neighborhood associations. The goal was to come up with creative ways to map problems in the community. This required critical reasoning on what features would be able to be displayed in map format (e.g. specific addresses mentioned versus tracts of the neighborhood and more abstract responses) versus what may have to be displayed in charts and graphs. Houses, blocks, and specific sections of sidewalks could be mapped by points, polygons, and lines, respectively (Fig. 4). Other more generalized responses (e.g. disinvestment without specific location details) were not as straightforward to map and were often better synthesized in graphical displays.

Old West End in Muncie, Indiana, was the neighborhood with the most responses returned. Thirty online responses and three paper versions were returned. With a very active neighborhood association, it greatly aided the enthusiasm for the survey. The map below shows strengths and weaknesses mapped in Old West End (Fig. 4).

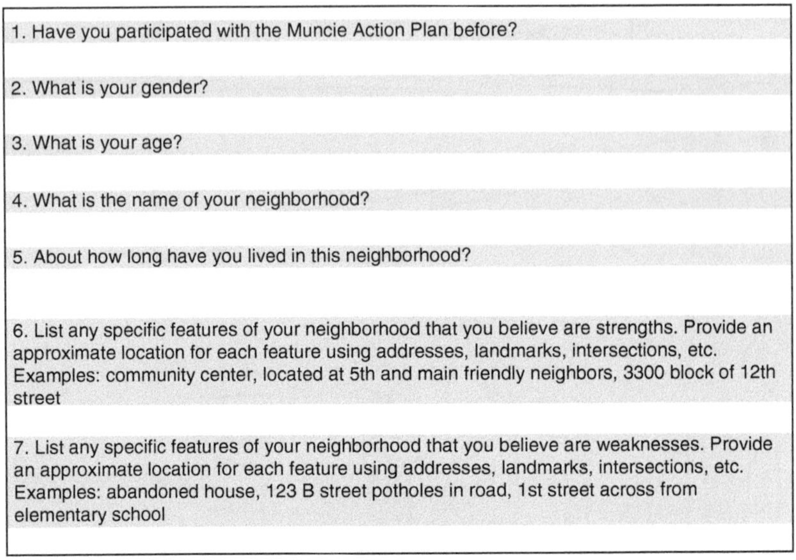
1. Have you participated with the Muncie Action Plan before?

2. What is your gender?

3. What is your age?

4. What is the name of your neighborhood?

5. About how long have you lived in this neighborhood?

6. List any specific features of your neighborhood that you believe are strengths. Provide an approximate location for each feature using addresses, landmarks, intersections, etc. Examples: community center, located at 5th and main friendly neighbors, 3300 block of 12th street

7. List any specific features of your neighborhood that you believe are weaknesses. Provide an approximate location for each feature using addresses, landmarks, intersections, etc. Examples: abandoned house, 123 B street potholes in road, 1st street across from elementary school

Fig. 3 The online survey format

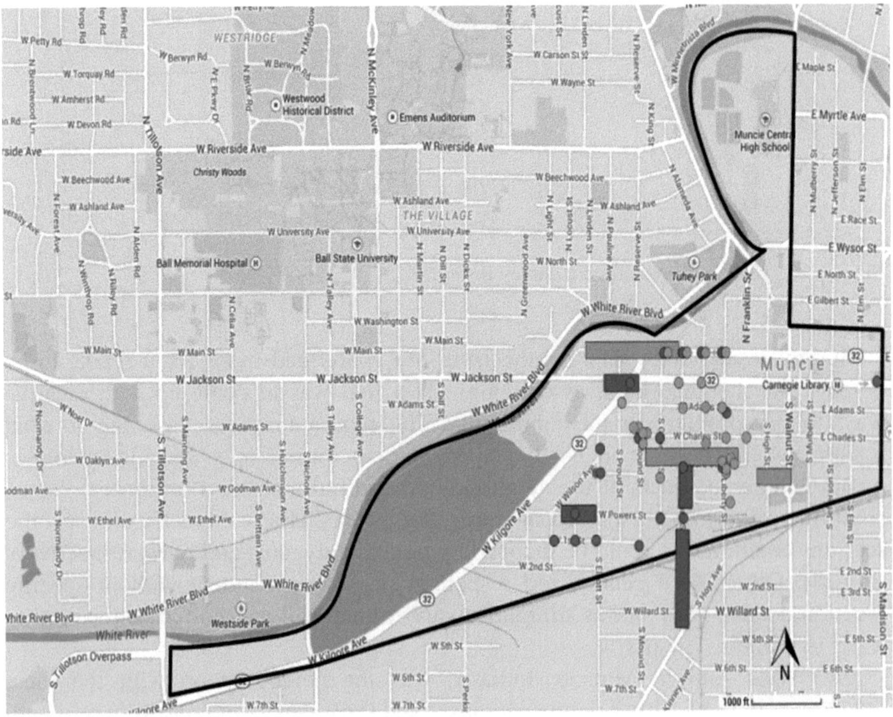

Fig. 4 Example of mapped results for Old West End Neighborhood

4 Results and Discussion

In general, the major concern commonly identified by residents across all the neighborhoods in the study had to do with the condition of sidewalks and roads in the neighborhoods. This usually came about from a perspective of safety of children or those who enjoy riding their bikes or jogging in the neighborhoods in terms of sidewalks. Disinvestment, reflected in abandoned residential properties, was another large concern throughout the various neighborhoods, particularly in older neighborhoods such as the Old West End. Interestingly, neighborhoods that were identified with a lot of negative issues in the original city-wide surveys showed more positive aspects when only residents of those neighborhoods were surveyed. Similarly, our surveys reveals smaller negative issues (e.g. sidewalks instead of widespread blight) when residents living in the neighborhood mentally mapped their own immediate surroundings as opposed to reflecting on the city as a whole. Many residents of the neighborhoods had particular features they took pride in, such as restored homes or pocket parks, that were often overlooked in the city-wide survey results.

Other concerns spanning across neighborhoods included large blighted tracts of land formerly occupied by industrial uses, lack of businesses nearby, and limited access to bus shelters. Road conditions were also commonly cited across neighborhoods. On the reverse, historic features, and proximity to Ball State University and the Minnetrista cultural center were commonly cited as strengths across neighborhoods.

Old West End consistently returned mentions of restoration of historic houses as a strength. Community members took a great deal of pride in the work that has gone into restoring residential properties in their neighborhood and continue to rally around these houses that have been restored. Historic landmarks were also consistently cited as strengths, alluding to the residents' pride in its history and location. Sidewalks were consistently mentioned as a weakness as some had cracks and crumbling pieces, which was also an overall note for the city as a whole when the Muncie Action Plan originally gathered large-scale responses.

Old West End, along with other neighborhoods such as Anthony Neighborhood, mentioned a lack of signage as a negative. With some roads being converted to one way streets, especially in downtown, there were specific intersections noted that needed indicative signage. Road conditions, with pot holes being specifically cited, were also raised in multiple neighborhoods. Lack of churches and presence of liquor stores was noted specifically as weaknesses in Forest Park, while Anthony residents specifically mentioned sewer drainage as an issue.

5 Conclusions

Surveys collected from meetings held throughout the city do have the ability to collect a large amount of results quickly, but may not always accurately display what the problems truly are in a community. By going through various neighborhoods and focusing solely on what is in that area, data can be obtained from residents on how to fix what is directly in their community. Mental mapping is a valuable research tool that helps survey participants visualize and communicate their thoughts in an easily accessible way [9, 10].

However, an important caveat of the changing preferences of society is that an online survey may generate more interest and participation. Online surveys gave the respondents the ability to finish the survey on their own time, which may have allowed them to make note of particular strengths and weaknesses. The downside to the online version of the survey is that it did not allow for mental mapping. In the future, with the increased use of tablets and similar, mental mapping may be able to synthesize more easily with online survey responses [11, 12].

The results of this survey and those like it will allow city officials to target specific areas and results with their resources. In a time when many cities face limited budgets, targeted responses that delineate right where bad potholes, lacking sidewalks, and missing signage in a city exist can be far more beneficial and solutions oriented than general responses of "bad roads" and "missing sidewalks". By going to individual neighborhoods we can utilize an important asset in the form of community members, who often know their area better than anyone else.

The goal for future research of this nature may be to combine the results and benefits of mental mapping with the changing public interest toward online surveys and mobile devices. With the rising use of smartphones and tablets, generating online surveys and data collection may be a better way to reach wider ranges of people and collect real time data within cities and neighborhoods.

Acknowledgments The authors wish to thank the numerous residents of Muncie that took part in this project. We also thank the Muncie Action Plan and various neighborhood association leaders for their support. This project was supported by the 2013 Provost Immersive Learning Grant at Ball State University.

References

1. Cope M, Elwood S (eds) (2009) Qualitative GIS: a mixed methods approach. Sage, Thousand Oaks
2. Henry-Nickie M, Kurban H, Green RD, Phoenix JA (2008) Leveling the playing field: enabling community-based organizations to utilize GIS for effective advocacy. URISA J 20 (2):33–41
3. Lynch K (1960) The image of the city. MIT Press, Cambridge, MA
4. Lynch K (1984) Good city form. MIT Press, Cambridge, MA
5. Gould P, White R (1993) Mental maps. Rutledge, New York
6. Elwood S (2009) Integrating participatory action research and GIS education: negotiating methodologies, politics and technologies. J Geogr High Educ 33(1):51–65
7. Fraser M (2012) Down and out in the new Middletowns. Dissent 59(1):27–34
8. Muncie action plan: vision and action plan for Muncie, IN. http://muncieactionplan.org/wp-content/uploads/2012/02/MAP_Report_FINAL.pdf
9 Carr S, Rodwin L et al. (1984) Kevin Lynch: designing the image of the city. J Am Plan Assoc 50(4):523–525

10. Hospers G-J (2010) Lynch's the image of the city after 50 years: city marketing lessons from an urban planning classic. Eur Plan Stud 18(12):2073–2081
11. Knox P, Marston S (2012) Human geography: places and regions in global context. Prentice Hall, Upper Saddle River
12. Hughes AL, Palen L (2009) Twitter adoption and use in mass convergence and emergency events. In: Proceedings of the 6th international ISCRAM conference, Gothenburg

The Ecological Impact of the Korean Saemaul (New Rural Community) Movement, 1970–1979

Chung Ho Kim

1 Introduction

Lester R. Brown (2008) argues in his book, *Plan B 3.0: Mobilizing to Save Civilization,* that South Korea is a reforestation model for the rest of the world based on the fact that South Korea was largely deforested only 50 years ago. He then points out why Korean reforestation was successful, as shown in the quotation below, making the claim that it is possible to reforest the earth based on the case of the successful rapid reforestation of South Korea:

> ***South Korea** is in many ways **a reforestation model for the rest of the world**. When the Korean War ended, half a century ago, the mountainous country was largely deforested. Beginning around 1960, under **the dedicated leadership of President Park Chung-Hee**, the South Korean government launched a **national reforestation effort**. Relying on the formation of **village cooperatives**, hundreds of thousands of people were mobilized to dig trenches and to create terraces for supporting trees on barren mountains.*
>
> *Se-Kyung Chong, researcher at the Korea Forest Research Institute, writes, "The result was a seemingly **miraculous rebirth of forests from barren land**." Today **forests cover 65 percent of the country, an area of roughly 6 million hectares**. While driving across South Korea in November 2000, it was gratifying for me to see the luxuriant stands of trees on mountains that a generation ago were bare. **We can reforest the earth!** (emphasis is mine)* [1]

This paper starts from the seemingly miraculous success of Korean reforestation by asking two basic questions: "How did it succeed?" and "Why did it succeed?" However, this paper raises a variety of fundamental questions:

- Was it a true success, and, if so, what were inherent conditions of the success?
- Were there any problems that arose during the process in spite of the success, and, if so, what were they?

C.H. Kim (✉)
Department of Urban Design and Planning, University of Washington, Seattle, WA, USA
e-mail: charisut@u.washington.edu

X. Chen, Q. Pan (eds.), *Building Resilient Cities in China: The Nexus between Planning and Science*, GeoJournal Library 113, DOI 10.1007/978-3-319-14145-9_10

Fig. 1 Miraculous rebirth of forests from barren lands [2]

- Were the problems general or specific?
- If we want to reproduce the success today, how can we realize it while overcoming many problems?

As a result, the following research examines the possibility of utilizing the Korean's strong, top-down movement as a measure to enhance social-ecological resilience [21]. For this purpose, the paper investigates KSM, which led to the success of Korean Reforestation, focusing on the relationship between a strong, top-down approach and its ecological impact. Thus, the research expects that it can shed light on viable approaches for sustainable development, especially in developing countries (Fig. 1).

2 Research Design

2.1 Research Problem and Questions

First of all, the research assumes that every society could have different approaches to increase their social-ecological resilience. In other words, every society could have different measures to achieve contemporary sustainable development. In fact, this basic assumption arises from the researcher's personal criticism of today's sustainable development.

Contemporary, worldwide problems, such as climate change and energy crisis, require a strong interest in sustainable development. However, their basic viewpoints and practical approaches have a bias towards standardized solutions based on engineering, which result in ignoring the inherent differences of culture and environment.

Thus, this research raises the question of whether or not the implementation of strong, top-down approaches in developing countries is effective in enhancing social-ecological resilience. For this question, the paper analyzes the extreme cases towards rapid modernization and urbanization in Korea: the Korean Saemaul Movement (1970–1979) and Korean Reforestation (1973–1978). Both cases were implemented under the authoritarian government led by President Park in the 1970s.

Interestingly, the historical reviews for the Korean Saemaul Movement (KSM) and Korean Reforestation are generally very positive. These movements have been regarded as the greatest influential governmental policies, which have contributed to national development throughout the history of the Republic of Korea [3–5]. In addition, they have become popular modernization models that have been referenced in the Third World [6, 7].

However, their ecological impacts have not been fully examined in comparison to interests regarding their economic and social impacts, even though they have fundamentally converted Korean society and nature from traditional villages with barren lands into new modern villages with forested lands. To sum up, the research problem is the 'Strong Korean Top-Down Approach and its Ecological Impact.'

2.2 Research Framework and Methodology

The framework of this paper undertakes four main steps. First of all, it identifies the background, nature, and historical progress of KSM as a strong, top-down approach through literature review and statistical data.

Secondly, the research analyzes the ecological influences of KSM with regard to resource management. For this, the paper interprets regime changes of resource management that were affected by KSM either directly or indirectly.

Third, the research focuses on the background, nature, and historical progress of Korean Reforestation as a strong, top-down approach through literature review and statistical data. Since Korean Reforestation was fundamentally rooted in KSM, the former can provide more specific and detailed information than the latter.

Finally, the paper investigates the ecological influences of Korean Reforestation, especially on a national-scale and global-scale. In other words, the research explores ambivalent impacts, such as national-scale success and global-scale consequences based on the world-system theory [8, 9].

3 Korean Saemaul Movement, 1970–1979

3.1 Brief Background on the Republic of Korea

Historically, the Joseon Dynasty (1392–1910) existed as a single, independent nation on the Korean Peninsula before Korea was under the Japanese colonial forces (1910–1945). During the Joseon Dynasty, neo-Confucianism was a national religion and ideological base for governance rules [20].

In addition, agriculture was a primary industry supporting Korean society and economy until President Park initiated the first Five-Year Economic Development Plan in 1962, which pursued rapid economic growth through industrialization. Meanwhile, Korea suffered from the Korean War (1950–1953) between South and North Korea, which completely devastated the country.

Geographically, the land area of South Korea is about 38,691 $mile^2$, or 100,210 km^2, which is similar to the state of Indiana in the United States. Mountainous land makes up about 65 % of the whole country, with relatively high mountains in the east area. The country has few natural resources, of which include coal, iron ore, and limestone.

Socially, South Korea has nearly 50 million people living on limited land with few natural resources. The population density is very high with approximately 1,271 people per square miles in 2010. Thus, the country is currently one of the most densely populated regions in the world.

Economically, the nation has one of the world's fastest growing economies. South Korea has impressively achieved rapid economic growth within the past 50 years. To be specific, South Korea was poorer than most of the African countries in the early 1960s. The Gross Domestic Product (GDP) per capita was only $82 at that time. However, the GDP per capita based on Purchasing Power Parity (PPP) reached $30,700 in 2011.

3.2 President Park, Chung-Hee, 1917–1979

Even today, President Park, Chung-Hee (1917–1979) is considered a very controversial figure based on his achievements as President. Some people argue strongly that he was a hero who led the Korean modernization and industrialization movement. Others condemn him as a dictator who suppressed Korean democracy and human rights until his assassination in 1979. According to numerous surveys, however, more than half of the respondents have stated that he was the greatest president throughout the history of Republic of Korea.[1] Furthermore, among all of the government policies that he implemented, KSM was considered to be the best.

[1] In President Park, Chung Hee's Internet Memorial [10], the results of numerous surveys are well summarized. Although the website is for the celebration of President Park's achievements, it provides reliable information.

Fig. 2 President Park, Chung-Hee (1917–1979) [10]

The personal life of Park, Chung-Hee is also very controversial. He was born the youngest son of the local collapsed gentry in 1917. He was later admitted to the Daegu Teacher's Gymnasium, a competitive high school for prospective primary teachers and, after graduating from the 5-year study in 1937, was a primary teacher for 3 years. He then voluntarily joined the Manchukuo Imperial Army Academy, completing his studies with top marks in 1942. He was selected for officer training at the Army Staff College in Japan, later graduating third in his class [19].

In 1945, after the end of the World War II, he went on to serve in the military of the Republic of Korea. He then became a leader of military coup on May 16, 1961, which allowed him to have absolute power for 19 years. However, he was fatally shot in 1979 by Kim, Jae-kyu, director of the Korean Central Intelligence Agency (KCIA), who had been his loyal subordinate for a long time. Presently, Park, Geun-Hye, his oldest daughter, is now the president of South Korea, taking office on February 25, 2013 (Fig. 2).

3.3 *Korean Saemaul Movement as a Top-Down Approach*

3.3.1 Background and Nature

In order to correctly understand the nature of KSM, it is necessary to comprehend the political climate of South Korea in the early 1970s. The local politics had been under the executive control of administrative elites in the central government since 1961 when the Park administration had absolute power [11].

Under those circumstances, the Park government proposed the Five-Year Economic Development Plans of 1962–1966 and 1967–1971, which generated remarkable economic growth that averaged 9.7 % annually for those years [12]. However, it also exacerbated the economic gap between the rural sector and urban sector. Boyer et al. (1991) describes this point clearly as shown in the quotation below:

> ***Growth in agriculture and rural areas**, however, **lagged far behind that in the industrial sector and urban South Korea**. The average rate of economic growth during **the first five-year plan** was **7.8 percent**, but only **5.3 percent for agriculture**. The agricultural sector worsened during **the second five-year plan**, when it grew at an annual rate of **only 2.5 percent, compared to 10.5 percent for industry**. On average a rural household in 1963, but **only 56 percent of the average household income** in 1969 (emphasis is mine).* [11]

Ironically, the rural sector in South Korea was very important to the Park government. This was because it was their political base, even though Park and his planners dreamed that the Republic of Korea would become an industrial society in the near future [13], a concept that was highly associated with Park's dictatorial government. Park held three consecutive terms of office as the fifth (1963–1967), sixth (1967–1971), and seventh (1971–1975) President. Moreover, Park was chosen by the people's direct election, even though he amended the constitution of the Republic of Korea.

However, the Park government thought that their political popularity was decreasing, as they had many political opponents. Thus, Park passed the new authoritarian constitution in 1972, legitimizing his dictatorship based on the people's indirect election. The affair was called 'October Yusin,' which was derived from the Meiji Yusin, or Meiji Restoration, the catalyst of Japanese modernization in the nineteenth century. In this context, Sorensen (2011) assessed the nature of KSM as one of Park's Yusin Period strategies as shown in the quotation below:

> *The New Village Movement, begun in 1971, was designed in part to shore up **Park's rural support**, and was central to **Yusin developmental strategy**. **The New Village Movement**, in fact, can be paired with **the Heavy and Chemical Industrialization Program** as one of the two legs of Park's Yusin Period development strategy. **Park was personally and deeply involved** in the drafting and implementation of both programs (emphasis is mine).* [13]

3.3.2 Historical Progress

To briefly summarize, KSM was a political initiative launched in 1970. It is often said that it aimed to modernize the rural economy and to change the traditional and unscientific mentality of the peasants [13, 14]. In addition, it was a strong, top-down leadership driven by President Park, Chung-Hee.

As shown in Fig. 3, President Park often used to write calligraphy in order to clearly present his mottos, which included modernization, national regeneration, and development and growth. In other words, he tried to emphasize spiritual revolution as well as economic growth. In fact, his core mottos were, "We will live better, too," and "We can do it." Therefore, KSM can be interpreted as a platform for realizing national agendas according to changing situations.

According to Boyer et al. (1991), the New Village Movement is divided into two major periods, 1970–1972 and 1973–1978 [11]. During the first period, the movement started from the pilot projects of central government distribution of 335 free

Fig. 3 Korean Saemaul movement [10]

bags of cement to each of South Korea's 33,267 villages. The villages were expected to use the cement for the ten government-designated village projects comprising the Program for Village Environmental Improvement.[2]

Surprisingly, it achieved huge results far exceeding initial expectations. For example, according to the government's evaluation in July 1971, the expenditure of the equivalent of US$11 million for the cement had yielded village improvements valued at US$32.6 million, nearly three times the government's estimate. As a result, the dramatic success allowed the Park government to pursue a more systematic and planned approach to KSM.

During the period of 1973–1978, the Park government decided to expand KSM to all villages, urban areas, factories, schools, and even the military. The government classified villages into three categories according to their stages of development: undeveloped (basic), developing (self-helping), and developed (self-sufficient) villages. The main goal of the classification was for all villages to become 'developed' by 1981. Table 1 indicates how KSM was actively expanded in the 1970s.

3.4 Ecological Impact of Korean Saemaul Movement

3.4.1 Regime Changes of Resource Management

Interestingly, President Park wanted to transform the 'Traditional Village' of those days into the 'Modern Village,' which had western-style houses with gabled roofs

[2] The Program for Village Environmental Improvement became the initial thrust of KSM, expanding from ten first-year target projects to twenty projects thereafter. First-year projects for each village included (1) reforestation of nearby terrain, (2) broadening village access roads, (3) repairing and improving village dikes, (4) preparing a village compost barn, (5) deepening the village pond, (6) repairing and maintaining the pond, (7) keeping the village, ditches, and gutters clean, (8) constructing a community well, (9) exterminating rats, and (10) establishing a village laundry facility [11].

Table 1 Expansion of Korean Saemaul movement [11]

Year	No. of participant villages	No. of man days (thousand)	No. of projects (thousand)	Total investment (billion won)
1971	33,267	7,200	385	12.8
1972	35,031	106,852	1,099	132.8
1977	36,557	137,193	2,463	463.5
1979	36,271	242,078	1,788	758.2

Fig. 4 Comparisons between a Traditional House and a KSM House [10]

Table 2 Regime changes of resource management

Resource		Traditional (natural-resource-based)	Modern (industrial-resource-based)
Building material	Wall	Soil, wood, straw	Cement, reinforcing bar
	Window	Traditional paper	Glass
	Roof	Straw and grass	Slate
Fuel	Cooking	Firewood	Coal, oil
	Heating	Firewood	Coal, oil, gas
Lighting		Kerosene lamp	Electricity
Water		Well	Well pump

and punching windows. President Park and his planners regarded the 'traditional village' as a negative place with passivity, stagnation, disease, and poverty.

Figure 4 clearly reveals the difference between the two villages. They each look completely different, especially in terms of the materials used and the images of the built environments.

As a result, KSM led to the regime changes of resource management as shown in Table 2, from natural resource-based materials to industrial resource-based ones. For example, building materials were changed from soil, paper, wood, and straw to cement, glass, and reinforcing bars. Fuels for heating or cooking were also changed from firewood to coal, oil, or gas.

4 Korean Reforestation, 1973–1978

4.1 *Korean Reforestation as a Top-Down Approach*

4.1.1 Background and Nature

As mentioned above, South Korea has a high population density and large mountainous areas comprising approximately 65 % of the whole country. In addition, the rainfall is concentrated during the summer months of June through September. The rainfall in July alone is nearly 28 % of the whole annual precipitation. The issue of the forest has historically been very important in Korea because it provides a variety of benefits: holding water, alleviating droughts, preventing floods and landslides, providing foundation for biodiversity, maintaining air quality, and so on.

4.1.2 Historical Progress

The reforestation process has continuously been pursued since Korea was deforested by Japanese exploitation, Korean War, reckless deforestation for firewood, lack of policing, and so on. Both Table 3 and Fig. 5 indicate how passionately the Park government reforested the country. The Park government exerted strong leadership during the periods of the Pre-Reforestation (1962–1972) and First Reforestation (1973–1978). However, the achievements of the Pre-Reforestation Period (1962–1972) were not as satisfactory as those of First Reforestation Period (1973–1978).

Table 3 Historical progress of Korean reforestation [15] (Unit: area –hectare, number of trees – million)[a]

Period	Pre reforest (1962–1972)	1st reforest (1973–1978)	2nd reforest (1979–1987)	Whole reforest (1962–1987)
Total area	149,902	179,962	107,319	142,099
Timber area	63,236	59,626	56,325	60,011
Rapid growth Tree area	10,607	59,467	47,184	34,544
Fruit tree area	20,182	25,599	3,146	15,535
Fuel tree area	55,732	34,638	0	31,572
Others area	145	632	664	437
Number of trees	420	493	213	365

[a]Numbers in the figure refer to annual averages during the period

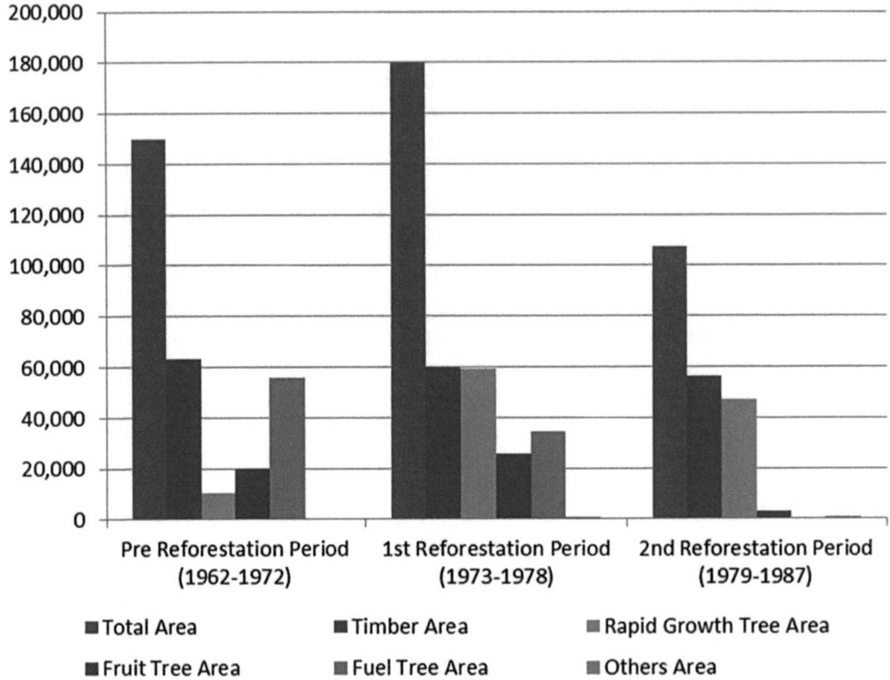

Fig. 5 Historical progress of Korean reforestation [15] (Unit: area –hectare, number of trees – million [Numbers in the figure refer to annual averages during the period])

To be specific, the Pre-Reforestation Period (1962–1972) had reforested areas and numerous trees similar to that of First Reforestation Period. However, the former focused more on fuel trees than the latter did. On top of that, most of the reforested trees in the 'Others Area' found in Table 3 were also used as firewood [16]. As a result, it was impossible that Korean reforestation took effect during the period of Pre-Reforestation (1962–1972).

On the other hand, the First Reforestation Period (1973–1978), which was very rapid and intensive, focused not on fuel trees, but on timbers or rapid growth trees. Furthermore, the 10-year plan was completed within just 6 years. This also implies that certain fundamental changes of forest utilization happened during this period, including the substitution of firewood for home use with fossil fuels.

4.2 Ecological Impact of Korean Reforestation

4.2.1 National-Scale Success

The success of Korean Reforestation is undisputed around the world. What were its inherent causes? Bae et al. (2006) argues that there were several reasons why Korean Reforestation succeeded [16].

First, President Park had powerful leadership in reforesting the whole country rapidly and fervently. He prioritized the issue of reforestation as the government's primary agenda. He monitored the progress continuously and thoroughly, mobilizing public officers as well as encouraging people.

Second, a variety of governmental ministries cooperated closely with each other in order to realize President Park's goal. Initially, the Ministry of Agriculture and Forest was in charge of the reforestation before the First Reforestation Period (1973–1978). However, since the result was not satisfactory, President Park ordered the Ministry of Home Affairs to undertake the task. The Ministry of Home Affairs was one of the most powerful government ministries at that time because it included all administrative public officers, such as the police and the prosecuting office. As a result, Korean reforestation was executed under the ministry's management and supervision.

Third, the citizens enthusiastically participated in the reforestation process [17]. Although the Park government was a fearsome authoritarian regime, the public actively joined the movement. In those days, Korea had two holidays for the reforestation: Tree-Planting Day (April 5) and Tree-Culturing Day (First Saturday of November). In addition, there was a specific period when the government motivated all people to plant trees (March 21–April 20). Even today, planting trees is thought to be important and meaningful to Koreans. For example, many leaders enjoy planting trees in important public places to commemorate taking up or retiring their posts.

Finally, the continuous economic growth in Korea in the 1970s was a very important factor that resulted in substantive effects on Korean reforestation. The Park government had enough funding to support the task of reforestation. Likewise, the citizens generally had increasing disposable incomes because it led to fundamental changes of life styles, so called from traditional to modern. However, the most important factor in Korean reforestation was the substitution of firewood for home use with fossil fuels such as coal, oil, or gas.

Table 4 shows the changes in cooking fuel consumption during the period from 1970 to 1990. In 1970, firewood and coal were the most important fuels for cooking. At that time, coal was mainly used in urban areas, while firewood was used in rural areas. However, the use of firewood as cooking fuel has decreased continuously, reaching 2.5 % of the entire consumption in 1990. Meanwhile, the use of coal hit its peak of 65.7 % in 1980, followed by a decrease to 10.3 % in 1990. This was because the use of gas became the dominant cooking fuel in South Korea.

4.2.2 Global-Scale Consequences

Along with national-scale success, Korean Reforestation created global-scale consequences. Currently, South Korea has a very low domestic self-sufficiency in timber, even though the figure has increased from 5.7 % in 2000 to 15 % in 2011 [18]. Hardwood, according to the 2011 statistical data, has 51.6 % domestic self-sufficiency, which reached 4.45 million m^3 [18].

Table 4 Changes of cooking fuels consumption [16] (Unit: number of households)

Year	1970	1975	1980	1985	1990
Coal	3,016,873	4,330,663	5,238,919	4,612,344	1,166,223
	(52.1 %)	(64.2 %)	(65.7 %)	(48.2 %)	(10.3 %)
Oil	37,907	58,481	200,619	782,345	253,297
	(0.7 %)	(0.9 %)	(2.5 %)	(8.2 %)	(2.2 %)
Gas	11,481	50,764	482,910	2,526,366	9,298,171
	(0.2 %)	(0.8 %)	(6.1 %)	(26.4 %)	(81.9 %)
Electricity	4,316	16,583	22,640	139,060	307,690
	(0.1 %)	(0.2 %)	(0.3 %)	(1.5 %)	(2.7 %)
Firewood	2,720,275	2,289,302	1,794,113	1,406,105	280,687
	(47.0 %)	(33.9 %)	(22.5 %)	(14.7 %)	(2.5 %)
Others	1,914	4,557	230,000	105,141	48,472
	(0.0 %)	(0.1 %)	(2.9 %)	(1.1 %)	(0.4 %)
Total	5,792,766	6,750,350	7,969,201	9,571,361	11,354,540
	(100.0 %)	(100.0 %)	(100.0 %)	(100.0 %)	(100.0 %)

Table 5 Korean foreign-dependence of energy and oil [18]

Year	1970	1980	1990	2000	2005
Dependence of energy (%)	47.5	73.5	87.9	97.2	96.8
Dependence of oil (%)	–	61.1	53.8	52.0	44.4
Cost of energy import (billion $)	–	6.59	10.93	37.58	66.7

In addition, South Korea has an extremely high degree of foreign dependence on energy. Table 5 shows the degree to which South Korea has depended on foreign energy, especially fossil fuels exported mainly from the Middle East. The dependence has continuously increased from 47.5 % in 1970 to 96.8 % in 2005. According to the 2011 statistical data, the total amount of energy import reached 262.6 million TOE (Ton of Oil Equivalent), which came to $121.6 billion of energy import [18].

As a result, the absolute foreign dependence on energy and resources has pushed South Korea into the rigidity trap, decreasing Korean social-ecological resilience. In fact, this phenomenon results from the distinct Korean economic structure, which is heavily dependent on international trades. Since South Korea has few natural resources and a small economic market, the Park government concentrated on export-centered economic growth. This key economic policy has been sustained since it was first established in the 1960s.

Interestingly, the energy imports reached 32 % of entire imports in 2010, which were equivalent to the total exports of Korea's main manufacturing industries, such as semiconductors, automobiles, and shipbuilding. In other words, Samsung Electronics, Hyundai & Kia Motors, Daewoo Shipbuilding are earning dollars to buy oil and gas from the Middle East.

5 Conclusion

This research investigated the ‘Strong Korean Top-Down Approach and its Ecological Impact’ with two extreme cases: the Korean Saemaul Movement and Korean Reforestation. For this research, the paper assumed that every society could have different approaches and measures to enhance their social-ecological resilience, as well as achieving contemporary sustainable development. The research, then, conducted a literature review and utilized statistical data under the research framework based on the theories of historical ecology. The research reached following conclusions.

First of all, KSM was a strong, top-down approach to modernize the rural economy and change the traditional and unscientific mentality of the peasants. Although the movement provided construction materials or support grants, it was basically a spiritual platform for national agendas to be implemented according to changing situations.

Second, KSM affected regime changes of resource management from natural resource-based materials into industrial resource-based materials.

Third, the national-scale success of Korean Reforestation resulted from the strong leadership of President Park, cross-sectoral support by various governmental ministries, the public’s passionate participation, continuous economic growth, and the substitution of firewood for home use with fossil fuels.

Fourth, the global-scale consequences of Korean Reforestation resulted in an extremely high foreign dependence of energy and resources, which have led South Korea to fall into the rigidity trap, decreasing the social-ecological resilience. In fact, this phenomenon results from the distinct Korean economic structure that is heavily dependent on international trades caused by few natural resources and small economic market.

Finally, it was meaningful to investigate the positive impact of the Korean strong, top-down approaches and to figure out their accompanying adverse effects according to both the national-scale and global-scale.

Acknowledgment The author would like to express appreciations to Professor Stevan Harrell at the University of Washington for substantive comments in his classes and Professor Daniel Abramson for his considerate encouragement.

Partial support for this research came from a Shanahan Endowment Fellowship and a Eunice Kennedy Shriver National Institute of Child Health and Human Development training grant, T32 HD007543, to the Center for Studies in Demography & Ecology at the University of Washington.

References

1. Brown LR (2008) Plan B 3.0: mobilizing to save civilization. W. W. Norton, New York
2. Kim J-C (2011) Korea: miracle country of reforestation success. The Korean Forestry News, Seoul

3. Burmeister LL (1987) The South Korean green revolution: induced or directed innovation? Econ Dev Cult Chang 35(4)
4. Moore M (1985) Mobilization and disillusion in rural Korea: the Saemaul movement in retrospect. Pac Aff 57(4):577–598
5. Park S (2009) Analysis of Saemaul Undong: a Korean rural development programme in the 1970s. Asia Pac Dev J 16(2):113
6. Dore RP, Mars Z, Brandt VSR (1981) Community development: comparative case studies in India, the Republic of Korea, Mexico and Tanzania. Croom Helm, London
7. Lo F-c, Song B-N (1979) The Saemaul Undong: the Korean way of rural transformation. United Nations Centre for Regional Development, Nagoya
8. Cumming GS, Cumming DHM, Redman CL (2006) Scale mismatches in social-ecological systems: causes, consequences, and solutions. Ecol Soc 11(1). http://www.ecologyandsociety.org/vol11/iss1/art14/
9. Abel T (2007) World-systems as complex human ecosystems. In: Hornborg A, Crumley C (eds) The world system and the Earth system: global socioenvironmental change and sustainability since the Neolithic. Left Coast Press, Walnut Creek, pp 56–73
10. President Park, Chung Hee's Internet Memorial. http://www.516.co.kr
11. Boyer WW, An P-m (1991) Rural development in South Korea: a sociopolitical analysis. University of Delaware Press, Newark
12. Korean Statistical Information Service. http://kosis.kr
13. Kim H-A, Sorensen CW (2011) Reassessing the park Chung Hee era, 1961–1979: development, political thought, democracy & cultural influence. University of Washington Press, Seattle
14. Oh M-S (1998) Peasant culture and modernization in Korea: cultural implications of Saemaul movement in the 1970s. Korea J 38(3):77–95
15. National Archive of Korea. http://theme.archives.go.kr/next/forest/viewMain.do
16. Bae J-S, Lee K-B (2006) Impacts of the substitution of firewood for home use on the forest greening after the 1945 liberation of Korea. J Korean For Soc 95(1):60–72 (in Korean)
17. Lee DK, Lee YK (2005) Roles of Saemaul Undong in reforestation and NGO activities for sustainable forest management in Korea. J Sustain For 20(4):1–16
18. Statistics of Energy and Resource. http://www.kier.re.kr/energy/statistics.jsp
19. Cyber President Park, Chung Hee. http://www.presidentpark.or.kr/presidentpark/pages/main.jsp
20. Encyclopedia of Korean Culture. http://encykorea.aks.ac.kr/
21. Walker BH, Salt D (2006) Resilience thinking: sustaining ecosystems and people in a changing world. Island Press, Washington, DC

The Analysis of Two Increases and Two Reductions Policy of Shanghai

Liao Yu-qing and Huang Jian-yun

1 Research on the Background of Two Increases and Two Reductions Policy

Since economic reform, Shanghai, the leading city in the Yangtze River Delta, has attracted a growing number of settlers by dint of its economic strength, economic growth speed and employment prospect (Fig. 1). However, as Fig. 1 shows, with the expansion of agglomeration effect, Shanghai is facing increasingly severe contradictions in population, social and environmental problems.

Shanghai Urban Master Planning (1999–2020) estimates that in 2020, the actual resident population in Shanghai will reach about 16 million, which was however exceeded in 2000. Figure 1 suggests that resident population in Shanghai soared by 1,988,400 from 1999 to 2003. By comparison, registered population showed a slow increase, growing by 286,500 in these 5 years. The major reason for the increase in Shanghai population proved to be the influx of non-native population, the ratio of which to total resident population rose year by year. In 1999, non-native population accounted for about 1/6 of total resident population while nearly 1/4 in 2003. At that time, urban center was overcrowded, and more importantly, non-native population, especially the large quantity of low-income groups, lacked the ability to afford the expensive apartment in Shanghai urban center, thus choosing to buy the apartments at a lower price in the suburbs. As the suburbs became flourishing, the original urban center kept going downhill. The worsening of living environment has capacitated more residents to settle in suburbs, which, as a result, caused the external expansion of built-up areas in Shanghai. Up to 2003, the administrative region had covered an area of 6,340.5 km^2. As Table 1 shows, compared with other

L. Yu-qing (✉) • H. Jian-yun
Department of Architecture, School of Naval architecture, Ocean & Civil Engineering, Shanghai Jiao Tong University (SJTU), Shanghai 200240, China
e-mail: liaoyq25@163.com; huangjianyun2004@163.com

X. Chen, Q. Pan (eds.), *Building Resilient Cities in China: The Nexus between Planning and Science*, GeoJournal Library 113, DOI 10.1007/978-3-319-14145-9_11

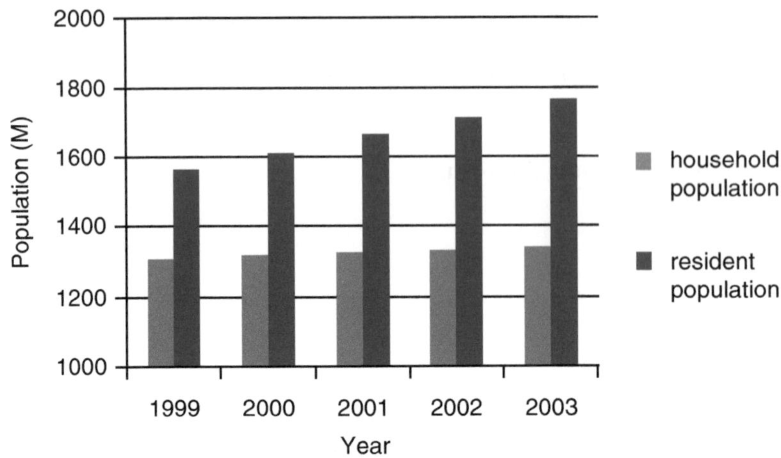

Fig. 1 The population statistics of Shanghai (1999–2003) (Source: The statistical yearbook of Shanghai (2000–2004))

Table 1 The population of cosmopolitan cities and Shanghai in 2003

International metropolis	Land area (km^2)	Population (m)	Population density (in h/km^2)	Statistical time	Scope definition
Shanghai	6,340.5	1,765.847	2,785	2003	Shanghai (16 districts and 1 county)
New York	1,214.4	808.57	6,658	2003	New York city (5 districts)
Tokyo	2,187	1,238.52	5,664	2003	Tokyo metropolitan
Paris	105	214.7	20,370	2003	Paris metropolitan
Hong Kong	1,104.3	673.85	6,102	2003	Hong Kong Island, Kowloon Peninsula, New Territories

Source: The statistical yearbook of Shanghai (2004) and the comparative study on land scale of the international metropolis, Beijing: China Building Industry Press, 2009 [1]

international metropolises, Shanghai held the largest population with the lowest population density in 2003.

Shanghai population faced imbalanced distribution in space. Population density of the inner-ring urban center was much higher than that of outer-ring suburb, and it seemed that the suburb with vast land but small population was still capable of offering sufficient land for development purpose. As a matter of fact, Shanghai has been highly exploited. Based on the comparisons about the proportion of construction land area to urban area, in 2005, the construction land area of Greater London was 1,596.2 km^2, taking up 23.7 % of the gross planning area of Greater London; the construction land area of Ile-de-France reached 2,723 km^2, accounting for about

23.4 % of gross area of Ile-de-France; HK construction land area only constituted 23.4 of gross land area in HK; however, in 2003, Shanghai construction land area represented nearly 30 % of gross urban area [1]. The constant growth of construction land area aimed primarily to cater to people's demand for houses. As population growth went beyond expectation and planning was divorced from practical development, there were a growing number of depressing skyscrapers and outdated supporting service facilities that constantly worsened urban environmental quality. This could be embodied in the comparison of land composition between Shanghai and New York in 2002. Taking green ratio for example, in 2002, entertainment land of New York covered an area of 157.08 km^2 (corresponding to the municipal green land contained in Shanghai construction land), constituting 25 % of total construction land, while Shanghai urban green land covered an area of 187.58 km^2, only representing 8.3 % of total construction land.

On December 3, 2002, Shanghai was granted the right to host World Expo 2010. In the meantime, in a new *Shanghai Urban Master Planning*, Shanghai Government declared that Shanghai would be built as an international metropolis with four centers, including international economic enter, financial center, trade center and shipping center. By dint of this opportunity, Shanghai expected to control population growth and urban expansion in Shanghai and improve urban environment by amending *Shanghai Town Planning Ordinance* published in 1995 as a new policy. Urban environmental reform placed focus on urban center, whose environmental problem was mainly reflected in overhigh floor area ratio, and excessively small open space and green land. Against this backdrop, Shanghai Government put forward *Two Increases and Two Reductions* Policy in 2003.

2 The Proposal of Two Increases and Two Reductions Policy

In October 2003, the fifth Planning Working Conference in Shanghai firstly made clear the thoughts of Two Increases and Two Reductions. Two increases referred to the increases of public green land and public space in urban center, while two reductions meant the reductions of floor area ratio and total building quantity. Regulations of Shanghai Municipality on Urban and Rural Planning published on January 1, 2011 emphasized again that the formulation and implementation of urban center planning should attach importance to the increase of public green land and public space as well as the control of total building quantity and high-rise buildings.

Shanghai government repeatedly stressed that Two Increases and Two Reductions Policy was only applicable to urban center because the problems of overhigh floor area ratio and overmany high-rise buildings were centralized in central city. Additionally, from 2003 to 2011, the specific contents of two reductions were modified from the reduction of floor area ratio and total building quantity to the control of total building quantity and high-rise buildings, which reflected that the execution of policy faced the problem of rigid uniformity, namely, the universality of policies and standards. The application of control implied that even in urban center, the non-core urban business areas or the major settlements of urban population with overlarge floor area ratio and building quantity should reduce the floor area ratio and building quantity to a larger degree in accordance with the actual condition, but the areas which were of proper or slightly small floor area ratio and building quantity as well as of development potential should maintain or slightly reduce the floor area ratio and building quantity. This would encourage the exploitation and then the development of these areas, and meanwhile, proper control was required besides the focus on development. Since the execution of this policy over 10 years ago, the contents of two increases have kept unchanged. This meant that it was a long-term and tough process from the execution of this policy to the realization of expected objectives, which could not be accomplished overnight.

Two Increases and Two Reductions Policy was executed to heighten the urban environmental quality of Shanghai, control the population growth speed in urban center, alleviate the social contradictions and conflicts, and control urban construction behavior, which was aimed at forging Shanghai as an international metropolis and a livable eco-city. It reflected the planning thought of "organic decentralization, people orientation, ecological priority, and smart growth", which has presented a useful solution for the existing problems in Shanghai urban society [2]. At present, China's urbanization level has surpassed 50 %. What is more, Shanghai has stepped into the post-industrial period, in which urban development has shifted from quantitative change to quality change. As a result, it is inappropriate to measure economic and social development level with GDP as before. Instead, urban construction needs to pay more attention to humanistic care, giving priority to the facilities that can improve people's life quality, which is the function that should be exerted by the Two Increases and Two Reductions Policy. In addition, in order to cope with the population growth which cannot be accurately estimated by means of conventional planning prediction, this policy can be executed to control population growth speed in the hope of mitigating the population pressure on Shanghai.

3 Feasibility Analysis on Two Increases and Two Reductions Policy

As the literal meaning shows, Two Increases and Two Reductions Policy was expected to address the problems concerning land development and environmental improvement. Besides high development intensity analyzed above, Shanghai land utilization status was also characterized by low land benefit. In 2002, Shanghai GDP stood at RMB 624 billion, with construction land GDP reaching RMB 221 million/km^2 [3]. Compared with other Chinese cities, Shanghai took the lead in construction land benefit, whereas it only constituted a fraction compared with metropolises in developed countries. For example, in 2006, New York created GCP of US \$ 47.8 billion, and construction land output value of US \$ 772 million/km^2 (RMB 6.15 billion/km^2 based on the exchange rate of US dollar to RMB in 2006), which evidently showed that there was a huge difference in construction land benefit between Shanghai and foreign developed regions. Another characteristic was that the growth speed of total land utilization quantity exceeded land requirements presented by economic growth and population growth. The output value of the secondary and tertiary industries increased by around 47 % from 1999 to 2002, whereas the built-up area by 100 % surprisingly. From the added values of the secondary and tertiary industries created by the every square kilometer of built-up area in 1999, 2000, and 2002, it can be seen that the figure declined from RMB 654 million/km^2 in 1999 to RMB 479 million/km^2, which suggested that since 1999, land utilization of urban built-up area has showed an uneconomical trend, and besides, urban economic growth has become more reliant on the extensive utilization of urban land, with total urban land utilization amount out of the control [3].

In terms of urban environment, in addition to the depressing skyscrapers and insufficient open space, the crowd in urban center, especially CBD section, proved to be one of the problems that resulted in traffic jam and parking difficulty. Two Increases and Two Reductions Policy produced the effects as below: controlling development intensity, tapping the potential of existing construction land, curbing the suburbanization process by comprehensively developing the land to raise the unit output and intensive land application, delivering more open space and green land, controlling the quantity and concentration degree of huge buildings to avoid the jam of pedestrian flow and traffic flow [4]. According to the analysis on the problems of urban development and the expected effect of the policy at that time, Two Increases and Two Reductions Policy was effective.

4 Defects of Two Increases and Two Reductions Policy

Government must make corresponding adjustments with the change of the times and the environment so as to suit the development and reform of the times. Otherwise, the result of policy implementation may go against the intention of policy formulation. Since the formulation and implementation over 10 years ago, Two Increases and Two Reductions Policy has indeed made some achievements. However, as an international metropolis with four centers, Shanghai has kept its population soaring in recent 10 years, which suggests that a growing number of people will settle down in Shanghai in the future, and meanwhile, urban center will still serve as the major carrier of resident population. The central aggregation of such population will be of positive significance for the improvement of urban center's vigor and reinforcement of its core position. On this basis, this policy will be no longer applicable to some regions. For example, core business areas, such as Xujiahui and Lujiazui, should enhance development intensity, attract settlers and raise popularity on the premise of assuring proper public space and green land.

Since full implementation, this policy has not only developed perfect laws and regulations and multilevel planning system, but also worked as the compulsory requirement for Shanghai to formulate regulatory planning. However, its implementation process handled all problems according to single standard without flexibility. Though this policy has defined the applicable scope (urban area) and clearly stipulated the upper limit of floor area ratio (floor area ratio of residence shall be no more than 2.5 and that of office building 4), single standard was still applied in different regions of one central urban area in the planning implementation process. There are eight Central urban districts in Shanghai such as Huangpu, Xujiahui and Changning, which take on different development status with different functional orientations. Thus, future development goals cannot be treated in the same manner. In view of this, Shanghai is expected to combine with local conditions in the future to flexibly handle the relationships among floor area ratio, total building quantity, high-rise building, green ratio, public space and other related factors. Additionally, Shanghai is required not only to encourage core areas to accommodate more population by regulating market supply and land development intensity, but also to improve environmental quality by means of land functional replacement, old town transformation, urban renewal in the backward old towns. Meanwhile, Shanghai needs to control population growth and alleviate the contradictions between the increase of resident population and the insufficiency of supporting service facilities.

Two Increases and Two Reductions Policy was originally intended to reduce floor area ratio and high-rise buildings and then the quantity of the residential buildings and commercial office facilities in central urban districts, which aimed to further dampen excessively rapid population growth by decreasing the resident and employed population. In fact, influenced by other factors, Shanghai keeps a rapid growth in actual resident and employed population every year. However, without corresponding adjustments for land supply policy, the pure reduction of urban land supply will inevitably lead to the tense and contradictory demand-supply relationship and finally the sharp rise of land price. Then, the relocation compensation cost of old towns will increase with land price. In the face of the increased land cost and development cost, real estate developers have to raise housing price. As a consequence, urban residents will be unable to afford the house and forced to settle down in suburbs. This result completely goes against the original intention of Two Increases and Two Reductions Policy.

5 The Implementation of Two Increases and Two Reductions Policy

Since the release of *Two Increases and Two Reductions* Policy in 2003, Shanghai Planning Bureau, coupled with planning bureaus in different districts, has classified the projects left by history. For example, in 2004, central urban districts have sorted out 376 projects left by history, including 326 development projects with high floor ratios, and besides, proper treatment was carried out for not only 4 construction

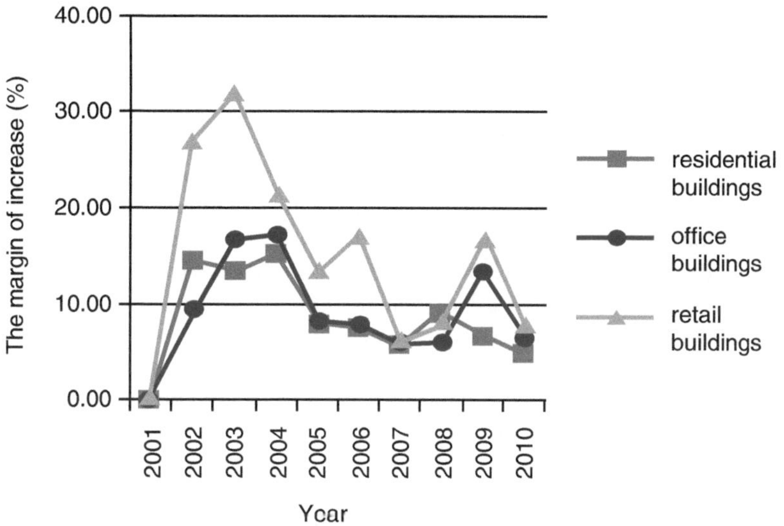

Fig. 2 Variations in categories of Shanghai's land area (2001–2010) (Source: The statistical yearbook of Shanghai (2002–2011))

projects with residential floor area ratio exceeding 2.5 and commercial office floor ratio surpassing 4, but also other construction projects inconsistent with planning requirements. After controlling, adjusting and optimizing development projects with high floor area ratio, the projects substantially decreasing such ratio constituted 85 %, with the total floor area falling by 3,685,000 km^2 and 15.3 % [5]. The figure below shows the increase of residential building area, commercial-residential shop area and office building area from 2001 to 2011.

According to the analysis in Fig. 2, the period from 2003 to 2004 witnessed the growth peak of residential building area, reaching 15.22 %, because the *Two Increases and Two Reductions* Policy was just put into practice without making good achievements. Additionally, this was closely associated with the impact of housing policy reform in 2003 on real estate market. Afterwards, the increase saw a decline from 2004 to 2005 and arrived at 4.84 % in 2010 at the lowest level, which reflected that *Two Increases and Two Reductions* Policy has achieved initial success. The floorage change of office building and store was also largely identical with minor differences. The major difference was that there was a huge increase during 2008 and 2009 because world expo was going to be held, because of which developers and investors increased development efforts.

In the past 10 years, *Two Increases and Two Reductions* Policy has made some achievements, but it was supposed to pay more attention to its disadvantages. In the future, how should this policy develop and improve itself to cater to the requirement of urban construction in new ages? With this question, the author primarily makes an analysis on three aspects including urban design, traffic impact assessment and resilient city.

In urban design, consideration should be given to the important factors, such as urban skyline, landmark building and building size. The changeful urban skyline and magnificent skyscrapers can provide visual shock and enjoyment for urban residents to enhance their identification with the city. Therefore, in the future implementation of this policy, single standard should not be adopted to impose constraints on the building size of construction projects in all blocks of central urban districts. Additionally, under the guidance of integrated urban design elements, all constructed projects need to enrich urban landscape and improve cultural influence. Besides, appropriate policy relaxation is required to attract developers with incentive mechanism of urban design so as to carry out the construction of public service facilities. At the same time, efforts should be spent in forging an intensive and high-dense urban center by increasing floor area ratio as well as height and quantity of high-rise buildings, so as to create central urban landscape with modern characteristics. For the purpose of controlling population and building quantity, the policy can implement dynamic equilibrium for the controlled quantity that is decomposed according to the formulation unit, and besides, the region which is positioned to be with low function should hinder the development with lower floor area ratio and control the quantity of high-rise buildings.

The urban construction activities, such as the renovation of old towns, and the development of new urban land, will be sure to cause the change of regional traffic flow, which will be turned into the important factor that influences commuting. The preliminary formulation of this policy failed to take into account this detail. Thus, full consideration should be given to the implementation, feedback and adjustment of this policy in the future. At present, traffic impact assessment in the countries such as U.S. is always deemed as the project development prerequisite of policy object, with developer inviting tenders from consulting company. Then, the consulting company will submit traffic impact assessment report to local government. If the development project fails to meet the stipulated service standards of traffic facilities, developers must amend their development schemes or assume the construction expense of the traffic facilities with the same influence so as not to impose burdens on periphery traffic facilities [6]. In the future, *Two Increases and Two Reductions* Policy should not consider the upper limit of floor area ratio and the quantity of high-rise buildings as the only factor influencing development intensity. Instead, it is supposed to bring in the traffic impact assessment factors of construction projects, comprehensively analyzing and considering local status, geographical condition, land utilization structure, traffic condition, planning purpose, project's functional orientation, traveling habit of the settled group, road network planning, traffic facility, etc. On this basis, *Two Increases and Two Reductions* Policy needs to present the legal control measures for development intensity based on scientific theory. This aims to ensure that under such development intensity, traffic facilities are able to meet the living and working requirements of residents in good commuting status in local region, without imposing pressures on the traffic facilities of surrounding area.

From the hurricane Katrina in 2005 to Ya'an Earthquake in 2013, it can be seen that traditional cities appear to be vulnerable to major natural disasters. Against this backdrop, overseas scholars have firstly presented the theory of resilient city. In their view, resilience refers to a system's ability to remove interference and maintain its basic structure and function. That is, when dealing with the external force arising from the changes of various factors, such as natural environment, social factor and political factor, city, society and people show some adaptability and self-resilience ability [7]. To a certain degree, this original intention is consistent with the *Two Increases and Two Reductions* Policy published before presentation of resilient city theory: the increased public space and urban green land can be applied to leisure in daily life. In case of any major natural disaster, they can be automatically turned into the shelter, distribution center of relief materials, place used by relief agency, etc. Meanwhile, the reduction of floor area rate for the purpose of controlling the excessive population growth can also better the operation status of infrastructure system, especially the ultimate load of urban traffic system. Then, in case of any disaster, there will be more flexible dispatching space and available equipment. The coincidence between *Two Increases and Two Reductions* Policy and resilient city theory suggests that in the future, this policy will still become effective. Under the guidance of resilient city theory, there is a need to make clear the objective and significance of increase and reduction, subdivide the scope, project, and quantity index of increase and reduction, which can ultimately contribute to the realization of flexible and healthy urban development, the high-efficient response to emergencies, and the diversified urban living environment.

6 Conclusion

Two Increases and Two Reductions Policy is an attempt made by Shanghai government to address the problems, such as the rapid growth of Shanghai population, and the worsening of urban environmental quality. As the inevitable choice for Shanghai, *Two Increases and Two Reductions* Policy is of far-reaching significance for Shanghai to construct a modern international metropolis and a livable eco-city. However, in combination with various factors, such as practical development of Shanghai, integrated urban design, traffic impact assessment, and resilient city, long-term efforts and explorations still need to be made, so as to scientifically increase and reduce floor area ratio, high-rise building, green ratio and public space and then realize the healthy urban development.

References

1. Shi Y, Zhihong P, Yongjian C (2009) The comparative study on land scale of the international metropolis. China Building Industry Press, Beijing
2. Monographic Conference (2012) The Shanghai municipal planning and land resources Bureau's work conference on the implementation of two increases and two reductions policy. Shanghai Urban Plann 5:71–73

3. Jing Dong (2005) The research on the characteristics of Shanghai land use amounts. In: Proceedings of the international conference on "The urban rational growth and land policy in China"
4. Liu Hui (2012) Shanghai construction land situation analysis and countermeasures of industrial land. Shanghai: Shanghai Land Resources
5. Guorong M (2004) Two increases and two reductions shows its achievements. Shanghai: Shanghai People's Congress Monthly
6. Pan Dongbo, Yangtianyi, Huang Tongyuan (2004) Traffic impact assessment system on urban construction project. J Chongqing Univ 2004(07):75–77
7. Jianming C, Hua G, Degeng W (2012) Studies on foreign resilient city. Progress Geogr 2012 (10):1245–1255

Facade Renovation and Utilizing Ways in the Historical Commercial Streets: Rigid Planning at the Compound Level Needed in Pingyao Ancient City in China

Chenchen Li, Jie Yin, and Yang Lei

1 Introduction

The World Relics Committee of UNESCO highly praised the Pingyao ancient city as the outstanding example of Chinese Han nationality cities during Ming and Qing Dynasties, remaining all features of such periods, Pingyao ancient city reveals a picture of unexpected cultural, social, economic and religious development in Chinese history.

The ancient city walls, streets, buildings and traditional customs has attracted over 2.6 million visitors annually since the Pingyao ancient city was on the World Cultural Heritage List. In successive years, tourism has become the fastest growing one among the Pingyao County's main economic indicators [1]. The historical "shi-glyph" commercial streets in the ancient city is the carrier of its tourism development and also protects the inheritance of the historical and cultural style and features. Along with the rapid development of the tourism commercialization, tourism-related shops on the historical "shi-glyph" commercial streets is on the increase, and the style and features of the streets have changed.

We went to the Pingyao ancient city in Shanxi in July 2012, and mainly studied the utilizing ways and the facade style and features of more than 400 shops along the "shi-glyph" commercial streets (Fig. 1), trying to explore the changing characteristics of the streets under the background of tourism commercialization. The

C. Li (✉) • Y. Lei
School of Urban Planning and Design, Shenzhen Graduate School of Peking University, Shenzhen 518055, China
e-mail: noseful@126.com

J. Yin
School of Urban Planning and Design, Shenzhen Graduate School of Peking University, Shenzhen 518055, China

School of Urban and Environmental Sciences of Peking University, Beijing 100871, China
e-mail: yinj@pkusz.edu.cn

X. Chen, Q. Pan (eds.), *Building Resilient Cities in China: The Nexus between Planning and Science*, GeoJournal Library 113, DOI 10.1007/978-3-319-14145-9_12

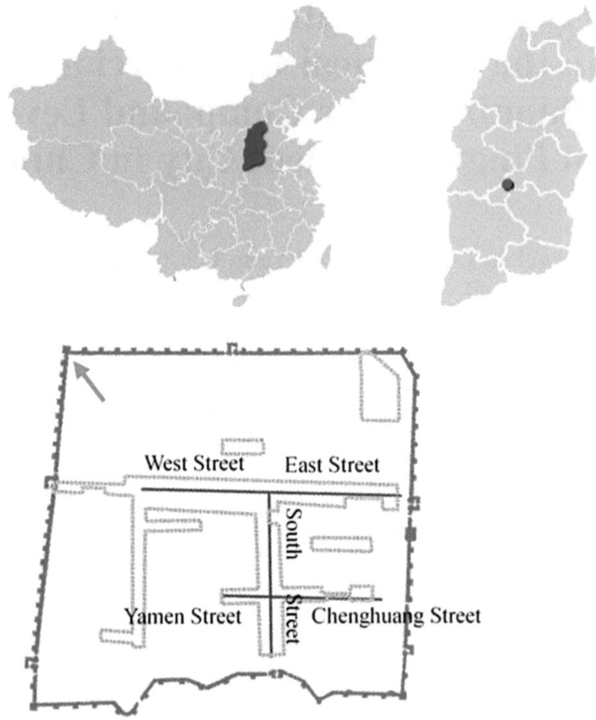

Fig. 1 Research scope in Pingyao ancient city of Shanxi province: the historical "shi-glyph" commercial streets

utilizing ways refer to the management content of the shops and its use of commercial space.

The "shi-glyph" commercial streets of Pingyao ancient city includes West Street, East Street, Yamen Street, Chenghuang Street and South Street. They are also the primary protected streets in Pingyao County Master Plan in the core protected areas in 2000 [2]. The "shi-glyph" commercial streets have been the most prosperous historical business district in the ancient city. Tourism-related shops and also the cultural relics and historical buildings, such as the Rishengchang, Qingxuguan temple, Shilou, Chenghuang temple, Confucian temple, Seat of the County Government and others contribute to the popularity of the streets.

2 Classfied Utilizing Ways

The four hundred and fifteen researched shops along the "shi-glyph" commercial streets are divided into five categories according to the utilizing ways: 1. Dining 2. Hostel 3. Tourism commodity 4. Tourism services 5. Non-tourism-related. 1–4 categories are tourism-related (Table 1). Tourism-commodity shops account for 67.0 % and constitute a vast majority. Hostels account for 14.5 % as the second.

Table 1 Classified utilizing ways of the shops along the streets and the quantity

Classification	Operating subdivided	Quantity
Dining	Restaurant, Snack shop	33
Hostel	Hostel	60
Tourism commodity	Handicraft shop, Specialty shop, Antique shop, Clothes shops, Tea shops, Supermarket, Video and book shop, Instrument shop, Creative products shop, Drug shop, Toy shop, Optical shop	278
Tourism services	Bar, Pedicure, Bicycle rental, Bicycle repair, 5D cinema	13
Non-tourism-related	Grain shop, Grocery store, Snack shop, Supermarket, Watch shop, Barber shop, Shroud shop, Living Furniture shop	31

Table 2 Newly increased tourism-related and non-tourism-related shops since 1997

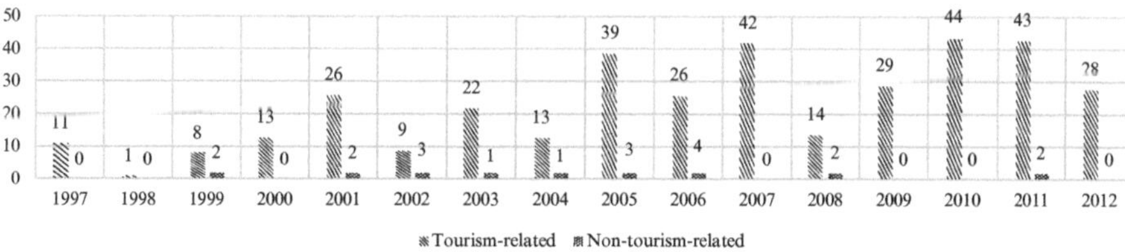

The tendency of tourism commercialization of the Pingyao ancient city becomes more obvious since it proved itself to be a world cultural heritage. Non-tourism-related shops only account for 7.5 % in 2012 and its proportion in total quantity is decreasing every year for the last 15 years (Table 2). Shops for tourists in the "shi-glyph" commercial streets are far more numerous than those for local residents.

The spatial distribution of utilizing ways of the shops along the "shi-glyph" commercial streets is shown in Fig. 2. They have a different distribution in the five streets (Table 3). Most shops only take advantage of the buildings by the streets that are part of a compound, is used for commerce. Others take full advantage of a whole compound, gathered in the West Street, South Street and Yamen Street (Fig. 2). The number of whole compounds used for commerce is 82.

3 Compound Façade Renovation Patterns

Based upon our research, there are five facade renovation patterns summarized for the compounds along the "shi-glyph" commercial streets (Table 4).

Pattern 1–3 have an extremely strong influence on the compound facade, directly perforating the wall or adding a new architecture, therefore they destroy the traditional style and features of the facade more. Over all these three patterns account for a small share of the total (3.8 %).

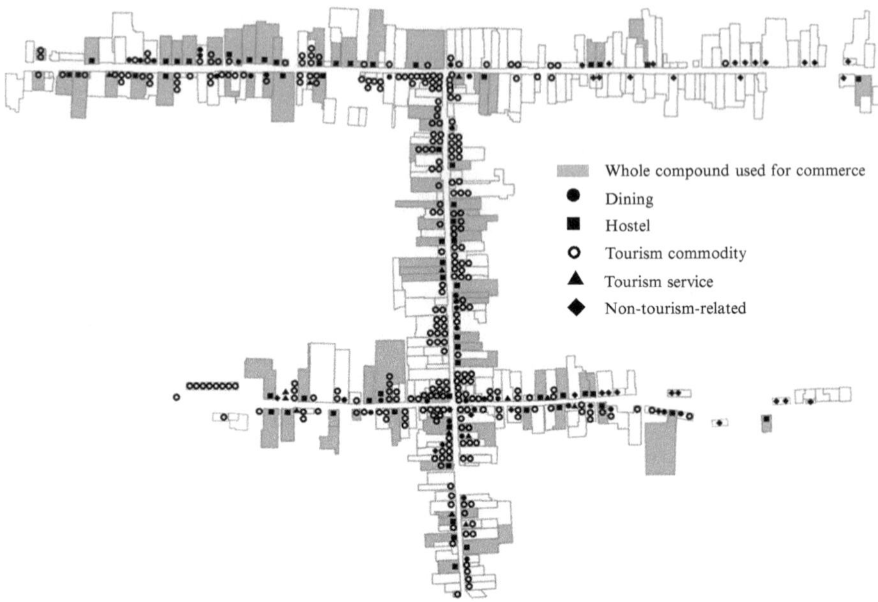

Fig. 2 Spatial distribution of the classified utilizing ways of the shops along the streets

Table 3 Quantity of the classified utilizing ways of the shops in the five streets

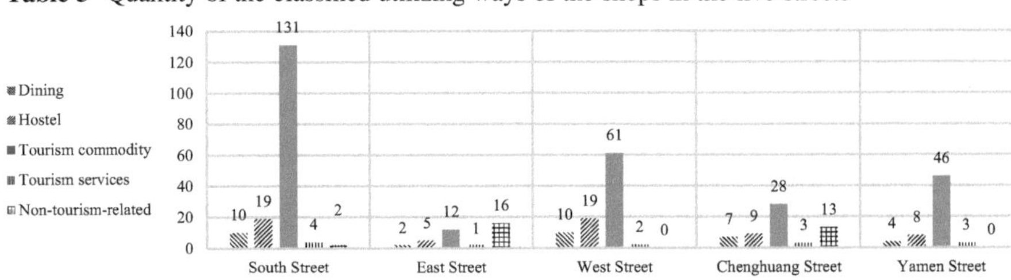

Pattern 4–5 renovate the form (this mainly refers to the doors and windows), the materials, the color and other aspects of the compound facade. They have a bit weaker influence on the compound facade compared to the former three patterns, but these two patterns account for a large number (Pattern 4 accounts for 59.1 % and Pattern 5 accounts for 37.1 %). From the point of view of the use of the commercial space, Pattern 4 is a complete utilization of compound facade along the streets. The shop with a completely utilized facade is managed by one shopkeeper and the shopkeeper renovates the facade relatively as a whole. It can better protect the traditional style and features of the ancient city streets. There is usually a change in Pattern 4 that the windows are renovated to doors and thus increases the accessibility of the shops (Fig. 3). Some other changes are the materials or color. Pattern 5 is the incomplete utilization of compound facade. The facade of a compound along

Table 4 Façade renovation patterns of the compounds along the streets

Pattern	Original Facade	Renovated Facade	Photograph	Quantity
Pattern 1	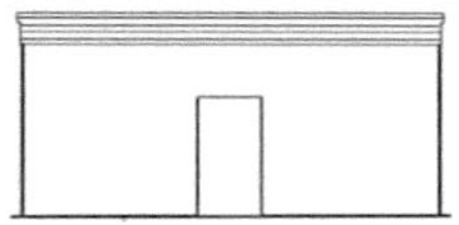 On the front elevation	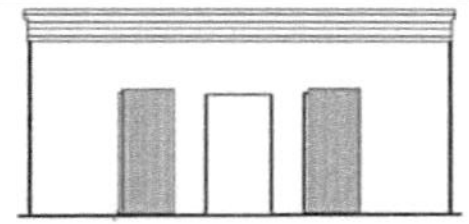 Perforating the wall		3
Pattern 2	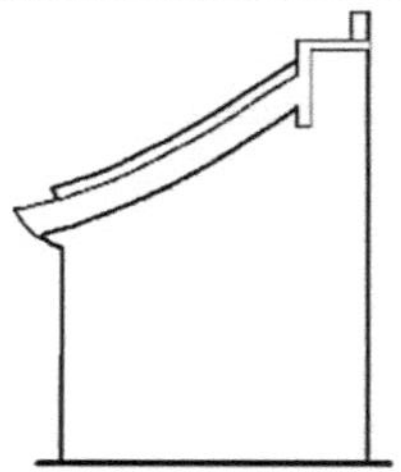 On the side elevation	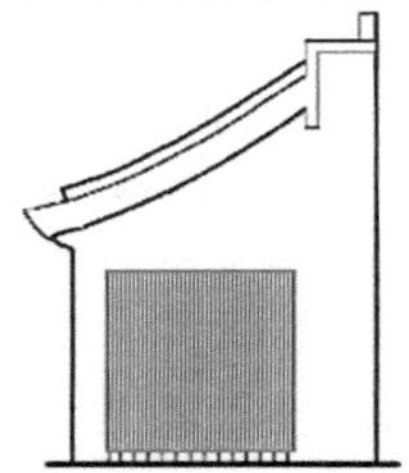 Perforating the wall		5
Pattern 3		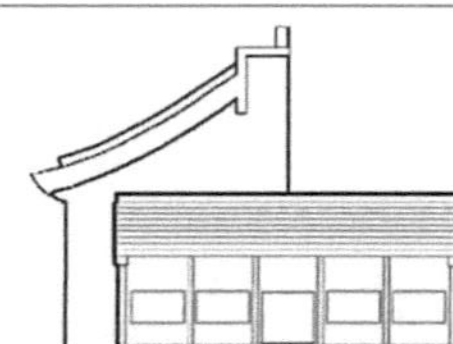 Additional architecture		1

(continued)

Table 4 (continued)

Pattern	Original Facade	Renovated Facade	Photograph	Quantity
Pattern 4	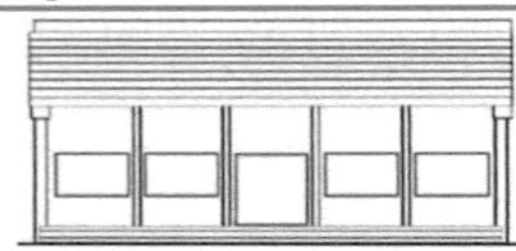 Facade complete utilization	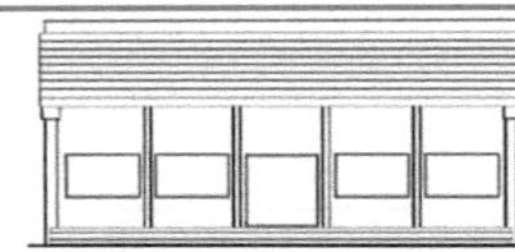 Changes of form, material, color and others	 	140
Pattern 5	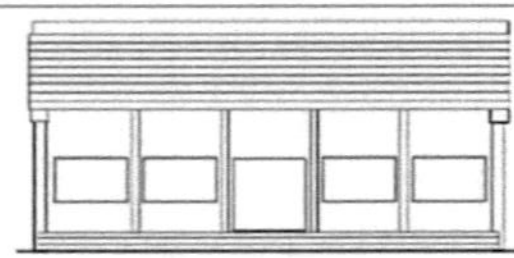 Facade incomplete utilization	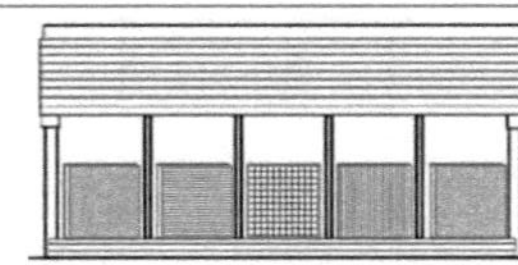 Facade subdivision		88

The original facades are made reference to The Ancient City and Local Style Houses in Pingyao, published by Tianjin University Press [3]

Fig. 3 Compound facade renovation along the streets in Pattern 4 (The first photography keeps the original facade form and expands the windows to show the inner space of the hostel. The second one changes the windows to doors on the facade)

Fig. 4 The facade subdivision of Pattern 5 (A complete facade is divided by several shops utilized in different ways. The first photograph shows that the facade of the compound along the South Street has three openings and each opening is occupied by one shop. The shops include a pedicure salon, a dining bar and a tourism-commodity shop)

the street is divided into several parts for several shops and shopkeepers have different renovating activities on the subdivided facade. Facade subdivision causes the traditional style and features of the streets to be disharmonious (Fig. 4).

Pattern 5 accounts for a large number and has a strong influence on the style and features of compound facade along the streets. We can see its wide distribution in West Street, Yamen Street, Chenghuang Street and South Street (Fig. 5). Only East Street has a cluster of non-tourism-related shops with almost no Pattern 5. There are also 19.5 % of the whole used compounds for commerce facade subdivided (Fig. 5) and the other whole used compounds have completely utilized facades. The destruction impact caused by its wide distribution in both the traditional "shi-glyph" commercial streets and the compounds is evidently severe.

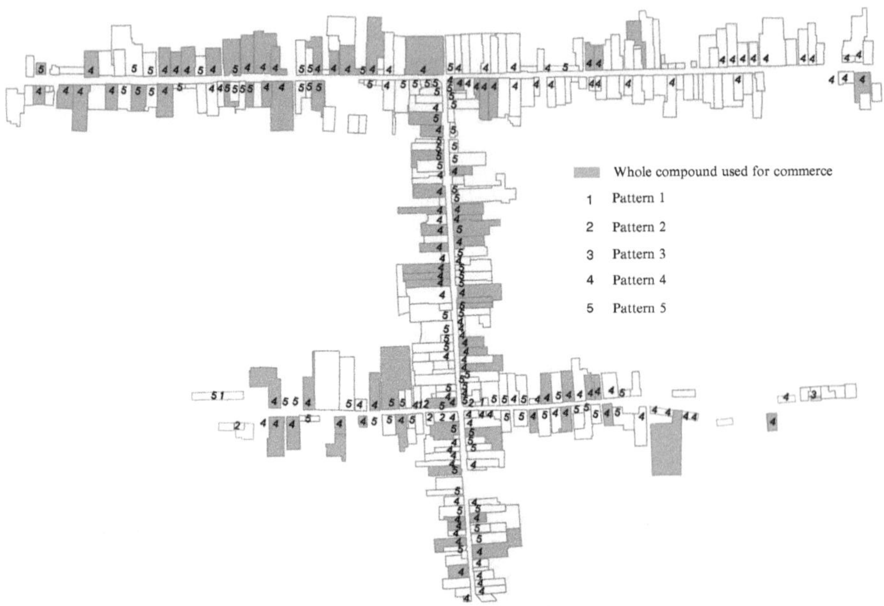

Fig. 5 Spatial distribution of the compound facade renovation Patterns along the streets

4 Relationship Between the Utilizing Ways and the Compound Façade Renovation Patterns

From the comparison between the spatial distribution of the facade renovation patterns (Fig. 5) and the utility patterns of the shops along the "shi-glyph" commercial streets (Fig. 2), we can see that the shops with different utility ways have different facade renovation patterns.

Table 5 shows that the Pattern 1, Pattern 2 and Pattern 3 which are set as the most destructive patterns to the traditional style and features of the streets are more for the tourism commodity shops. There are totally 216 shops distributed in the 88 compounds that are facade subdivided. Every compound has more than two shops in it in average. In the meantime, 82.4 % of the 216 shops are tourism commodity shops which is also the main utilizing way to the facade subdivision. Paying close attention to the tourism commodity shops and its facade renovation has become a very important point to the protection of the traditional style and features of the streets. Dining shops account for 19.3 % and are the second most utility way using facade subdivision. The facade renovation of hostels and non-tourism-related shops primarily use the Pattern 4, which treats the compound facade as a whole. Furthermore, the renovating activities can protect the traditional streets better. The tourism service shops, which are the minority utilization way, use the Pattern 5 in its facade renovation.

Table 5 Corresponding quantity between utilizing ways and compounds façade renovation patterns

Pattern	Total quantity (compound)	Utilizing ways and quantity (shop)				
		Dining	*Hostel*	*Tourism commodity*	*Tourism services*	*Non-tourism-related*
Pattern 1	4	—	—	3	—	—
Pattern 2	5	1	—	4	—	—
Pattern 3	1	—	—	1	—	—
Pattern 4	140	10	36	58	3	14
Pattern 5	88	22	7	178	6	3

5 Discussion

The protection and inheritance of the historical and cultural style and features of the Pingyao ancient city is the motivating force behind the tourism and commerce. The"shi-glyph" commercial streets have become an important space, balancing the old city's protection and the tourism development.

Shanxi Pingyao Ancient City Protection Regulation (1999) [4], Pingyao County Master Plan (2000) [2] and other relevant plans take the traditional "shi-glyph" commercial streets as the primary protected area and streets. The architectural appearance is strictly protected. The form, materials and color of the facade may not be arbitrarily changed. Disharmonious construction should be removed or renovated in order to restore the historical style and features. These are all rigid rules trying to preserve the traditional style and features at the street level. The relevant plans also permit the diversity of the utilizing ways, such as hostel, dining, handicraft shop, museum, traditional entertainment, and tourism-related product making. However, the style and features of the facade along the streets has been greatly influenced by the tourism commercialization at the compound level in fact. The facade subdivision of the compounds and the renovation of the facade form, materials and color have especially been well beyond the framework of the relevant protection and utilization.

Formulating limited rules of the relevant protection plans of the Pingyao ancient city to protect the traditional style and features only at the street level is obviously not enough. All suffer from the absence of more detailed control measures to the facade renovation at the compound level along the streets. The diversity of the utilizing ways is mainly reflected in an architecture or a compound. This is a very important reason that plans now cannot control the damage to the facades in the tourism development process. Shouldn't the planning make new adjustments to respond the new problem if we need to?

In the near future, we need to add the corresponding rigid rules at the compound level to control and guide the diversity of the utilizing ways. There can be multiple utilizing ways using the same architectural space of the compounds along the streets together, but the flexibility of this utilizing way cannot go beyond the framework for the relevant protection and utilization. It requires rigid planning at the compound level to keep the flexible changes in utilizing ways of the shops within bounds and to ensure the integration of the style and features of the traditional streets and facade.

6 Conclusions

This study shows that there are different facade renovation patterns of the compounds along the "shi-glyph" commercial streets in Pingyao ancient city. The figure of the spatial distribution of classified utilizing ways of the shops along the streets makes it clear to identify the relationship between the facade renovation patterns and the utilizing ways. Facade subdivision caused by several different utilizing ways of shops has a wide and strong influence on the style and features of the compound facade along the streets. Comparing the complete utilization and subdivision of the façade, the former can protect and inherit the historical and cultural style and features better than the latter.

Discussions concerning rigid planning at the compound level must be made to keep the flexible changes in utilizing ways of shops and its facade subdivision activities within bounds. Recently, the relevant protection plans for the Pingyao ancient city concerning the protection of the traditional historical facade and space at the street level is not effective. The changes of the style and features of the compound facade along the streets have proven this well. The diversity of the utilizing ways are mainly reflected in an architecture or a compound. Multiple utilization ways using the same compound or architectural space along the streets together is acceptable, but it needs more detailed and rigid rules at the compound level to be under restraint. How planning will control and guide these facade changes in the future and find the balance between the protection of the ancient city and the tourism development shall need additional research efforts.

Acknowledgments The authors are grateful to all the researchers from Peking University for their in-depth investigation and images acquisition of the shops along the "shi-glyph" commercial streets in Pingyao ancient city. They are Xu Chunhui, Liu Fengyou, Chen Guo, Song Zheng, Wang Pei and Zhou Yanlv.

References

1. Wang L (2009) Study on tourism development strategic planning of Pingyao ancient city[D]. University of Architecture and Technology, Xi'an
2. Shanxi (2000) Academy of urban and rural planning and design. Pingyao County Master Plan, Xi'an
3. Song K (2000) The ancient city and local style houses in Pingyao [M]. Tianjin University, Tianjin
4. Shanxi Provincial People's Congress (1999) Shanxi Pingyao ancient city protection regulation. Xi'an

Method Exploration of Sensitive Public Space System Planning in Urban Newly Areas on the Viewpoint of Systematology

A Case Study of Urban Design of Bali Lake District in Jiujiang City, Jiangxi, China

Mengchen Li and Ying Lin

1 Foreword

With the development society, changing times and innovation theory, the category of urban design as the initial simple city shape environmental design has gradually expanded to a wider form from the city to the realm of human behavior psychology, from the material space to the social culture, from the beautiful and comfortable to the natural ecology, from the functional design to sustainable development. As a result, nowadays urban design has developed into a rich connotation and denotation of the concept, which is associated not only with urban planning, architecture, also closely related to many other fields and disciplines [1, 2].

According to the general law of science, the greater range of system engineering is, the more difficult the operation will be completed [3]. This article selects the sensitive public space system as the research object in the integrated urban design, appropriately controls a range of systems engineering, from the perspective of system theory, explores the optimization design method of sensitive public space system, then extends to the whole integrated urban design process. Based on this, we explore for generality work ideas and methods for such a comprehensive systems engineering.

M. Li
Wuhan Land Use and Urban Spatial Planning Research Center, Wuhan, China
e-mail: 183185363@qq.com

Y. Lin (✉)
School of Architecture and Urban Planning, Huazhong University of Science and Technology, Wuhan, China
e-mail: updocly@foxmail.com

X. Chen, Q. Pan (eds.), *Building Resilient Cities in China: The Nexus between Planning and Science*, GeoJournal Library 113, DOI 10.1007/978-3-319-14145-9_13

2 Systematology of Sensitive Public Space Planning

2.1 Introduction of Systematology

Systems theory is made up of African American Austria biologist Ludwig Von Bertalanffy created in 1948 [4]. Taking system as the research object, systems science mainly from system perspective such as the system integration, internal relation, the relationship between system and external environment, the system dynamic development deliberates the operation law of systems. Systematic thinking model is brought up in system science, a new scientific way of thinking to solve the multi-factor complex, dynamic and changeable organized system based the people provides a holistic, comprehensive, hierarchy, dynamic and open principle [5]. To deal with the problem of complex integrated system optimization design, the introduction of systematic theory is of great help.

2.2 Systemic Integrated Optimization Mechanism

"Based on the perspective of systematics, the essence of the whole optimization system is under the condition of the system is a new force within the system characteristics of subsystems or elements of change, and also the consequences of its associated changes and restructuring so as to form to adapt to new work force (optimization) system [6]." Ludwig Von Bertalanffy said. Therefore, the system changes and the generation of the new system are based on the system property change of the internal subsystem or system elements. Structure change between each subsystem or system elements will make system change. Lastly the model of "analysis-transformation-stability" systemic optimization mechanism will be evolved [7, 8]. Based on this model, it can be generated out of the process and the method of "system construction" [9]. Through the method of "system construction", not only sensitive public space subsystems can be organized, but also the whole system can be generated, namely the overall structure of urban design. Here, we introduce the system construction of sensitive public space planning and the generation of subsystems.

3 System Construction of Sensitive Public Space Planning and Design

From the above description, according to the "analysis-transformation-stability" systemic optimization mechanism model, the whole system of sensitive public space will be generated after several steps. Firstly, we summarize it as evaluation subsystem, namely the analysis step, grasp of the inherent conditions and restrict factors of internal areas, and recognition of the external condition of master planning; Next is guidance subsystem, namely the conversion steps, integration of the elements of master planning and the status quo, and then optimization by combining their own characteristics and the internal correlation between each other with the formulation of the planning principles, target positioning and design concept, which covers and sums up the sensitive external public space planning and design of organization structure and inner spiritual kernel; Then generated by guiding them to consolidate sensitive public space planning and design achievement perfect management system, namely steady steps, that can be summarized as control subsystem.

In the end, the interrelated and interaction subsystems form an overall system of sensitive public space planning and design in integrated urban design, namely the "evaluation – guide – control".

4 Case Study of Urban Design of Bali Lake District

4.1 Introduction of the Case

4.1.1 Origin of the Case

The rise of Bali Lake district is the development strategy guidance of Jiujiang city, is the common goal of each planning and design level; Public service facilities in advance and landscape ecological network construction will be key decisions of inner urban to Bali Lake district. Under the guidance of the strategic goals, Bali Lake district has carried on the micro level of urban design and landscape design in a number of projects, and now, in urgent need of holistic and systemic consideration and refactoring these urban design projects are, for the future development of Jiujiang city. Therefore, integrated urban design is imperative, as shown in Figs. 1 and 2.

4.1.2 Research Framework

Based on this case, through empirical investigation and design participation, theoretically supported by system construction, research framework of "evaluation subsystem-guide subsystem-control subsystem" can be formed to explore the method of sensitive public space system planning and design, as shown in Fig. 3.

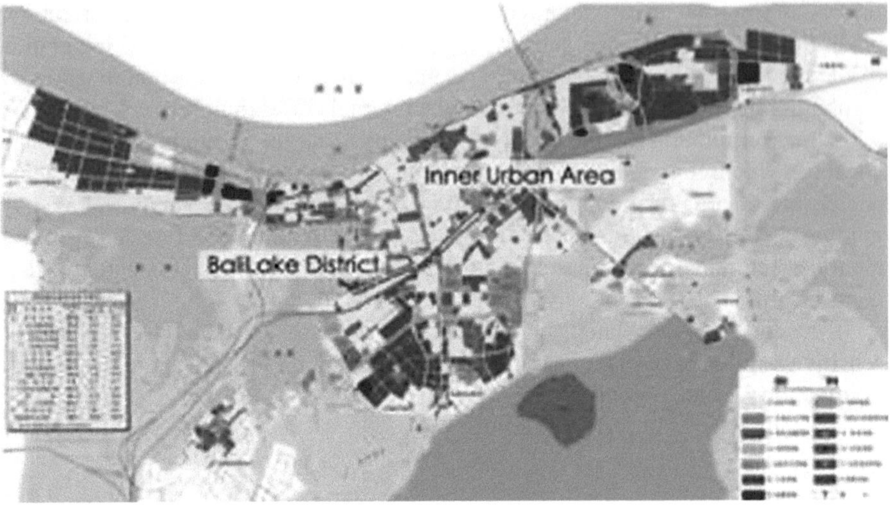

Fig. 1 Development relation

Fig. 2 Newly area rising

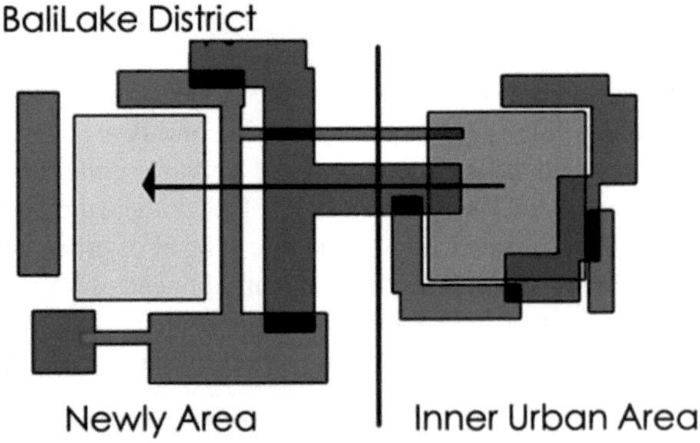

4.2 Evaluation System – Analysis of Background and Present Situation

In the evaluation subsystem, the grasp of the inherent conditions and restrict factors of internal areas, and recognition of the external condition of master planning, we explore its inner and outer factors providing the basis for the next step, integrated optimization design.

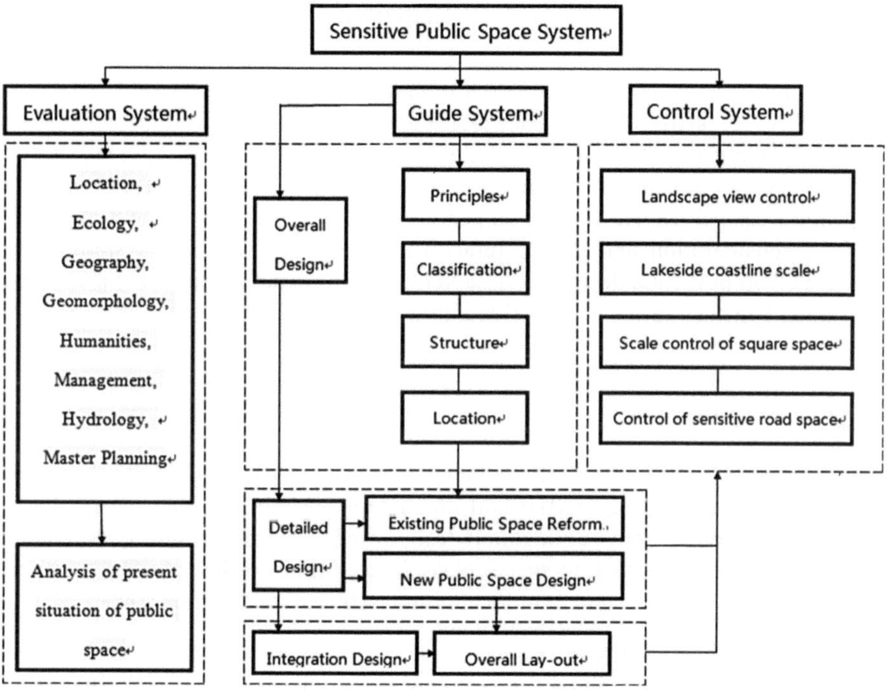

Fig. 3 Research framework

4.2.1 Overall Background

Based on considerations of external factors on the project, from location, ecology, geography, geomorphology, humanities, management, hydrology, master planning, analyze the external factors affecting sensitive public space system.

4.2.2 Present Situation

Consider sensitive public space within the system factors, carries on the analysis of present situation of public space, such as through the aspects of landscape planning area present situation analysis, evaluate its good natural landscape protection, lakeside landscape looked like interface, landscape type is missing, plot internal landscape system has yet to be perfect, there is insufficient space on the viewing public, etc [10, 11].

4.3 Guide System Systemic – Integrated Optimization Design

In the guide system, through the integration of design level and design activity, to the quality of sensitive public space environment can be improved from the macro level systematically. At design levels, integrate the framework of "overall design-detailed design-integration design". On design activities, different types of sensitive public spaces such as the ribbon park, commercial plaza are being designed by overall consideration and design refactoring. Through the guide system of the integrated optimization design, cover and outlines external organizational structure and inner spiritual kernel of sensitive public space system.

4.3.1 Overall Design

Sensitive public space refers to the urban public space which can be felt urban space environment quality, city image and characteristic, behavior etc. easily. In the process of system overall optimization design, we clearly confirm the classification principles of sensitive public space, classify according to the function, form the planning structure of system, and locate important sensitive public space.

In this case, bellows is the principles.

Characteristic principle: show urban architectural characteristics and space environment, improve the image of the city.

Perception principle: easy recognition and awareness.

Accessibility principle: the accessibility of behavior and the line of sight accessibility is good, while being used or viewing frequency;

Difference principle: elements of public space in both space and time level have unique functional or aesthetic significance.

Based on these principles, then the structure is formed, as shown in Fig. 4:

One circle: Around the Bali Lake circle;

Two zones: West ecological zone and southeast ecological zone;

Three Belt: Shili river ecological landscape belt, Jieliu river ecological landscape belt, Jiaotan Lake landscape belt;

Six Road: Around Lake east road, Victory avenue, Changhong west avenue, Qianjing west road, Yongning road, Shayan road;

Nine Nodes: Financial plaza, Changhong zone commercial center, Shengli Park, Boat dock, Relics Park, South cluster commercial center, Bali Lake Park, Sports Park, Administration center.

4.3.2 Detailed Design

This part emphasizes the integration of design activities, through specific design principle and overall structure in the overall design process; we do detailed design

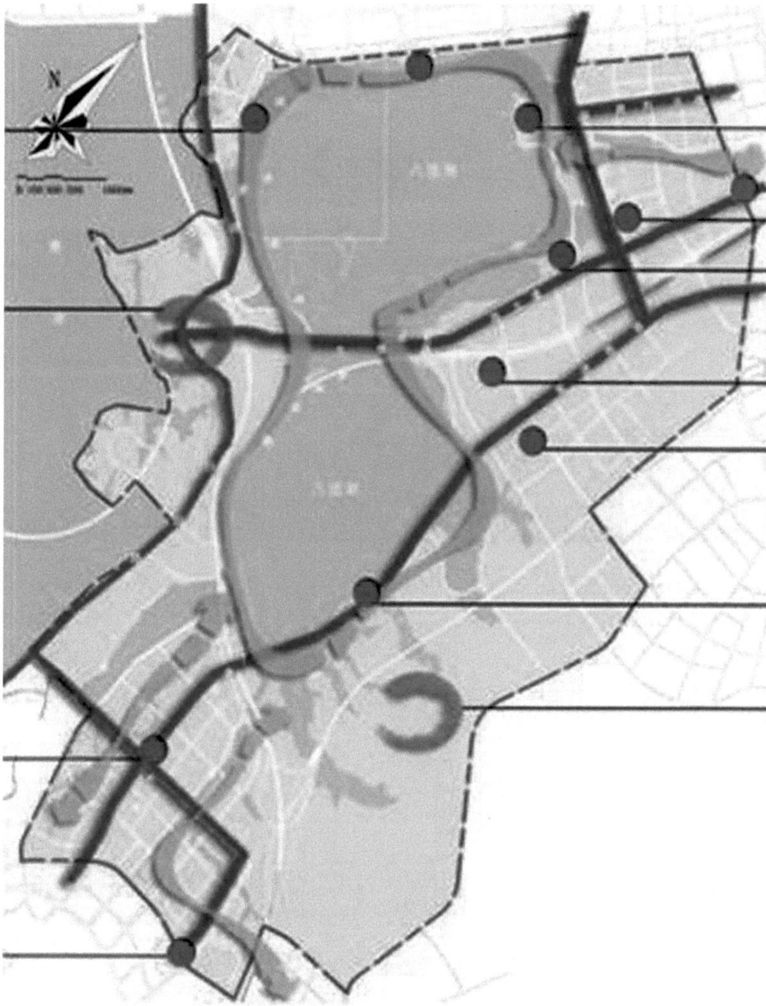

Fig. 4 Planning structure

of the existing or not sensitive public space to implement design essentials on overall level, as shown in Figs. 5 and 6.

Here is the sample of south group commercial center.

1. Design concept – Value achievement

Function value: satisfy people leisure, entertainment, shopping, business, trade and other activities;

Urban value: circle enclosing square fully expresses the city space form cluster, strongly controls the area center.

Fig. 5 South group commercial center I

Fig. 6 South group commercial center II

Cultural values: provide a self-presentation and ideas exchange 3d platform, reflecting the users' various potential positive behavior consciousness.

Aesthetic value: through the formation of appropriate scale size, good visual function, multi-layered composite landscape and delicate space environment, create aesthetic landscape matching people's psychological and physiological activities.

2. Guidance of the square architecture interface

The interface length of underlying business architecture should be appropriate, to create a friendly and pleasant street building space; Building interface length must not exceed 120 m length. Suggestion is given priority to high-rise building size with square and circular tower, that building length shall not exceed 40 m interface, and try to avoid plate level settings.

4.3.3 Integration Design

In a systematic, validity, organic, uniqueness, intuitive principle, we are doing integrated optimization design of sensitive public space system in Bali Lake district, to form a integration, coordination, systematic sensitive public space, as shown in Fig. 7.

4.4 Control System – Construction Management Control

In implementation of the integrated urban design activities, emphasizing the design control level is very necessary to form a consistent and sustainable design of control system, which can make the management agencies can guide and influence the planning and design work for the project development and urban promotion from overall to local, from macroscopic to microcosmic, to realize the overall goal of urban design.

In above two systems, the system as a whole makes arrangement and design to material elements of sensitive public space. In design implementation management process, which needs to guide and standardize the sensitive public space optimization design, the control system of construction and management of sensitive public space should be formed.

- Landscape view control for the important public space. View analyze to the important landscape through different angle and different distance, in order to control the buildings heights, as shown in Fig. 8.
- Lakeside coastline scale. We make continuity interface control for the lakeside shore plot, in order to improve the recognition of places and promoting interaction of construction and spatial. Mainly is control for the lakeside construction

Fig. 7 Integration design

interface and lakeside skyline. At the same time, we make indicators and patterned guide and control for the waterfront space, as shown in Fig. 9.

- Sensitive road space control. We make indicators and patterned control for buildings interface of sensitive road space, as shown in Fig. 10.
- Scale control of square space. Some requirement can be proposed to square space from the space dimension and construction interface scale elements. Firstly we encode for sensitive public space including road, lakeside and square plot, as shown in Fig. 11, and then we do detailed control in these code spaces.

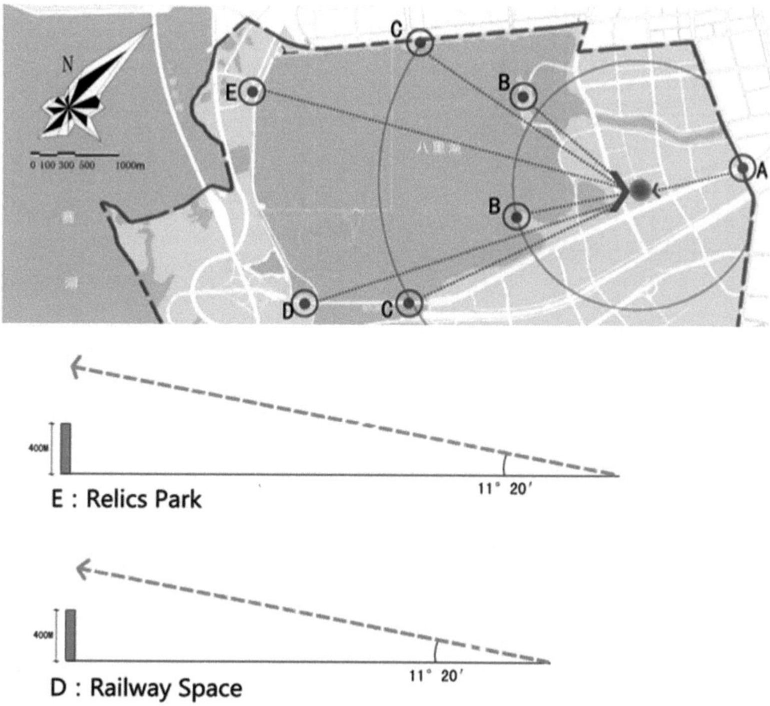

Fig. 8 View control

5 Conclusion

The super-large scale object of design work, both being in different periods of present situation, features and with different nature of the project design activities; not only in consideration of space structure in the macro level, but also with the balance of each construction key-point in micro-level, which involves fields being from pure urban design development to history, culture, ecological environment, urban management and some more categories, has become a comprehensive system engineering [12, 13].

Based on system science inquiry, With Urban Design of Bali Lake District in Jiujiang City as an Example, through system construction of sensitive public space system planning and design method, the purpose is to provide promotion significance of methodology for creating a macro integrated structure of urban form and space environment, in order to achieve the urban design goals, that is constructing ideal and good urban space environment, maintaining and creating urban space characteristics and quality.

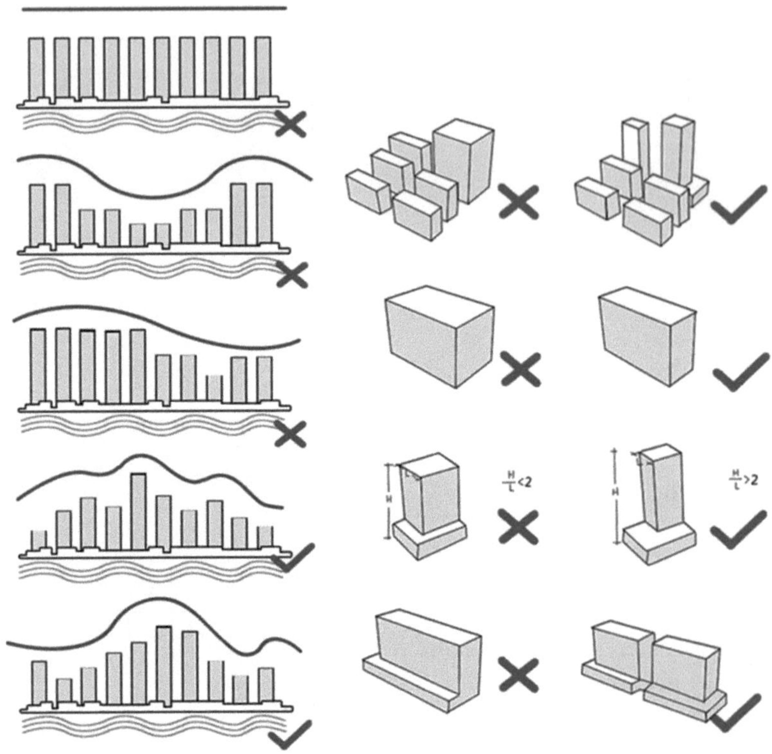

Fig. 9 Lakeside skyline control

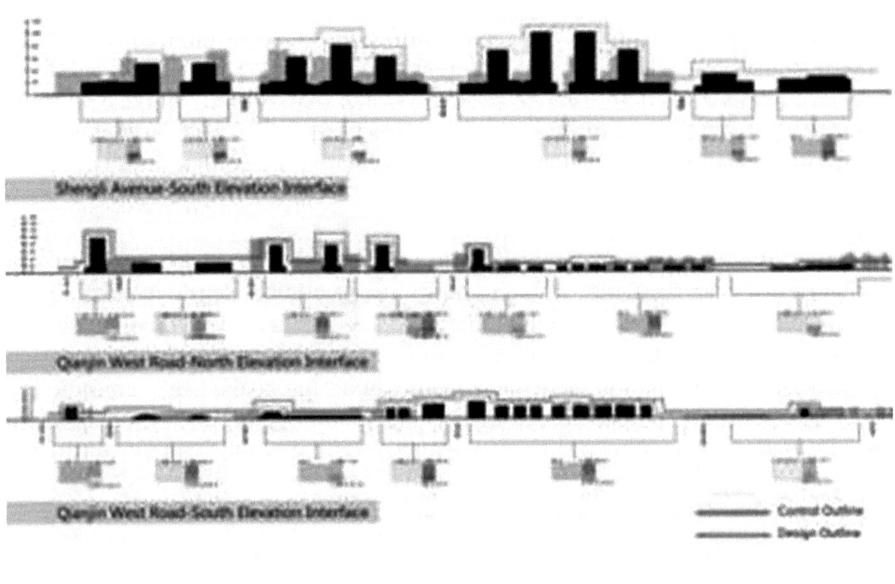

Fig. 10 Interface control of road space

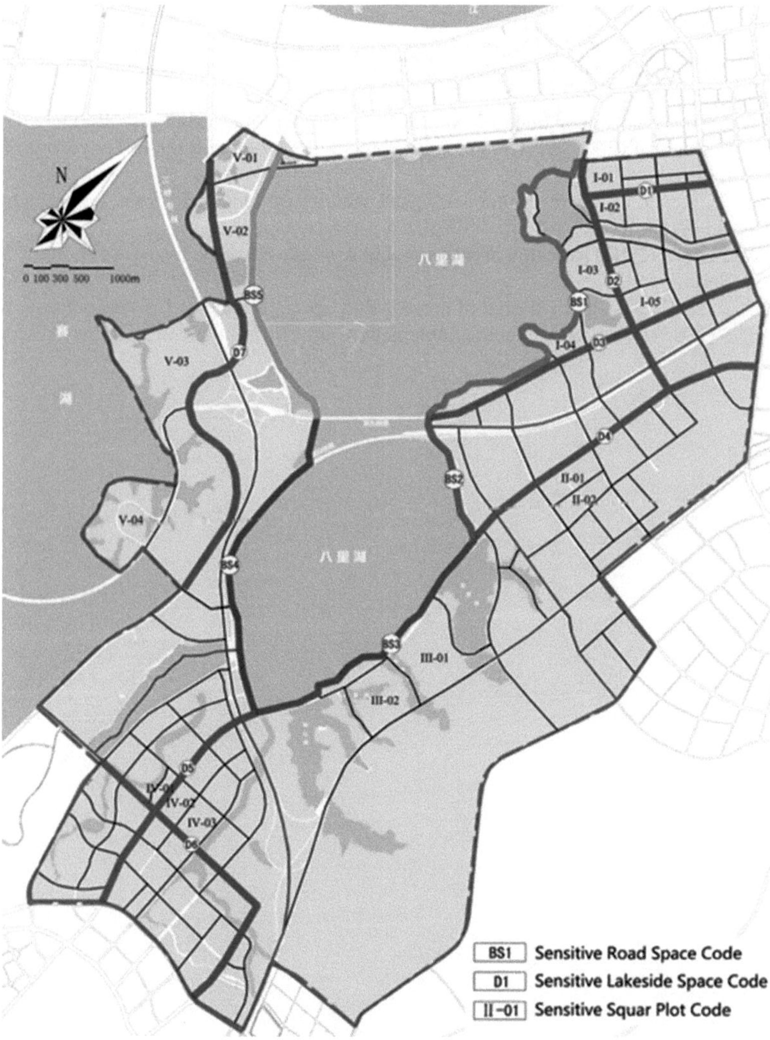

Fig. 11 Square plot code for control

References

1. Yu B (2008) Extraordinary urban design-idea system detailed. China Architecture & Building Press, Beijing
2. Lu J (2005) Urban design mechanism and creative experience. Southeast University Press, Nanjing
3. Chen B (1991) Systems engineering of urban planning. Tongji University Press, Shanghai
4. von Bertalanffy L (1950) An outline of general system theory. Br J Philos Sci 1(2):134–165
5. Von Bertalanffy L (1972) The history and status of general system theory. Acad Manage J 15 (4):407–426

6. Von Bertalanffy L (1950) The theory of open systems in physics and biology. Science 111(2872):23–29
7. Chang S (2004) Systematic science method overview. China University of Political Science and Law Press, Beijing
8. Chen H (2004) System engineering introduction. Higher Education Press, Beijing
9. Chen Y, Zhong J (2006) Systematology and method overview. Renmin University of China Press, Beijing
10. Zhang Y, Ren Z (2005) Regional ecological security on landscape scale. J Northwest Univ 35(6):815–818
11. Su W, Yang Y (2007) Research of urban spatial structure based on landscape ecology. Science Press, Beijing
12. Wang P (2002) Systematic design of urban public space. Southeast University Press, Nanjing
13. Wang S (2005) Urban design facing implement. China Architecture & Building Press, Beijing

Analysis of Framework and Optimization of the Urban Natural Landscape System (UNLS)

Case Research of Eco-efficiency of UNLS in Tianjin, China

Pengbo Li, Jun Wu, Yan Jiang, and Lei Meng

1 Introduction

With the accelerated process of urbanization in China, the rapid urban construction has caused serious damage to the ecological environment of the cities. Although more attention has been paid in recent years to the reconstruction and management of the urban environment during and after the construction process, there are still problems to be solved when planning to construct urban green space systems to enhance the function of the urban ecological environment [1, 2]. This is the key study of urban landscape planning, designing, ecology and other disciplines.

Much scientific research has confirmed that an urban green system improves the function of the urban ecological environment. Many analyses and researches have been done on green land factors [3]. This research includes analyses of the urban green land system [4], valued ecological function and construction indicators of urban green systems.[5], These analyses have revealed some indicators of urban green systems [3, 6, 7], such as: per capita public green area, urban green land rate, green coverage, green quantity and green seeing rate. These analyses have promoted the improvement of the ecological functions of urban green space.

However, with the development of Urban Science and Ecology, urban–rural integration theory and the concept of three-dimensional urban greening, ideas and theories of landscape pattern have gradually seeped into every aspect of urban construction [8]. Urban green space has gradually been absorbed into the peri-urban natural environment [9].

Supported by the Natural Science Foundation of Tianjin (No. 08JCYBJC05100), the Art Science Foundation of Tianjin (No. A12068)

P. Li (✉) • J. Wu • Y. Jiang • L. Meng
Department of Architecture, Tianjin Chengjian University, Tianjin, China
e-mail: pengbotj@gmail.com

X. Chen, Q. Pan (eds.), *Building Resilient Cities in China: The Nexus between Planning and Science*, GeoJournal Library 113, DOI 10.1007/978-3-319-14145-9_14

The traditional urban green land system has limited the urban built-up area, which increasingly cannot adapt to the needs of urban ecological construction. Green space in urban areas is no longer just green land [8]. Some urban land is covered with pavement, water and other constructed elements, and three-dimensional afforestation and roof afforestation have been gradually built up [10].

So, the green land system should evolve into an urban natural landscape system, including the urban open and built-up surroundings. In this open space one can use the hardscape and soft nature landscape.

The urban natural landscape system is the meso-scale system, including the urban green land system of built-up areas (this is the urban green land green aforementioned) and certain suburban environments. It would introduce the suburban natural landscape's ecological effects into urban center through the construction of urban–rural integration [11]. The function of ecological protection and improvement of the UNLS can be explained by optimization methods. And the built-up area of the suburban environment can be integrated with the protective film for the urban environment and give the urban green space a healthier and organic quality [12]. The quality of the urban eco-environment will be improved and a pleasant living environment for urban residents will be created.

2 Structure and Framework of UNLS

2.1 Structure Elements of UNLS

The suburban and built-up areas of urban regions are two parts of the UNLS. There are different ways to understand these elements.

Most of the urban landscapes are artificial scapes, in which little persists from nature. So there are two types of urban landscape according to their properties: artificial landscape and retained pure natural landscape. The artificial landscape in urban areas is informed by and sourced from its origin in nature. The artificial landscape is more concerned about a beautiful visual effect, although the ecological role is also important. In retained pure natural landscape, on the one hand, ecological value is more important; on the other hand, it is also valued for its associated history.

If the classification is according to foundation and appearance of the landscape, there are two types of the urban landscape: landscape and waterscape [13]. Landscape includes both the land covered with vegetation and that covered with hardened paving. Waterscape is the area in which water is the main landscape factor. Waterscape includes wetlands, lake landscape, river landscape etc. Landscapes and waterscape are the majority of roles in the urban landscape in the eco-environment. Their aesthetic value is also an important part. In a modern landscape, more and more landscapes are paved with hardened materials. Hard-paved landscapes weaken the ecological function of the landscape, but they also have special features and carry out the function of corridors. So, according to those special features, hard-paved landscapes should be listed as a kind of landscape.

According to the shape of the urban landscape and the role in the improvement of urban ecological function, there are four types of urban landscapes: massive green space, scattered green space, green corridor across the urban area, and suburban green space around the urban area.

1. *Massive green space*

There have been large areas of green land in urban areas, such as: urban park, district park, waterfront park and semi- preserved natural park. These parks range from several to tens of hectares in area. They typically feature rich vegetation and varied types of space for people to stay, play and do many activities. So these massive green spaces have higher eco-efficiency and are the most important space in urban areas.

2. *Scattered green space*

Scattered green space is the small green space distributed in urban areas. Such areas are dozens to hundreds of square meters; the larger green spaces are thousands of square meters to several hectares.

They spread to the sides of street, in yards belonging to building units or enterprises and in residential districts, etc. The scattered green space is the foundation of the urban green space and is the ecological matrix of the urban landscape.

3. *Green corridor*

In urban area, the road, street, and river are the continuing factors crossing the entire urban area or district. They are often decorated by vegetation and some hardened paving. That builds on the green corridor to contact the massive green space with the scattered green space, integrating them to form the whole urban landscape [9]. Because of the characteristic of line, the green corridor has the function of exchange of information, energy and material [14]. This is the most important function for the urban ecological environment.

4. *Suburban green space around urban areas*

On the outskirts of urban area, there always is the natural environment and it forms a green belt surrounding the urban core. These natural environments construct the protection layer of the urban environment. They protect the urban area from the bad effects of high wind, air pollution and sand-storms, etc. They also provide fresh air, water, and suitable conditions for urban greenspace to enhance the quality of the urban ecological environment.

2.2 Function of the UNLS

The four elements of the UNLS interact and achieve a virtuous cycle in the urban ecological environment, the function of which is the self-healing of the urban ecology. The **massive green space** generally has the biggest area in urban area;, its ecological function is great, and it can greatly improve the temperature, humidity and pollution in the air, such as PM2.5, and other conditions of urban areas. **Scattered green space** has smaller areas and lower levels of function, but it has the great quantity of overall area.

With the function of superimposed scattered green space combined with the massive green space function, the ecological function of the green space is greatly enhanced. Suburban green space surrounds urban and provides a virtuous outdoor environment for urban residents. It and transports good material into the urban area, such as fresh and cool air, and transports bad material out of the urban area, such as pollution and hot air. In the transport process, the **green corridor** plays the key role. It is the only channel to connect the green circle with the urban area and urban landscape. Its most important function is the exchange of ecological elements between the inner and outer urban environment [13]. Therefore, the function of the UNLS is complex. It is the superposition of function of the various elements.

2.3 Optimized Framework of UNLS

According to the analysis above, the framework of the UNLS should take the four elements into account and integrate together. Therefore, the framework should be: massive green space combines with scattered green space which promote each other; all elements connect between internal and external sides of the urban area, and the ecological cycle should be achieved.

To optimize the framework of the UNLS, it should be considered in the urban planning [13], such as:

1. The quantity, area and position of the massive green space and scattered green space are the most important factors. They are the foundation of the urban ecology. The land should be reserved to develop the massive green space and scattered green space in urban planning and renovation。
2. The suburban green belt and peri-urban natural environment are the protective layers of the urban environment. Suburban green belt should have a certain width and structure. In order to construct the layer of the urban environment, some key sections should have different structures (Fig. 1).
3. Several green channels should be reserved and planned in the urban landscape planning process. They should be located according to the position of massive green space and suburban natural environment, and their design should be coordinated with that of the urban master plan, including watersheds, wind direction, etc.

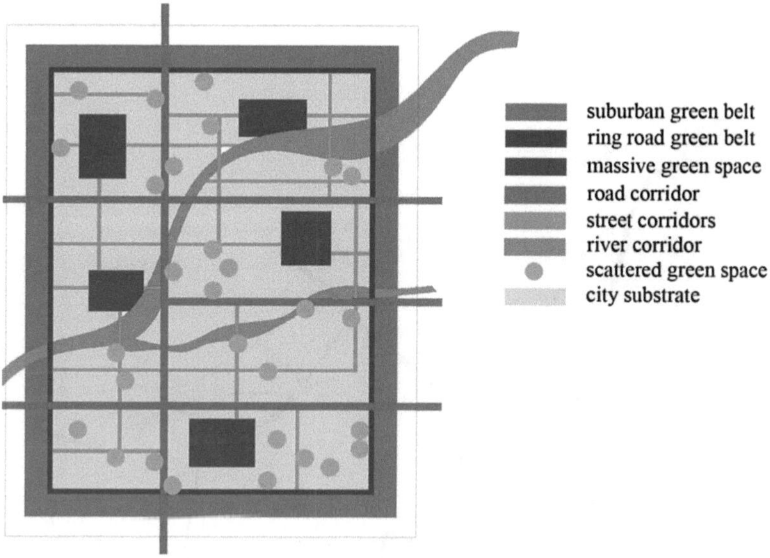

Fig. 1 Structure model of UNLS

3 Research of Tianjin UNLS

3.1 Analysis of Structure Elements

At the beginning of 2011, there were approximately 4,109 ha of massive green space in Tianjin. There are 28 parks in central urban areas. Only the Water Park area contains more than 1,000 acres. The area of the parks, such as the Zoo, Beining Park, Activity Center of Younger, Xigu Park, Chonghong Park and Nancuiping Park, are more than 300 acres. The others are not more than 100 acres.

There are 912.5 ha of protectedgreen space in Tianjin's urban green space system. The protected green space inside the urban area it accounts for 5.7 % of the urban green space; scattered green space is 6449.3 ha. There are 345 main roads in the city, and green rate of the roads is 21.5 %.

3.2 Overall Layout of Urban Green Space

The massive green spaces are mainly located in the south and north of the urban area. The south's massive green spaces are aesthetically more pleasant than the north's and there are many lands that have never been greener.

The center of the urban area is the old urban district; there is only some little urban space. There is no massive green space; the area of the green space is not more than 2,000 m^2 (Fig. 2).

Fig. 2 Layout situation of Tianjin green space

River and street are the main line elements of the urban core. So, river landscape, belt and road green spaces are the main green corridor in the urban core. The rivers of Tianjin are the Haihe water system. There are five rivers in the Haihe water system: South Channel, North Channel, Ziya River, Daqing River, and Yongding River. In the second layer of the water system, there are 19 rivers and 6 artificial rivers including New Ziya River, Machangjian river, Duliujian River, New Yongding River, Chaobai River and New Huanxiang River. There are many streets in Tianjin including surrounding roads, a quick road, middle quick road and other streets distributed in city districts. Some green spaces on streets are street green land.

3.3 Assessment of Eco-efficiency of Tianjin UNLS

In order to analyze the eco-effection of the UNLS, the research used temperature as the indicator to discuss the relationship between the green space area and the temperature effect..

When the urban green space area is 1–2 ha, it has a certain humidification effect, but the cooling effect is not obvious. When the green space area is 3 ha, the river effects of cooling and humidification are obvious; when the green space area is 5 ha, the cooling and humidification effects are very obvious; when the green area is greater than 5 ha, the cooling and humidification effects are extremely obvious and uniform. Some research indicates that when the green space area is more than 3 ha, it is capable of stable internal space humidity conditions, and makes a greater contribution to the surrounding environment [15].

3.4 Analysis and Assessment of Green Space Temperature Buffer

1. *Buffer analysis of temperature effects in Tianjin UNLS*

According to the cooling effect of green space (Table 1) [15], the research analyzed the green space buffer, based on the size of Tianjin's urban green area, to determine the zone of the buffer (Table 2).

2. *Greenland buffer evaluation*

From the buffer analysis diagram of the Tianjin UNLS, it can be seen that the third buffer range accounts for about 30 % of the urban area. It is mainly in the urban area's southwest side, north, east side of the city. And in the center of the urban area, the green space buffer area is rare. It is mainly due to large area of green space in the heart of the urban area, which is smaller, scattered, and does not make up a system. Therefore, the ecological effects of green space are smaller, the heat island effect is stronger, and compared to the edge of the city in the summer central urban temperature is higher than that in the edge district.

The River green belt in Tianjin includes the Haihe and Ziya Rivers, bringing the cool air from the suburbs. Especially, on both sides of the river in recent years in Tianjin, the rebuilding engineers have extended the big area of green space. It developed by widening and adjusting the structure of the river belt green space. It enhanced the corridor function of the belt with combining the ecological benefits of river green space and water. But for the reason of the narrow green space in the road and street belt, there is little effect on the urban eco-environment. So, efforts should be focused on the building of the road and street green belt in Tianjin UNLS.

From the analysis of the buffer, the overall layout appears that there are fewer green spaces in urban center, but it is great in edge district. And in the center of the

Table 1 cooling effect and radiation range of green space (about 14:00 in summer) [15]

Green space	Green space area (ha)	Temperature reduced (m)	Temperature difference (°C)
Shinjuku Imperial Garden	58.2	100–350	0.3–0.2
Koishikawa Botanical Garden	16.1	100–250	0.3–2.3
Zenpukuji River green	17.7	100–200	0.5–1.0
Grove of or? park	0.05	5	0.5

Table 2 Buffer of green space according to the size

Types of green space	Size of green space		Range of buffer (m)			Remark
			First buffer	*Second buffer*	*Third buffer*	
Massive green space (ha)	1	Areas more than 50	100	200	300	Green space includes water-scape, green land, paving land, etc.
	2	Areas more than 20 less than 50	50	100	200	
	3	Areas more than 5 less than 20	20	50	100	
	4	Areas more than 3 less than 5	20	40	60	
	5	Areas more than 0.5 less than 3	5	10	20	
Corridor green space (m)	1	Width more than 50 m	100	250	350	River width is included
	2	Width more than 35 less than 50	60	130	200	
	3	Width more than 15 less than 35	20	50	100	
	4	Width more than 5 less than 15	5	15	30	

urban area, the green spaces have not connected with each other. That upset the balance of the green space system. Though, the indicator of the green space of urban can achieve the standard to a certain extent. But the effect on the urban environment could not reach the requirements of comfort and livability for the whole city (Fig. 3).

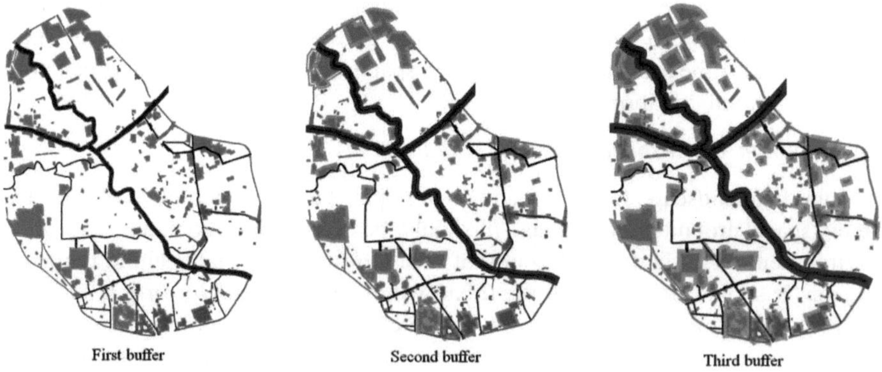

Fig. 3 Buffer analysis of temperature of Tianjin urban green space

3. *Optimization of Tianjin UNLS*

(a) *Overall framework*

Massive green space is one of the main green space types of urban ecological environment improvement. According to the characteristics of the Tianjin urban green space, the three layers system of urban parks should be improved in the central city of Tianjin to form an urban center with massive green space; urban parks, district parks and community parks. Urban parks are the urban ecological security. They should meet the needs of all residents of urban areas, from sightseeing to leisure, communication and other activities. District parks are the assistants of urban security; the vegetation-scape should be emphasized and the plant communities should be paid more attention. Community parks are the ecological protection auxiliary area. They serve the community for morning exercise, dancing, singing, playing chess and other leisure activities. Both the hardened paving and plant landscapes should be emphasized. The three level layouts of green space are the green elements of the urban area, and they are the key to the urban eco-environment.

Scattered green space is the common green space in the urban green space. It is the result of "see seam, insert green" in urban construction. Single scattered green spaces have small areas. But unity is strength. Therefore, its overall ecological effect is very important [16]. The layout of the scattered green space should be made according to the layout of massive green space and corridor green space to determine the form and plant configuration. Scattered green space is the supplement to massive green space.

Corridor green space should take the main road and main river of the urban area asprimary arteries, and take the other 1 type green space asauxiliary. The green space width of the subject should more than 30 m to form the communication corridors of between urban and suburban ecological environment.

Focusing on the problems of high fragmentation, layout imbalance and weak service functions, it should be solved through the green corridor construction to improve the green space connection and to dissolve isolated green space and to penetrate the scattered green space into residential areas, and municipal unit areas [17]. With the corridor green space connection, the scattered green space could be formed as a system layout: a large green space for the nuclear, small green space for the assistance, street green space and small garden for green point and small green corridor for connection.

(b) *Suburban green space around urban areas*

Suburban landscape forest belt and country parks are the main types of suburban green space around urban areas. It should be a corridor protecting the urban environment to mitigate the hazards of dust storms, industrial pollution and soil erosion. It should be built with the urban winds in mind, and to form a barrier to harmful climate and a corridor to favorable climate. The country parks should be built according to urban massive green space and corridor green space, to strengthen the liquidity of the inside and outside of the urban green space system. It should also dissipate bad quality air from the urban areas, and purify and import the good quality air to urban areas, to improve the urban ecological environment.

4 Conclusions

The urban natural landscape system is the basis of the urban ecological environment. Its structure should include massive green space, scattered green space, corridor green space and suburban green belt encircling urban areas (including country parks). Massive green space in urban areas should account for a large area to be set aside in urban planning, and to ensure proper size and location, so that it can meet its ecological effects and improve urban eco-environment.

Scattered green space is the basic green space in urban areas. It supplements and strengthens the ecological functions of the massive green space. It should be rationally distributed in urban areas, and distributed across the city, so that every corner of the urban area will be full of green, and the urban ecological environment will be improved.

Corridor green space is the only way of communication between urban and suburban ecological environments, and the way of exchange between the different green spaces [9]. More attention must be paid to the building and strengthening of corridor green space. It should be planned as a reasonable construction with width, structure and vegetation to organically link the internal artificial environment and the external natural environment, to communicate and exchange environmental resources and improve the ecological environment to get self-circulation and eco-balance.

The suburban green space around the urban area is the protection of the urban environment, and also is the transition of the urban artificial natural environment to the outside natural environment [18]. To strengthen the planning and construction of the structure and function of the suburban green belt could provide an excellent natural resource base for the artificial natural environment of urban areas and integrate it into natural environment to improve the quality of urban eco-environment.

Of all the elements of UNLS, the planning and construction of corridor green space is the weakness in modern urban landscape planning in China. In recent years, the planning and construction of urban landscapes in the river landscape and avenue landscape areas has made significant development. But most are based on urban beautification, on the reason of urban scape and image. The function of ecological improvement of the green space has not been studied previously. Therefore, plan and construct urban corridor green space in accordance with its ecological function, communication function, etc. To link the point (massive green space), surface (scattered green space) and ring (suburban green belt ring urban) spaces and enhance the ecological function of UNLS, blend the UNLS into the natural environment.

Acknowledgments This work received backing from the Natural Science Foundation of Tianjin and the Art Science Foundation of Tianjin. The author would like to thank colleagues on our team for their valuable comments and other help.

References

1. Gong ZL (2011) Wan. fractal features analysis of green spaces on rural–urban fringe in Guangzhou, China. In: 4th international workshop on chaos-fractals theories and applications, IWCFTA 2011. IEEE Computer Society, Hangzhou, 19–21 Oct 2011
2. Minghua Huang, Yang Wang, Xiaonan Shi (2012) The reconstitution of landscape pattern at master planning level in Ankang city[J]. Adv Mater Res P1724–1731
3. Jianxun Rui, Beiqi Shi, Di Shen, Weiqin Yao (2013) Shanghai urban green landscape model system based on MapServer. SPIE, Guangzhou
4. Kong F, Yin H, Nakagoshi N, Zong Y (2010) Urban green space network development for biodiversity conservation: identification based on graph theory and gravity modeling. Landsc Urban Plan 95(1–2):16–27
5. Meyer BC, Wolf T, Grabaum R (2012) A multifunctional assessment method for compromise optimisation of linear landscape elements. Ecol Indica 22:53–63
6. Kabbes KC, Windhager S (2010) Sustainable site initiative – protecting and restoring site ecosystem services. In: World environmental and water resources congress 2010: challenges of change. American Society of Civil Engineers, Providence, 16–20 May 2010
7. Kazemi F, Beecham S, Gibbs J (2011) Streetscape biodiversity and the role of bioretention swales in an Australian urban environment. Landsc Urban Plan 101(2):139–148
8. Acosta A, Carranza ML, Giancola M (2005) Landscape change and ecosystem classification in a municipal district of a small city (Isernia, Central Italy). Environ Monit Assess 108(1–3):323–335
9. Walmsley A (2006) Greenways: multiplying and diversifying in the 21st century. Landsc Urban Plan 76(1–4):252–290
10. Amati M, Yokohari M (2006) Temporal changes and local variations in the functions of London's green belt. Landsc Urban Plan 75(1–2):125–142
11. Ayad HM (2002) A strategy for an integrated ecological and urban landscape development: the urban natural network. Alexandria Eng J 41(5):849–866
12. Bryant MM (2006) Urban landscape conservation and the role of ecological greenways at local and metropolitan scales. Landsc Urban Plan 76(1–4):23–44
13. Gill SE, Handley JF, Ennos AR et al (2008) Characterising the urban environment of UK cities and towns: a template for landscape planning. Landsc Urban Plan 87(3):210–222

14. Fang Chen, Junmei Tang, Zheng Niu (2008) Estimating the impact of urbanization on LAI and fPAR in the Baltimore-Washington corridor area[J]. Can J Remote Sens 34(2):326–337
15. Fei Wu SL, Liu J (2006) Effects of urban green space with different vegetation quantity on temperature and humidity. Chinese Landsc Archit 22(7):56–60
16. Shapiro N, Valentine B (2010) Green street retrofit in an urban transportation infrastructure. 2010 green streets and highways conference: stormwater and environmental planning and design, pp 190–200
17. Li F, Wang R, Paulussen J, Liu X (2005) Comprehensive concept planning of urban greening based on ecological principles: a case study in Beijing, China. Landsc Urban Plan 72(4):325–336
18. Ji F, Chu J (2012) A study of the designing and planning of the urban green space landscape – Based on the service function of the green ecosystem. In: 2nd international conference on civil engineering, architecture and building materials, CEABM 2012. Trans Tech Publications, Yantai, 25–27 May 2012

Study on the Indicator System of Eco-regional Development Planning for Shanghai Chongming Island

Mingxing Liu and Gang Zeng

1 Introduction

In the traditional process of industrialization, regional development has experienced the process of 'pollution first, treatment later'. Western developed countries recognized early the importance of the ecological environment, energy conservation, industrial upgrading, consumption patterns changes, transformation of the mode of growth, sustainable development, and recycling economy. Ecological economic research has been carried out extensively on regional ecological environment construction practice, and there has been a large number of influential regional ecological development reports completed [1–3]. The ecological crisis that occurred in developed countries in the 1960s prompted many scholars to explore new regional development models and the formation of a regional building theoretical system based on the philosophy of sustainable development including the concepts of circular economy, ecological economy, clean production, and industrial ecology [4–7]. In addition, many foreign scholars have studied construction of an evaluation index system of ecological zones [8, 9]. The World Conservation Union (IUCN) released a "sustainability barometer", the Index of Sustainable Economic Welfare (ISEW), the Genuine Progress Indicator (GPI), ecosystem service indicators system (Ecosystem Services), Energy analysis indicators (Energy Analysis) and others. Among them, the index systems of the United Nations Commission on Sustainable Development (UNCSD), due to strong internal logic, have been widely adopted; SEER of eco-environmental accounting system design also has a decisive impact. These theoretical achievements are basically based on the Western developed countries, the highly developed level of development of productive forces and mature modern market economic system characteristics. They studied human,

M. Liu (✉) • G. Zeng
Department of Urban and Regional Economy, East China Normal University (ECNU), Shanghai 200062, China
e-mail: Starlmx1010@163.com; 13301659976@163.com

X. Chen, Q. Pan (eds.), *Building Resilient Cities in China: The Nexus between Planning and Science*, GeoJournal Library 113, DOI 10.1007/978-3-319-14145-9_15

economic, social, ecological, environmental, management and other aspects of harmonious relations in the post-industrial society which are not fully compliant with China's national conditions.

Chinese early theoretical study of the development of eco-regions is mainly an introduction of foreign theory. Since the 1990s, the focus of Chinese academic research can be divided into five stages: sustainable development [10–12], growth pattern [13], ecological economy [14, 15], circular economy [16–19], and ecological civilization construction [20–22]. Research trends are noticeably changed from the "idea" to "build" and from "singularity" to "comprehensive". Few studies were based on the ecological civilization regional development mode.

2 Background and Definition

Chongming Eco-island (CEI) is an alluvial island of approximately 1,270 km^2 located in the mouth of the Yangtze River. The archipelago of CEI is governed by Shanghai municipality and has the lowest population density in the province. The Shanghai Municipal Government established it as China's first and only eco-island. This presents an exciting proposal, however one that must be approached with thoughtful consideration for the agricultural and natural environments of Chongming. The island plays an important role in the greater prosperity and ecological health of the region with an emphasis placed upon the 'International Importance Wetlands Convention' recognized Dongtan Wetlands located on the island's eastern point.

This research inquiry into China's less developed regions' rapid development seeks to examine the eco-island development model based on the characteristics of developing countries.

3 Objectives and Principles

3.1 Objectives

In order to provide a scientific basis for government decision-making and to promote the process of the construction of CEI, we need to set up a scientific indicator system with three major functions:(1) To guide the direction of CEI's construction; (2) To standardize the practice of CEI's construction, and (3) To control the process of CEI's construction.

3.2 Principles

1. *Advancement*: A leading regional development model of China and the developing countries
2. *Scientific*: Based on the complex ecosystem theory, to achieve rapid development: (a) Evaluation of the current situation, problem diagnosis and guide construction; (b) Attention to the quality of regional economic development, achieve rapid development
3. *Interoperability*: Fitting with the status quo and future development needs of Chongming: (a) To emphasize the characteristics of Chongming; (b) meet the needs of enrichment and rapid development.

3.3 The Alternative Index Set of Assessment Subjects

Sustainable development is the key point of the construction in Chongming. Based on this, the selection of indicator should be learned from the existing index system of sustainable development. At present, the most influential index system is the Driving force-State-Response (DSR) proposed by UNCSD. The index system was established in 1996. Driving DSR has 100 and 34 indicators based on social, economic, environmental and institutional systems. There is a strong logical relationship among the indicators, and particularly the causal relationship between degraded environment and human destruction. In addition, the framework of the index system can be learned from the economic accounting system in comprehensive environment proposed by OECD and UNSD, the evaluation of sustainable progress in Europe proposed by the European statistical bureau, and the index system of sustainability proposed by the Chinese Academy of Sciences.

In addition, the establishment of the index system can be learned from Human Development Indicators (based on society) proposed by UNDP, National Wealth, Sustainable Economic Welfare (based on economy) proposed by Daly and Cobb, Environmental Sustainability (based on environment) proposed by Scientific Committee on Environment, and Ecological Services (based on ecology) proposed by Constanza and Lubcheno.

4 Working Thoughts and Theoretical Basis

4.1 Working Thoughts

1. *Three-dimensional scales*: (a) Time scale: Taking the demands of different development stages; (b) Spatial scale: Concerned about the feedback and the link of the ecological island, eco-village and township, and ecological projects;

(c) Management Scale: Paying great attention to the labor and interaction's division among the city, county, and town governments and grass-rooted organizations
2. *Three perspectives*: (a) Global vision; (b) Regional strategy; (c) Chongming characteristics.
3. *Three steps:*(a) Diagnosis, assessment of the status quo and integrated indicators; (b) Construction and optimization indicator system; (c) International test and comparative analysis of indicator system.

4.2 Theoretical Basis

We mainly evaluated various types of indicator systems in different spatial scales (global, national, regional/city), similar geographical conditions (Island), and similar development goals (ecological construction, sustainable development, etc.), analyzed their theoretical basis and methods, and studied their applicability, main features and enlightenments on CEI construction. On this basis, we studied and identified CEI construction indicator system's theory and model.

1. *Economics Discipline* (Externality Theory): It reflects mainstream economics ability to address the idea of sustainable development and is also ecological compensation's major theoretical basis. However, it lacks enough integration of non-mainstream economics such as the economics of property rights, resource economics, environmental economics, etc., and the ecological environment was only in support and to "burden" position.
2. *Ecology (foreign) Discipline* (Ecological Redundancy Theory, Ecological Niche Theory, Development Niche Theory): Redundancy theory in the mechanism of phytocoenosium stability was deeper, more universal than the old ones. Niche theory was introduced to the social science research system (development), and stated to concern ecology, economics, and sociology but did not place enough emphasis on the importance of the development of high-intensity areas of human activity.
3. *System Science Discipline:* (Cybernetics, Dissipative Structures, Coordination Theory) It focused on the constraints and interaction among population, resources and environment and the importance of people-oriented management control role, but all of the elements' logical relationship lacked an in-depth discussion.
4. *Sociology Discipline* (Functionalism, Conflict Theory): It pays attention to social problems, but not enough to the improvement of the ecological environment functions and the economic growth efficiency.
5. *Geography Discipline* (Theory of human-nature, Theory of human-nature harmonization): It consists of the regional ecological construction, the idea of sustainable development, and an emphasis on comprehensive balance. However, this discipline was not close enough to the development needs of different regions and different stages.

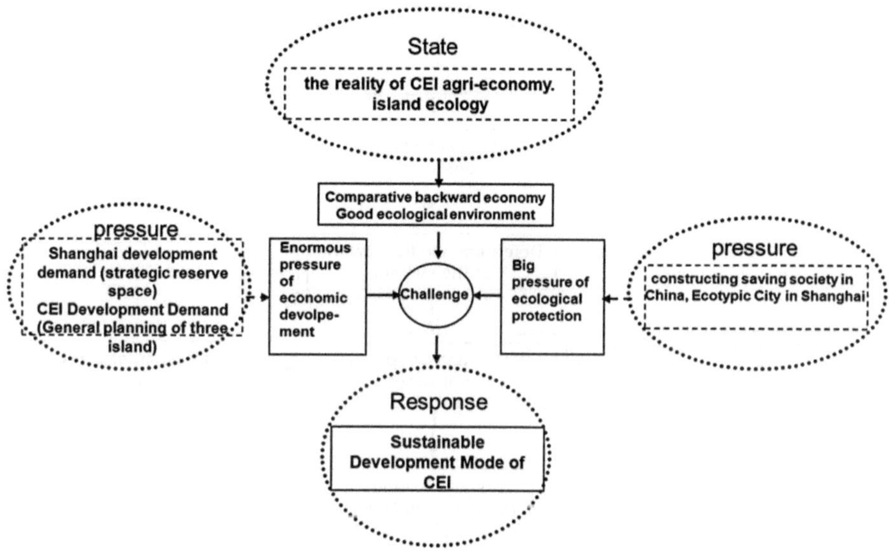

Fig. 1 The analysis diagram of CEI construction based on PSR model

6. *Ecology (domestic) Discipline* (Social, economical and natural compound ecological system): It reveals the intrinsic nature of the association among the legal system, the natural environment, resources, and human behavior; but it is less empirical analysis and yet to be tested in practice (Fig. 1).

4.3 Research Route

Among the related indicator systems and models, the following are most typical and popular: sustainable development indicator system; Pressure-State-Response Model (PSR); Environment-friendly Society indicator system; Ecological Provinces, Municipalities (County) Construction Indicator System; Socio-economic and natural compound ecological indicator system; Well-off Society Building Indicator System; professional sector indicator system; and others such as the human development indicator, the happiness and harmony indicator, urban competitiveness indicator, etc. (Fig. 2).

On the whole, they can be divided into three categories: the priority of social justice, the priority of economic growth, the priority of eco-efficiency.

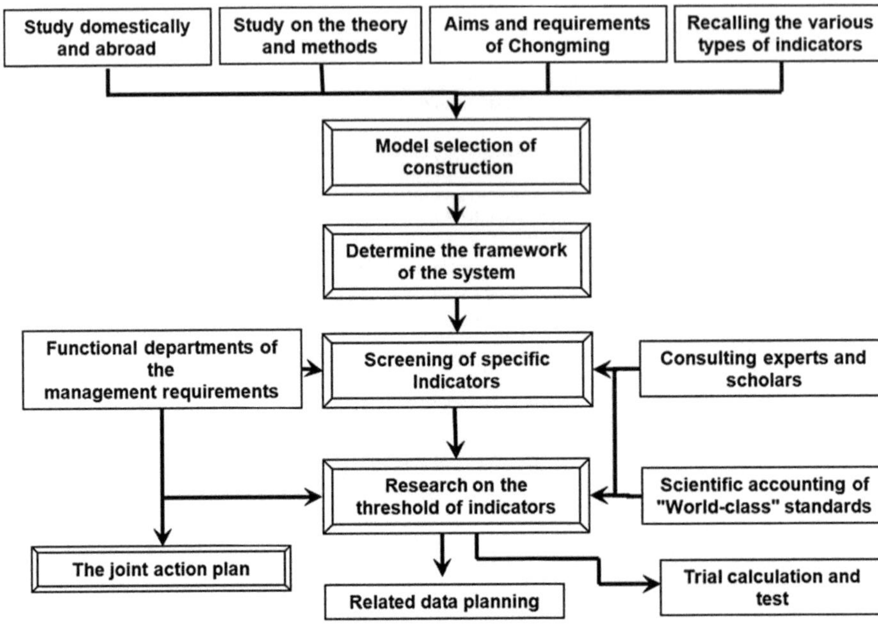

Fig. 2 Research route

4.4 System Framework Design:"5.3.X" Construction

The hierarchical structure that is used for the construction system of CEI is the index system of "Five, Three, X".

Level One (A) refers to thematic areas. According to the compound ecosystem theory, the index system can be classified as social harmony, economic development, friendly environment, ecological health and scientific management. By using specified indexes, it can reflect the overall process of the construction of CEI and meet the demand of the policy and macro-control set by the government.

Level Two (B) refers to assessment subjects. According to the PSR (Pressure-State-Response) theory, each area is composed of three assessment subjects. By using specified indexes, it can meet the demand of supervision and goal-seeking by government departments.

5 Indicator System Framework

Level Three (C) refers to specific indicators. Specific indicators (numbers are uncertain) can reflect the key content of assessment subjects. According to the standard of the world ecological island, the current domestic and international

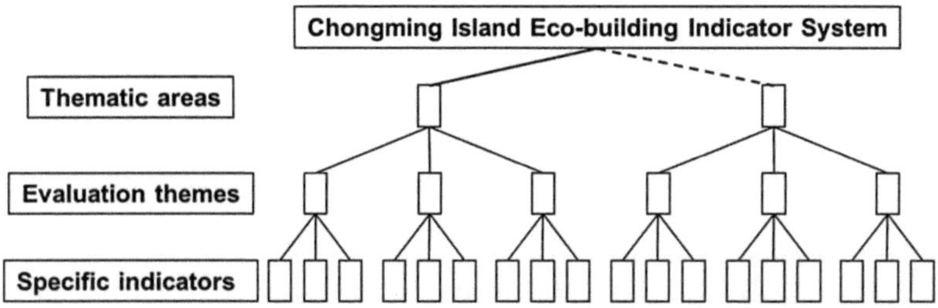

The first level	5	Social Harmony, Economic Development, Environment Friendly, Ecological Healthy, Management Scientific
The second level	3	design "three" evaluation themes in each area, corresponding to the respective logical relationship among "PSR"
The third level	X	Choose the specific indicators reflect the core content of the evaluation theme

Fig. 3 System frame design: 5-3-X structure

research, and the characteristics of Chongming, we have concisely proposed the index set composed of the twenty-four key indicators in which each subject corresponds to X indicators of the index set. Finally, specific indicators can be reflected in values, which meet the need for standardization and quantization of the construction in ecological island (Fig. 3).

5.1 The Establishment of Thematic Areas

According to the compound ecosystem theory, the five elements theory of Chinese tradition, the planning goals and functional orientation of CEI; we have designed the first grade indexes in society, economy, environment, ecology and management, which reflect the overall process of the construction of CEI. By using specified indexes, it can meet the demand of the policy and macro-control set by the government.

The basic functional domains of the five thematic areas can be respectively defined as the organic combination and interaction of the five thematic areas which have covered the main functional domains of the construction of CEI, formed the triangle framework of compound interaction, and maintained the stability and sustainability of the construction of CEI (Table 1).

Table 1 Indicators of the system

Area A and code	Theme B and code	Core indicators C and code
Social harmony A1	Social security/B1	Intact rate of urban Lifeline/C1
	Life quality/B2	Local surveyed unemployment rate/C2
	Social progress/B3	Expenditure per capita of the social development/C3
Economic development A2	Industrial model/B4	Output value ratio of organic/green agricultural/C4
		Modern services value-added account for the proportion of GDP/C5
	Economic performance/B5	The productivity of Unit area/C6
	Resource efficiency/B6	Energy consumption per GDP/C7
Environment friendly A3	Environmental pressure/B7	Amount of the Main Contamination/C8
		Land Development Intensity/C9
	Environmental Quality/B8	Days of API (Air pollution Index) I/C10
		Proportion of river water quality to reach III category and above/C11
		Nemerow Index of soil/C12
	Environmental Protection/B9	Emission stable standard rate of enterprise/C13
		Sewage treatment rate/C14
		Proportion of clean energy use/C15
Ecological health A4	Ecological safety/B10	Natural calamity loss rate/C16
		Species diversity/C17
	Ecological health/B11	The retention rate of natural wetlands/C18
		Standard rate of drinking water sources/C19
	Ecological protection/B12	Forest cover by percent/C20
		Public green land per capita/C21
Management scientific A5	Managerial capability/B13	Rate of ecological model village/C22
	Management mechanism/B14	Public administration scores of environmental performance evaluation/C23
	Public participation/B15	Public satisfaction rate of the environment/C24

5.1.1 Social Harmony

The functional domain in the society area is harmony which is the final goal of the construction of CEI. Based on human-orientation of the scientific development concept, the key goal of the construction of CEI is to improve the public's quality of life, promote the quality of livelihood, and optimize the human settlements. Therefore, the value orientation in social harmony is to achieve the goal of social harmony, promote the quality of living and maintain the sustainability based on the overall improvement in the public's quality of life.

5.1.2 Economic Development

The functional domain in economy area is development which is the important part of the construction of CEI. According to the general regulation, we must optimize industrial structure, promote modern ecological agriculture, develop advanced manufacturing and modern service, transform economic growth mode, and achieve sound and rapid development. Therefore, the value orientation in economic development is to develop environment-friendly economy by means of the establishment of elimination mechanisms in industries; maintain and increase the value of the economic benefits by means of the increased investment in natural capital; and establish the double heights in eco-technologies and eco-industries by means of the introduction of hi-tech industries and the creation of an ecological brand in Chongming.

5.1.3 System Science Discipline

The functional domain in environment area is friendship, which is the nature basis of the construction of CEI. Environment is the basic factor in human existence and development. A friendly environment cannot only reflect human environment-friendly activities, but also a social development-friendly inorganic environment. Therefore, the value orientation in a friendly environment is to reduce the environmental stress caused by social and economic activities, maintain the sustainability of the environmental quality, and achieve the friendly interaction between people and the environment by means of the rational utilization of environmental capacity.

5.1.4 Ecological Health

The functional domain in ecology area is health which is criterion of the construction of CEI. Ecological protection is the key concept of the construction. Leaving enough ecological space and improving the value of ecosystem services proposed by the general regulation has emphasized the harmony between human and nature. Therefore, the value orientation in ecological health is to reflect the priority ecosystem services in the compound ecosystem, maintain the positive derivation in natural ecosystem, and protect the ecological security.

5.1.5 Scientific Management

The functional domain in management area is science which is the strategic core of the construction of CEI. Management has the leading character, subjectivity and regulation on the eco-construction strategy. The construction of CEI is the government oriented strategic action which is the dominating function over any other factors. Therefore, the value orientation in scientific management is to establish an effective service-oriented government, improve scientific management capability and mechanism, and play the regulatory role on the practice of the construction.

5.2 The Hierarchical Design of Assessment Subjects

Based on the functional domains and value orientation in the five thematic areas, we have proposed the following fifteen assessment subjects in order to reflect the construction process and effect. By using specified indexes, this proposal can meet the demand of supervision and goal-seeking by government departments.

5.3 The Screening Analysis of Key Indicators

Based on the theoretical summary of domestic and international eco-construction and sustainable development, we have proposed the alternative index set. Considering the current situation in Chongming, we have further proposed the 24 key indicators.

5.4 Threshold and Target Identification Basis

International organizations recommend the indicators of sustainable development threshold for developed countries and the Eco-region's ecological development index.

The relevant indicator threshold was issued by the Chinese Construction Department (CCD), and Chinese Environmental Protection Department (CEPD) and targets the Chongming development plan.

6 Indicator System Framework

6.1 Chongming Indicator Calculation

Index indicator $=100-(|\text{threshold-status quo}|/\text{threshold}) \times 100$
Theme indicator $=\sum$Index weight $\times$ Index indicator
Area indicator $=\sum$theme weight $\times$ theme indicator
Composite indicator $=\sum$area weight $\times$ area indicator

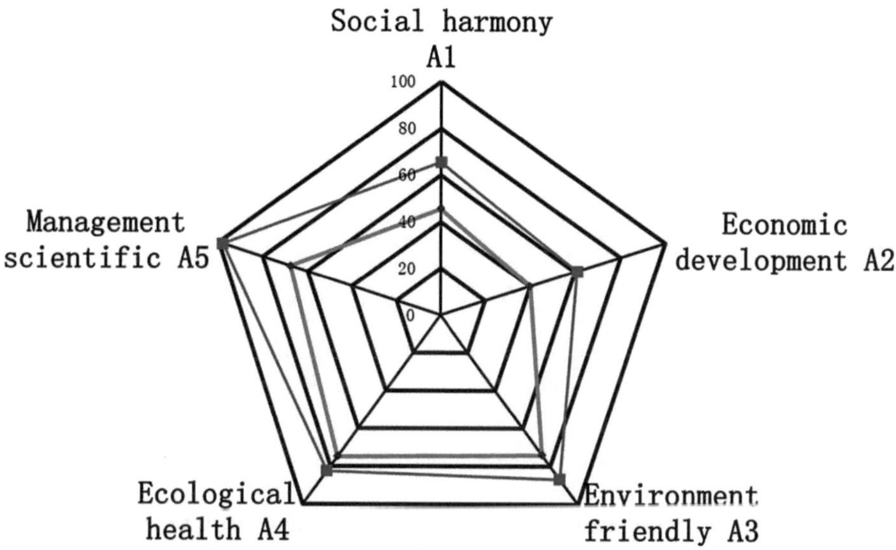

Fig. 4 Diagram for the relevance of the five thematic areas (2008, 2012)

6.2 Chongming Area Indicator and Strategic Position

We mainly evaluated various types of indicator systems in different spatial scales (global, national, regional/city), similar geographical conditions (Island), and similar development goals (ecological construction, sustainable development, etc.); analyzed their theoretical basis and methods; and studied their applicability, main features and enlightenments on CEI construction. On this basis, we studied and identified CEI construction indicator system's theory and model (Fig. 4).

1. *The status quo:*(a) Have the basis of building eco-island, but still a long way to a world-class ecological island; (b) The degree of ecological civilization and environment friendly score high; (c) Management science and economic development makes short boards for Chongming world-class ecological island building.
2. *Ecology (foreign) Discipline*: (a) Target: Maintain the health of the beautiful ecological environment at present; (b) Measures: Develop modern industries, transit management systems; (c) Focus: Improve the quality of residents' livelihood.

6.3 Chongming Indicator of the Theme and the Course of Action

Good social security, social progress need to improve; Poor economic performance; Good environmental quality, environmental protection not enough; Good

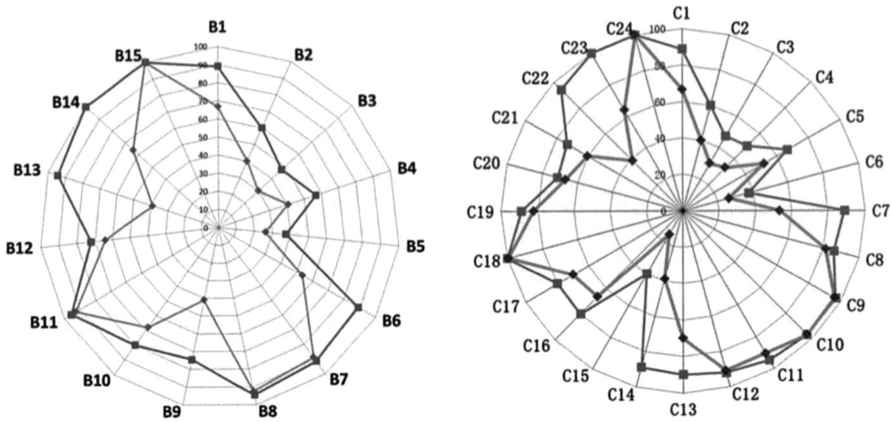

Fig. 5 Chongming indicator of the theme and the course of action (2008, 2012)

ecological health, ecological security needs to be strengthened; Good community participation, the management capacity to be improved (Fig. 5).

7 Conclusion

CEI is a secondary complex systems, a comprehensive experience of developed and developing countries. Its theoretical basis is the integration of the theory of complex ecosystems and the traditional Chinese theory of five elements. The power of CEI's construction comes from the combination of external support and internal potential, but external support is the key. It is difficult to get the present value and meet the specific demands of the different region which are the key issues of this world-class ecological island research. The recommendations are as follows preliminarily build the eco-island construction technology system, and exert a demonstrative and leading effect of science and technology. Establish the dynamic optimization system of "governance-monitory-feedback", give a scientific evaluation on CEI construction through the cooperation with UNEP on international evaluation. The popularity and influence of CEI construction will grow continuously at home and abroad by international cooperation, international forum organization and other interested parties.

References

1. Common (2005) Ecological economics. Cambridge University Press, Cambridge, UK
2. Lawn (2007) Front issues in ecological economics. Edward Elgar, Cheltenham

3. Ashford NA (2002) Reflections on the first decade of the journal of cleaner production. J Clean Prod 10(2):101–102
4. Azbar N (2004) Upgrading an existing treatment system to adopt cleaner production principals. J Clean Prod 12(7):789–795
5. Lambert AJD, Boons FA (2002) Eco-industrial parks: stimulating sustainable development in mixed industrial parks. J Clean Prod 22(8):471–484
6. MeGrath C (2001) Waste minimisation in practice. Resour Conserv Recycl 32:227–238
7. Odum HT, Odum B (2003) Concepts and methods of ecological engineering. Ecol Eng 20:339–361
8. Button K (2002) City management and urban environmental indicators. Ecol Econ 40:217–233
9. Fichtner W, Frankv M, Rent O (2004) Inter-firm energy supply concepts: an option for cleaner energy production. J Clean Prod 12:891–899
10. Li Feng, Liu Xu-Sheng, Hu Dan, Wang Ru-Song (2007) Evaluation method and its application for urban sustainable development. Acta Ecol Sin 27(11):4793–4802
11. Gu Chaolin (1994) On research direction of Chinese urban continued development. City Plan Rev 6:1–8
12. Zhang Chimu, Luo Ling (2005) On sustainable development of regional economy. Ecol Econ 12:61–64
13. Fu Yunsheng (2007) Changes of resources restrain and balanced development of districtive economy—inspection on the foundation of generalized resources and resources agglomeration. Acad Mon 39(11):73–80
14. Chen Liang, Wang Ru-song, Wang Zhi-li (2007) Niche assessment of China provincial social-economic-natural complex ecosystems in 2003. Chinese J Appl Ecol 18(8):1794–1800
15. Zeng Gang (2009) A new regional development model and approach based on ecological civilization. J Yunnan Norm Univ 41(5):33–43
16. Zhu Dajian, Zhu Yuan (2013) Deepen theoretical studies of circular economy under the background of eco-civilization. Bull Chinese Acad Sci 28(2):207–218
17. Li Ning, Ding Si-bao, Zhao Wei (2011) Study on spatial differentiation and optimization of circular economy development in China. China Popul Resour Environ 21(5):157–163
18. Yang Hao, Zhu Dong-yuan (2008) Recycling economic synthetically target system research in our country. Ecol Econ 06:52–55
19. Zhang Yongkai Chen Runyang (2012) The index system evaluation of the circular economy development and empirical analysis—take Gansu Province for example. J Ind Technol Econ 224(6):81–86
20. Song lin-fei (2007) Theory and practice of ecological civilization. Soc Sci Nanjing 12:3–9
21. Wu Yin (2000) Connotation and characteristics of ecological civilization. Ecol Econ 02:38–40
22. XU Chun (2008) Ecological civilization construction and human beings' all round development. J Guan Norm 44(4):10–13

A New Thinking of the Urbanization Route in Forest Ecological Function Regions: The Case of Shen-Nongjia

Liu Yun, Chang Lili, and Liu Ying

1 Introduction

Since twenty-first Century, urbanization has been the main theme of China's economical development. The government has focused all its resources to ensure the harmonious development of such urbanization. For instance, within the Major Function Oriented Region Planning, the Chinese government required that large-scale and high strength urbanization be limited during the national land development to key Ecological Function Regions (EFRs). Therefore, the coordination of harmonious development among urbanization, industrialization and ecology becomes an important issue.

In the National Major Function Region Planning of China (2010–2020), the functional orientation of national key EFRs is to protect the important national ecology security region, the demonstration area of harmonious coexisting between human and nature. The total area of this is about 386 million square kilometers, which takes up 40.2 % of the national land area. By the end of 2010, the population of this region has reached 1,100 million, which takes up 8.5 of the Chinese population. The national key ecological area is divided into five resource and

L. Yun (✉)
School of Urban Design, WuHan University, Wuhan 430000, China
e-mail: cloudleo@126.com

C. Lili
School of Architecture and Urban Planning, HuaZhong University of Science and Technology, Wuhan 430000, China
e-mail: 357498261@qq.com

L. Ying
Land Resources and Planning Bureau, East Lake High-tech Development Zone, Wuhan 430000, China
e-mail: 7948067@qq.com

X. Chen, Q. Pan (eds.), *Building Resilient Cities in China: The Nexus between Planning and Science*, GeoJournal Library 113, DOI 10.1007/978-3-319-14145-9_16

Fig. 1 The distribution of forest ecological function regions in China (Drawn by the author according to the Fig. 11, the National Major Function Region Planning of China, 2010–2020)

morphological categories: forest, steppe, wetlands, desert and hills. Among these, there are 11 forest ecological function zones in China (Fig. 1).

This paper points out the inadequacy of research into the urbanization of national key ecological functions; A good majority of the research available focuses on Gannan Yellow River water supplying ecological function zones. Wang Shu and Li Wei [1] analyzed its rural and urban status quo from the aspects of economy and ecology and discussed its developing strategies; Wang Shu et al. [2] analyzed the harmonious developing progress of Gannan ecological function zones based on coordinated degree and development degree mode, and discussed its developing strategies; Wang Wenhao discussed the necessity and selectivity of herdsmen settlement in this region and proposed corresponding suggestions [3]. Meanwhile, this "new thinking thesis" goes deep inside the forest function zones and discovers that Qinba, Wulin, Dabie Mountain areas, which have larger populations than others, are the regions where building activities most frequently take place. Consequently, the urbanization in these areas has been of a larger scale, and correspondingly, the factors that lead to their urbanization are more complicated. This paper focuses on Shen-Nongjia, which is located in Qinba is a representative case during the discussion.

2 The Features of Forest Ecological Function Regions' Urbanization

In the aspect of boosting urbanization and urban and rural construction, compared with general areas, forest ecological function zones have several features.

2.1 High Ratio of no Developing Zones and Prominent Ecological Service

Through the overlaying analysis of the forest ecological function zones and national prohibition exploitation zones (Fig. 2), it is easy to see that the forest ecological function zones have covered many no-developing zones such as nature reserves, national forest parks and national places of interest. Limitations such as vegetation, terrain, ecological protection requirements all dramatically restrict developable space available. This phenomenon is quite common in the local area, and is a special feature that differs from general administrative regions. No developing area mainly points to no urbanization planning during the national territory planning, which means no matter how large the scale of the forest ecological function zone is, the possibility of its urbanization is limited. Shen-Nongjia, for example, already has a national forest park, a national geopark, a national wetland park and a provincial tourist resort. Although they overlap each other, the area of national geopark within it is as large as 1,700 km^2, which takes up 52.86 % of the total forest area. This means that half of the territory is already prohibited from any construction for the purpose of protecting the vegetation, reducing deforestation, enhancing the effects of grain for green projects, and slowing the deterioration of the mountain's ecological environment. Consequently, industrial activities can merely focus on the suitable industries, characteristic industries and service industries. The range of choices for construction is severely restricted.

2.2 Low Population Density and Inadequate Space for Urban and Rural Construction

The population density of this area is far lower than the national average (Fig. 3). Shen-Nongjia reached a population of 80.1 thousand by 2010. The population density, however, is a mere 25 per square kilometer for a land mass of 3215.84 km^2. However, the available land for construction in this area (including settlements mining, land for transportation, and land for water conservancy construction) takes up a mere 0.51 % of the total national territory; unused land (including desert, wetlands, sand, saline-alkali land, bare land and exposed rock and shingle land) and arable land take up 4.67 % and 2.2 % respectively. The

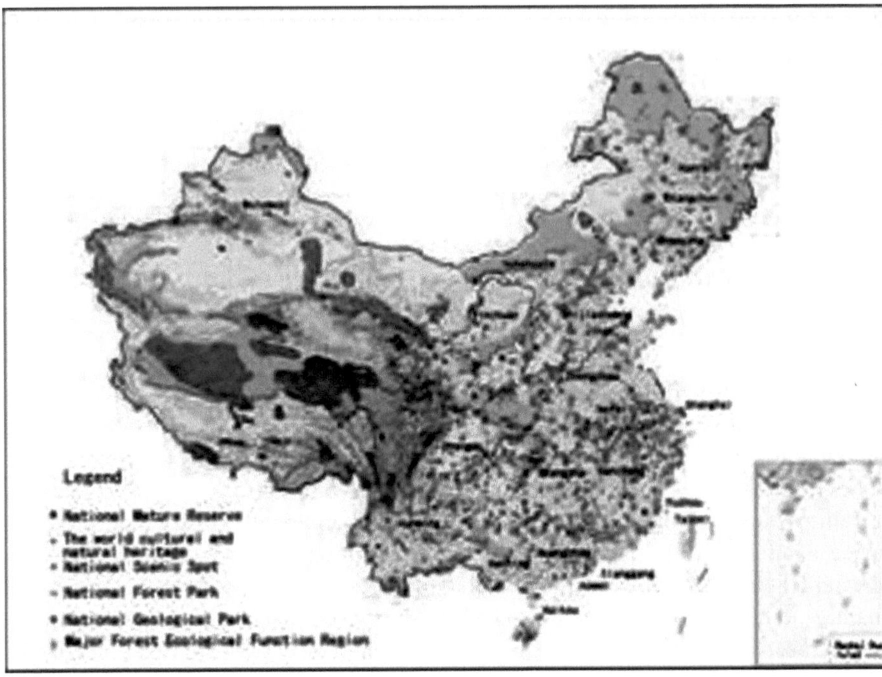

Fig. 2 The spatial relationship between forest ecological function regions and national prohibition exploitation zones (Drawn by the author according to the Fig. 12, the National Major Function Region Planning of China, 2010–2020)

distributive structure features little arable land, with fewer plains but more slopes. This means that only 8 % of the land is available for urban and rural construction, irrespective of the development condition of constructible space or farmland protection.

The feature displayed in the graph indicates that the human-and-land balance is still being maintained within the local area. Namely, the spatial capacity, population size and the scale of construction land are still very small. Based on this premise, it seems more reasonable to focus on the quality of its urbanization and the reasonableness of the relationships between urban and rural areas, rather than simple increasing of the number.

2.3 *The Obstacles That the Urbanization of Forest Ecological Function Regions Are Confronted*

From the point of view of urbanization policies for key ecological function zones, the concentration of industry and population in the EFR seems to be a necessary choice. However, this path usually dies young because it fails to achieve the

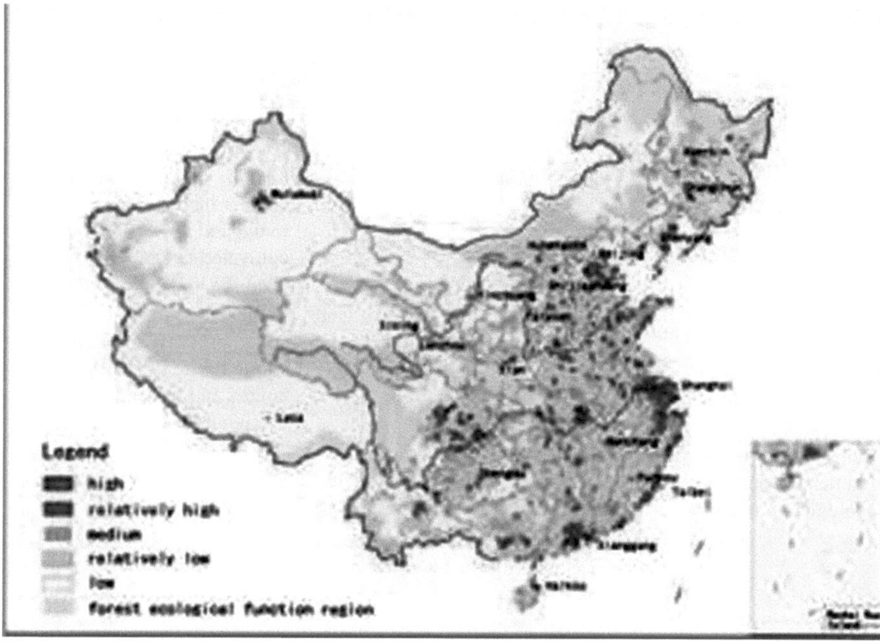

Fig. 3 The schematic diagram of population density in forest ecological function regions (Drawn by the author according to the Fig. 18, the National Major Function Region Planning of China, 2010–2020)

requited economy of scale. In correspondence to the current requirements of the urbanization policies, areas of this category face the obstacles listed in Table 1.

First, in the aspect of land urbanization, it is impossible to meet the current demands of urban and rural spatial development and protection. Furthermore, in the aspect of industrial development, the manufacturing and processing of agricultural and forestall products, tourism, and even leisure agriculture could be seen as better choices. However, the manufacturing and processing of the agricultural and forestall products needs large plantations to fulfill its raw material supply. Tourism and leisure agriculture also depend largely on the natural resources existing in the suburban areas. This means the development of suburban area has become an important premise. Thirdly, during the precess of population urbanization, the rising cost of urbanization away from home has become the uppermost obstacle for urbanization. Meanwhile, the top-bottom home urbanization proposes a higher requirement for the developmental level of centers.

In sum, although the policies of urbanization of this area have abandoned the beaten path of conventionality-industrialization driving urbanization, the thought which restricts the development of rural area is still implied. But the formation of current obstacles has much to do with the development of most rural areas. Therefore, changing approaches and conventionality of developing industries and expanding towns is needed if the urbanization of an area of this category should be promoted.

Table 1 The current obstacles of the urbanization of forest ecological function regions

Policy requirements	Solutions to the urbanization	Current obstacles subhead
Spatial urbanization: overall protection; focused development	The focused distribution of resources in urban areas	Only focusing, no expanding can not reach the best size effect
	Foothold development	Lacking going-to-scale development land, the cost of environmental protection and cultivated land compensation is on the high side
	Reduce the constructional land for villages	The demand for village construction is large, settlements' strong desire of building new houses and lacking awareness of saving land coexisting
Industry orientation; environmental friendliness; ecological protection	Develop the production and precessing industry of agriculture, forestry and animal husbandry	The development deficiencies of modern agriculture and forestry within the rural areas
	Develop tourism and leisure agriculture	The insufficient development and utilization of local natural resources in rural areas
	Develop other services	The quality and quantity level of labors is relatively low
Population urbanization; Reducing quantity and improving quality; Transferring and concentrating	Releasing the population pressure within the area by urbanization offsite	The rising integrated cost of urbanization offsite
	Transfer onto the towns in the region	The towns are underdevelopment, and the attraction of them is weak

3 A New Thinking of Forest Ecological Function Regions' Urbanization Way

3.1 The Proposal of the New Thinking of Urbanizing Development

It is advocated that an area of this category try to concentrate on the development of rural areas in order to indirectly boost the urbanizing development, starting with new village construction and green industries This new "village encircles city" line of thinking derives from the analysis below: for the current policies to proceed into the urbanization development, the logic relationship could be demonstrated in Fig. 4. In this figure, the dash-line represents the current obstacles; the bold-line demonstrates the possible solutions. Viewed form the graph, it is obvious that the development of rural areas is the support if urbanization is underway. First, judging from the secondary industry, the most prominent developable candidate within the

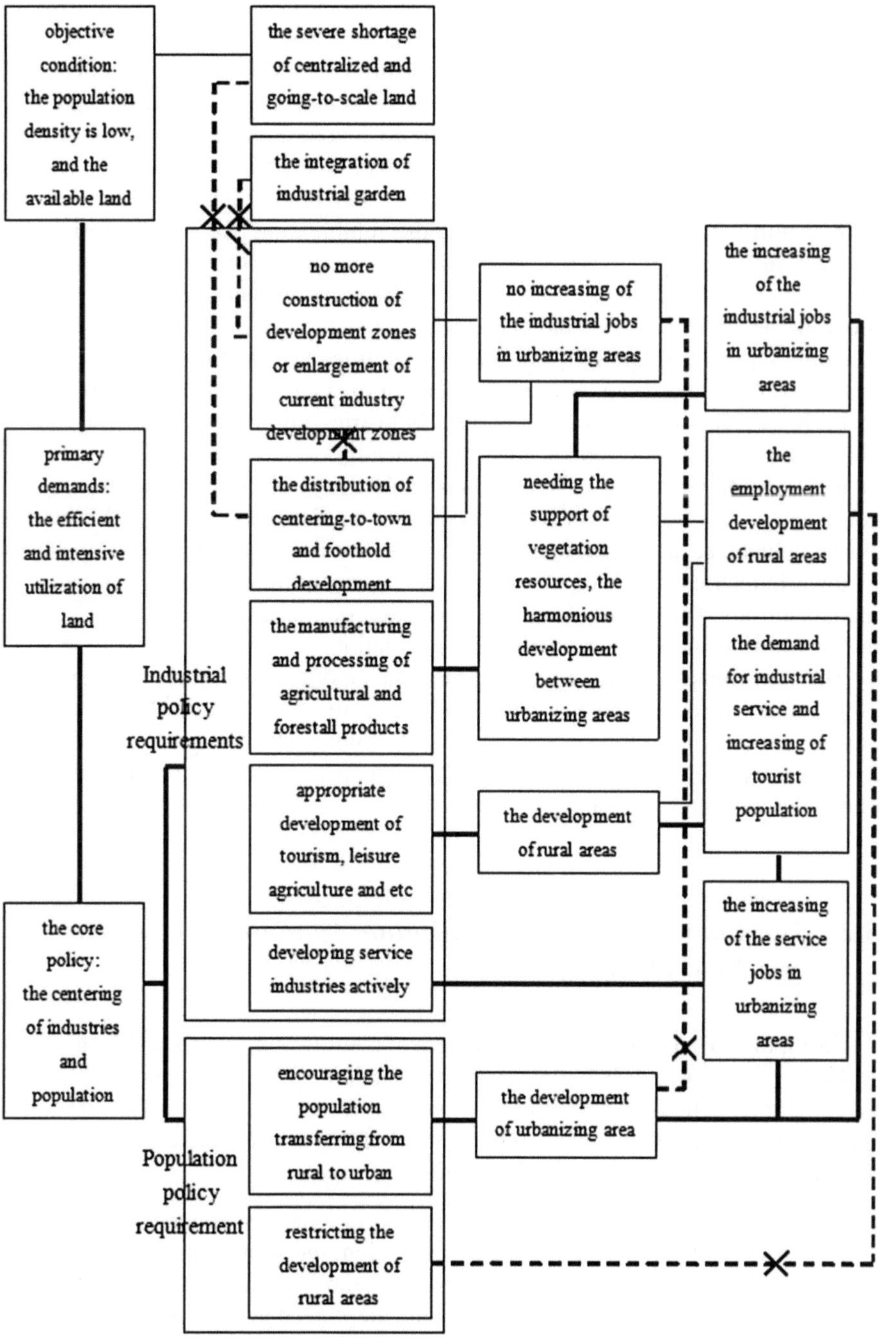

Fig. 4 The analysis of the solution to the urbanization in forest ecological function regions

area of this category should be agricultural production and processing of this category should be agricultural production and processing of this type needs large plantation to support its development, requiring the development of modern crop farming and manufacturing. It is the duty of urban areas to execute intensive processing, packing and marketing, so as to increase the employment and the capacity in urban areas. In other aspects, tourism and modern agriculture are also reasonable choices for the development of this area, both of which need the support of the natural resources in rural areas and the combination between the former and relevant infrastructure as well as the service facility. Both need planning and construction focused towards the feature spots, farms and growing regions in order to start. Furthermore, qualified service capacities for the rural areas, sufficient scientific knowledge and efficient practical abilities for local transactions are also necessary. Once these can be realized, local tourism and modern agriculture should feel a boost. The demand for the service industries further paves the way for the population centering-to -town, and consequently, a circle is formed. In Shen-Nongjia, the development of tourism has dramatically boosted the urban and rural construction, contributing to the formation of tourist towns such as Mu-yu and Jiu-hu, which see an annual increase in tourist population and the rapid expansion of town industries including service, leisure agriculture, and agricultural product manufacturing. This in turn boosts urbanization dramatically, and pushes forward the centering-to -town of industries and population.

This new thinking takes the driving mechanism of such areas of the urbanization process into account, and has played a role in the core conditions of resources advantages in urbanization precess. Table 2 provides the different characters and advantages of new thinking in the overall properties, theories and central points, compared with present policy.

3.2 The Coordination of Urbanization from the Perspective of a New Thinking

All in all, the construction and development of the rural areas proposed in this thesis is the key factor which influences the sustainable healthy development of urbanization progress should feature the development of rural areas as the key solution to the sustainable development of urbanization of the above-stated areas (Table 3). This urbanization can be categorized into three aspects:

Spatial Policy— appropriate concentration and rational dispersion. For the current industries, instead of one-size-fit concentration, the substandard ones should be concentrated into the urban areas. However, the standardized places, no matter rural or urban areas could be taken into the consideration of reservation, the premise of which certainly lies in the appropriateness of industry type. In case of urban and rural settlement construction, no blind emphasis on the extreme core development of town-centering should be advocated. Instead, the formation of a

Table 2 The comparison between the new and the old thought of the urbanization development in forest ecological function regions

Compare item	Current thinking	New thinking
Overall features	The subjective requirements, regarded the leading, ignore the driving mechanism of urbanization	Based on objective condition, the core advantages will be strengthened, and will play an important part in the progress of urbanization
Supporting theory	The traditional theory of urbanization, lay particular stress on "people in town", that is, the amount of urban population growth	The practical effect of local rural areas in the progress of urbanization should be accentuated. Namely, both the increase of "population in the town" and the improvement of the quality of "urbanizees in the countryside"
The kernel	Space: foothold development which emphasizes the decreasing of rural settlement space	Attaching importance to the development of construction of village and town system adhering to the local conditions to develop rural areas, instead of decreasing the land scale taken by villages
	Industries: emphasizing the development restrictions	Focusing on the relevance among the developments of industries and its influence towards the spatial construction
	Population: urbanization away from home	Local urbanization, which emphasize paving the way for the progress of centering-to-town of population

correspondent village system should be the key element, the construction of new village enhanced practically, and the ratio of which balanced. Through the integration of villages, the intensive land utilization should be correspondingly improved. Although the purpose of new village construction is not to emphasize the absolute centralization of the population and industries in towns, it still belongs to thinking of the relative centralization development, not mentioning staying the same and keeping dispersive.

Industrial policy—Agriculture-centralization and service for urban areas. Rural areas are unique for developing modern agricultural and forestall industries. Based on this advantage, the production chain should be extended, the industrial cost advantage formed within the limited scale through the combination of industry and agriculture, and consequently the uneconomic aftermath led by the limitation of remitting the industrial going-to scale development. Meanwhile, new agricultural industries which include the modern plantation, aquaculture, ecological leisure agriculture landscape creative agriculture, ecological leisure agriculture and landscape creative agriculture should be boosted. The author advocates that the new village construction being regarded the key point within the progress of urbanization in the areas stated above have its unique basement. Through combing the new village system, the new village

Table 3 The sustainable and healthy development of the urbanization of forest ecological function zones

Policy requirements	Policy adjustment	The coordination of urbanization
Spatial urbanization; overall protection; focused development	Appropriate concentration; rational dispersion	Settling the threshold for the scale of industrial garden areas; on blind emphasis on the centering-to-town; reasonable development of village system
		No unique emphasis on the enlargement of towns and cities; boosting new village construction and the intensive utilization of village land
Industrial guidance; environmental friendliness; ecological protection	Agriculture-centralization and service for urban areas	Pushing forward the development of modern agricultural and forestall industries
		Pushing forward the creative agriculture, new agriculture and new village construction
		Attracting tourist population depending on the unique natural ecological conditions
Population urbanization; reducing quantity and improving quality; transferring and concentrating	Appropriate concentration and co-increasing of quantity and quality	Encouraging local non-agriculture based on the village system planning
		Boosting urban service industry through the rapid development of rural areas

construction will combine with the local tourist industry more tightly, so as to attract a larger tourist population. Consequently, the demand for agricultural service and tourist populating service will advance side by side and thus stimulate the development of service industries. Finally, the new village construction will boost the development of urban service industry, and further promote the economy development of developing rural regions within the areas stated above.

Population policy—moderate concentration, quality and quantity rise together. Through the new village construction, make sure that a part of rural people can live in the rural parts, while the other part transfers into the town; this is essential to healthy urbanization.

As is shown in Zhu Nong's [4] studies, the rewarding impact of education in non-agricultural activity is usually better than that of in the agricultural. Educated family members have more possibilities to be attracted by the non-agricultural activities.

Consequently in the new village construction, human can improve their skills and educate the surplus labor that comes from farmers. They then have higher chances of getting a job in the city rather than live their lives with back-breaking farm work, transferring into the second and third industry. In this way, the urbanization will be realized earlier as long as people increase the investment in rural education and attach more importance to it.

In addition, relying on the new rural construction, local non-farm phenomenon will become more common in this region. On-farm should be regarded as an important part of the level of urbanization, which is a vital indicator of the quantity and quality of this class of regional urbanization.

4 Conclusion

The urbanization of forest ecological function zones has its own particularity. Chinese government has also proposed a series of developmental requirements and thoughts. According to the thesis, current urbanization has many obstacles in its way, and consequently goes against the sustainable development within the areas stated above to some extent. Changing approaches and conventionality of developing industries and expanding towns is needed if the urbanization of the area of this category is to be promoted. It could be a reasonable choice to concentrate on developing rural areas, so as to push forward the progress of urbanization, following the strategy of "village encircles city" from the perspective of new village construction and green industries.

Acknowledgment The study is supported by The National Natural Science Funds (No: 5118199) and The Specialized Scientific Research Fund of The Ministry of Education Doctoral Program (No:20100142120021).

References

1. Wang Shu, Li Wei (2009) The study on the urban and rural problems and development countermeasures of important water supply ecological function region within gannan, Yellow River. J Heilong Vocat Inst Ecol Eng 23:1–3
2. Wang Shu, Shi Peiji, Li Wei (2010) The study on the coordinated development between urban and rural areas of important water supply ecological function region within gannan, Yellow River. Territ Nat Resour Stud 35:50–52
3. Wang Wenhao (2011) The study on the herdsmen settlement and the countermeasures of important water supply ecological function region within gannan, Yellow River. J Anim Sci Vet Med 30:44–46
4. Zhu Nong (2003) The role of education within the rural family production activities and income in China. Chinese J Popul Sci 16:17–26

Quantitative Study of Housing Price Based on Huff Model and Hedonic Method

Yuan Li, Lang He, Junfeng Jiao, and Guoqiang Shen

1 Introduction

Housing, which is one of the most basic human needs, is a hot economic and social issues related to people's livelihood. Many factors influence the housing price and many researchers try to reveal the distribution rules on spatial variation in housing price. Some people [1–4] used GIS spatial interpolation method to get spatial distribution map and contour map of housing price. Some people [5] took Lanzhou as a case study and constructed the SD-GIS space-time simulation model to study the space-time distribution changes of residential price. Some [6] considered that space in the real estate market is important and official averages do not take into account the spatial correlation of housing price. They proposed the Kriging mean method to estimate mean housing price. Some [7] hold the view that absolute location must be viewed as interacting with other determinants of housing value. They presented an interactive variables approach and tested its ability to explain price variation in an urban residential housing market. Some [8, 9] established the hedonic price method, studied the relationships between different traffic convenient degree and house price. Some [10] constructed Hedonic Model and analyzed the

Y. Li (✉)
Department of Urban Planning, Tongji University, Shanghai, China

Department of Urban Planning, Xiamen University, Xiamen, Fujian, China
e-mail: yako79@gmail.com

L. He
Urban-Rural Planning & Design Institute of Guizhou Province, Guiyang, China

J. Jiao
School of Architecture, The University of Texas at Austin, Austin, TX 78759, USA

G. Shen
Department of Regional and City Planning, University of Oklahoma, 660 Parrington Oval, Norman, OK 73019, USA

X. Chen, Q. Pan (eds.), *Building Resilient Cities in China: The Nexus between Planning and Science*, GeoJournal Library 113, DOI 10.1007/978-3-319-14145-9_17

degrees how urban public service accessibility influences housing price. Some [11] revealed the relationship between airport noise level and real estate value. Some [12] detected how public school quality influence real estate markets in Howard County, Maryland. Some [13] used hedonic price modeling to derive different models of property price from which the contribution of the characteristics of the residential environment were estimated. It was demonstrated that an important factor was the accessibility to employment opportunities. Some [14, 15] took Knoxville TN as an example and focused on the relationship between neighborhood environment and housing price. They suggested that vacant land should be reasoning to promote its value when the housing price around it rises.

From the literature review, we found that there was a lack of comprehensive and systematic researches about the spatial distribution in house price. Basically, geography scholars focused on spatial analysis and less quantitative description about causes. Meanwhile, real estate and city economics researchers mainly focused on the application of Hedonic method to study the external environment influence in housing price [16]. Besides, in the previous studies, the distance variables are measured by air distance rather than road network distance, and this is not in conformity with the actual situation. Using Huff model, our study considered commercial center size and people's shopping preferences based on road network. Finally, through constructing Hedonic model, we quantitatively analyze the characterization factors' impact on housing price.

The remainder of this paper is organized as follows. Section 2 introduces study area and data. Section 3 discusses the analysis models used in this study. Section 4 presents the influencing factors we used. Sections 5 and 6 gives the results of analysis. Finally, discussion and concluding remarks are presented in Sects. 7 and 8.

2 Study Area and Data

Xiamen City is located in China's southeast coast, and in southeast of Fujian province. Back to Quanzhou plain and Zhangzhou plain, Xiamen is close to Taiwan Strait. As one of fifteen sub-provincial cities in China (one of five municipalities directly under the central government and one of the first batch of four special economic zones in China), it play the role of the regional financial services center between Taiwan and mainland, international shipping center in southeast China, trade center to Taiwan for mainland, the modern international port and scenic tourist city, etc. Taking the ordinary commercial housing in built up area as study object, we combine with GIS technology, Hedonic and Huff model to study the housing price's spatial pattern and influencing factors.

The sample data is mainly from SouFun real estate website (http://www.soufun.cn/) and real estate advertising on newspapers. We collected 106 housing samples with completed information. Besides, we used the basic GIS data from the government to establish a GIS database of Xiamen. The data contain road network, public

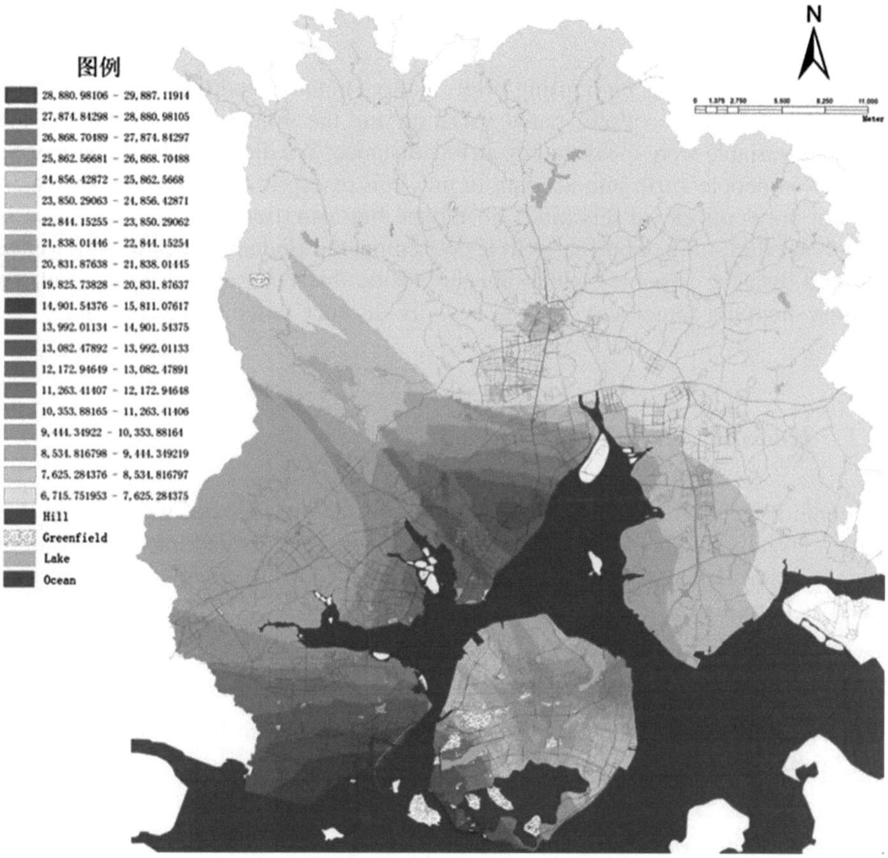

Fig. 1 Spatial distribution of commodity housing price in main urban areas of Xiamen

service and house sample and so on. With ordinary Kriging interpolation method, we performed spatial interpolation analysis of sample data.

As shown in Fig. 1, the housing price in Xiamen city generally showed a strong single-center distribution pattern. The price in island area was a gradient decreasing process from the edge of the island to the inside area. The housing price of outside area was much lower than Xiamen island. The highest price was around Zhongshan Road area. This area was the earliest commercial center in Xiamen city, where was a place close to the Gulangyu scenic zone and people gathered there in a long history. Besides, it formed a housing price sub-center around Yuandang lake and train station. Administrative Centre and external traffic hub were located in these two areas where transport was convenient, scenery was beautiful and public service was good. In the outside island region, both Haicang District and Jimei District had a relatively higher level housing price than Tong'an District and Xiang'an District.

3 Analysis Model

In the previous studies on housing price, authors mainly used GIS technology and Hedonic method to explore spatial variation and the influencing factors, and the distance variable were measured by airline distance. We didn't take road network's impact on people's trip into account in previous research. However, in this paper, we not only considered this kinds of impact, but also used Huff model, which is widely used in study of business circle. We compared Hedonic method's difference for two distance variables which are the airline distance between house and the closest shopping center and the road network distance based on Huff model.

3.1 *Hedonic Method*

Hedonic price method, also known as utility valuation method, was used in this paper. Usually, factors affecting the housing price were divided into three categories: location (L), structure (S) and neighborhood (N). The price could be formulated as Formula (3):

$$P = f(L, S, N) \tag{1}$$

A Hedonic price model often adopts following three basic function forms in the empirical research.

3.1.1 Linear Model

$$P = a_0 + \sum a_i c_i + \varepsilon \tag{2}$$

As shown in Formula (2), in which P represents the average residential price, a_0 is constant, c_i represents characteristic variable of order i, a_i is the characteristic price of variable i and ε is error.

3.1.2 Logarithmic Linear Model

$$InP = a_0 + \sum Ina_i c_i + \sum a_j c_j + \varepsilon \tag{3}$$

As shown in Formula (3), *i*ndependent and dependent variables adopt a logarithmic form. In which c_i is Continuous variable, c_j is neighborhood attributes which can't take the logarithm. It is 0–1 variable.

3.1.3 Semi Logarithmic Linear Model

$$InP = a_0 + \sum a_i c_i + \varepsilon \tag{4}$$

As shown in Formula (4), dependent variables adopt logarithmic form while the independent variable adopts a linear form.

3.2 Huff Model

Huff model, which was firstly constructed by David Huff in 1963, is a random probability model of gravity that considered the complicated road network in city. It defined the random probability as the ratio of urban service facility's attraction and the sum of all similar facilities' attractions. It aim to measure the probability that a consumer in a specific location go to a retail store. The probability could be formulated as Formula (5).

$$P_{ij} = \left(S_j/T_{ij}{}^n\right)/\sum \left(S_j/T_{ij}{}^n\right) \tag{5}$$

As shown in Formula (5), in which P_{ij} represents the probability customer in location i go to store j, S_j is the size of store j, and $j = 1, 2, 3 \cdots\cdots, n$; T_{ij} is the cost for location i to store j. n is effect index. If T_{ij} means time, then $n = 1$. In the condition that T_{ij} means distance, then $n = 2$.

4 Influencing Factors

We selected 21 factors or variables for our study. The variables were selected on the base of literature review and the availability of data. We classified the variables into three types: location characteristic (L), dwelling district characteristic (S) and neighborhood characteristic (N). The characteristic variables are shown as in Table 1.

5 Result of Huff

The variables X_1, X_2, X_3 and X_4 are measured by the tool of Network Analyst in ArcGIS 10.0. But X_1 and X_2 are based on Huff model (Formula 5). On the condition that the ratio of supermarkets or shopping centers' size (area) and the distance that dwelling district from them based on road network is highest, the corresponding

Table 1 Characteristic variables of houses and their meanings

Category	Characteristic variable	Description and measurement of variable
Location	ShoppingMDistHuff (X_1)	The distance between shopping malls and dwelling districts based on Huff model(m)
	SuperMDistHuff (X_2)	The distance between supermarkets and dwelling districts based on Huff model(m)
	DistCityCenter (X_3)	The closest distance between city center and dwelling districts based on road network(m)
	DistHospital (X_4)	The closest distance between hospitals and dwelling districts based on road network(m)
	NextToOcean (X_5)	The closest air distance from dwelling districts to ocean (m)
	LocationDistrict (X_6)	If dwelling district is on Xiamen Island (0 or 1)
Surrounding environment	PublicTransportservice (X_7)	If there are bus stops within 500 m (0 or 1)
	NextToPShool (X_8)	If there is a primary school within 800 m (0 or 1)
	NextToSSchool (X_9)	If there is a secondary school within 1,000 m (0 or 1)
	NextToKeyPShool (X_{10})	If there is a key primary school within 800 m (0 or 1)
	NextToKeySSchool (X_{11})	If there is a key secondary school within 1,000 m (0 or 1)
	KGService (X_{12})	If there is a kindergarten within 1,000 m (0 or 1)
	GreenFieldEffect (X_{13})	If there is green field within 500 m (0 or 1)
	BankService (X_{14})	If there is a bank within 500 m (0 or 1)
	VillageEffect (X_{15})	If there is a village within 500 m (0 or 1)
	RailwayEffect (X_{16})	If in the 500 m buffer area of railway (0 or 1)
Dwelling district characteristics	Parking (X_{17})	Parking space
	Households (X_{18})	Households
	GreenRate (X_{19})	Green ratio (%)
	FAR (X_{20})	Plot ratio
	PropertyManagementFee (X_{21})	Property management fee (Yuan/month $\cdot$ m^2)

distance are selected as the value of X_1 and X_2. The distance and attraction probability is shown as Tables 2 and 3.

According to the computing results in Tables 2 and 3, we built the O-D relationships between residential areas and shopping malls, also residential areas and supermarkets. The trip lines for shopping based on Huff model are shown as Figs. 2 and 3.

Table 2 The distance between dwelling district and shopping mall based on huff model

Dwelling districts	Shopping mall	Shopping mall's area	Distance	Attraction index	Attraction probability
1	Huijing	60,000	**4,827.99**	12.43	0.16
2	Mingfa	340,000	**14,214.80**	23.92	0.18
3	Mingfa	340,000	**1,560.40**	217.89	0.41
4	Mingfa	340,000	**3,298.87**	103.07	0.25
5	Minnan	200,000	**8,948.09**	22.35	0.17
6	Mingfa	340,000	**9,563.55**	35.55	0.19
7	Mingfa	340,000	**18,531.01**	18.35	0.18
...	...	...	...	...	...
104	Mingfa	340,000	**31,449.42**	10.81	0.17
105	Mingfa	340,000	**20,852.25**	16.31	0.15
106	Mingfa	340,000	**19,806.29**	17.17	0.17

Table 3 The distance between dwelling district and supermarket based on huff model

Dwelling districts	Supermarket	Supermarket's area	Distance	Attraction index	Attraction probability
1	Runzhou	18,000	16,678.60	1.08	0.10
2	Lehai	20,000	4,137.92	4.83	0.17
3	Mingfa carrefour	18,000	1,609.93	11.18	0.16
4	Darunfa	28,000	3,670.36	7.63	0.14
5	Darunfa	28,000	14,311.99	1.96	0.10
6	Darunfa	28,000	9,532.44	2.94	0.11
7	Darunfa	28,000	17,558.05	1.59	0.09
...	...	...	...	...	...
104	Runzhou	18,000	5,502.39	3.27	0.23
105	Darunfa	28,000	21,042.84	1.33	0.09
106	Darunfa	28,000	19,998.81	1.40	0.10

6 Result of Hedonic

The standard method for estimating hedonic price function was parameter method. First, we determined the functional form. Then, we needed to select the appropriate estimation method. Last, we needed to fit the sample data. In order to facilitate the estimation of parameters and interpretation, we made the hypothesis that the relationship between price characteristics was linear. Many researches adopt standard simple functions, such as a linear model, semi logarithmic model and logarithmic model, etc. The reason for selecting these function form was that the traditional least square estimation methods was convenient to statistical inference and hypothesis testing. In this paper, we estimated the housing characteristic price or marginal price through the model, then took the corresponding market analysis.

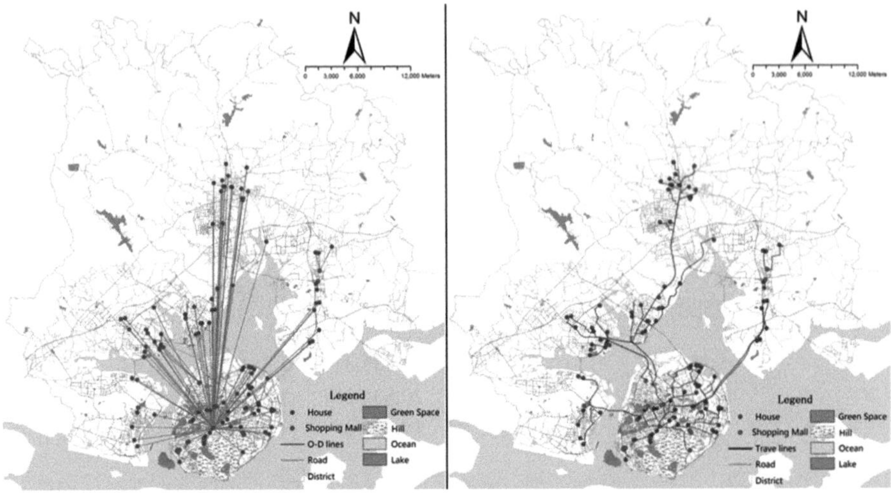

Fig. 2 The O-D lines of dwelling-shopping mall and travel lines based on maximum attraction probability

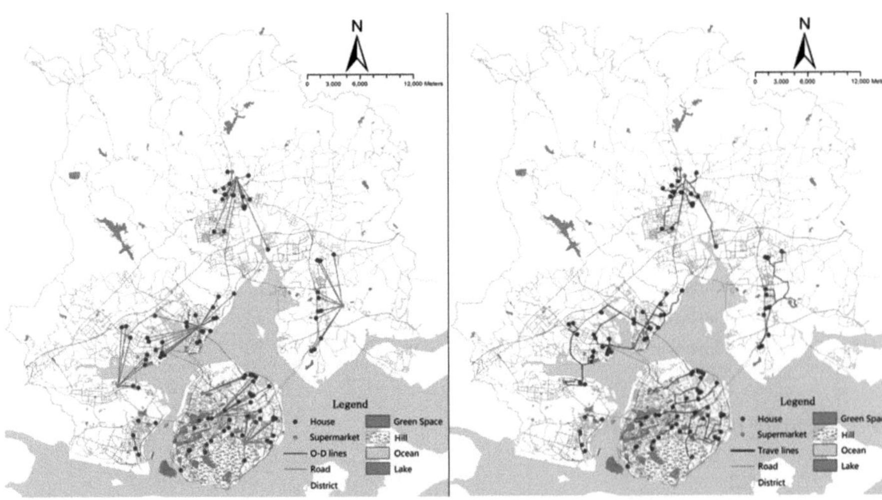

Fig. 3 The O-D lines of dwelling-supermarket and travel lines based on maximum attraction probability

In Economic theory, there were no explicit rules to specify which function forms and feature variables hedonic adopted for hedonic housing modeling. Thus, this study compared the three kinds of functional forms in order to select a suitable model. Meanwhile, we monitored the co-linearity between variables by variance inflation factor (VIF).

Table 4 showed the parameter comparison of three functions. Linear model multiple correlation coefficient R was 0.901. It demonstrated that the fitting

Table 4 Comparison of three functions

Function type	R	R2	Adjusted R2
Linear model	0.928	0.861	0.854
Semi-logarithmic model	0.927	0.860	0.853
Log-linear model	0.919	0.845	0.840

precision of linear model was better than the other two. Thus, we choose linear model to study the relationships between housing price and housing characteristics. According to the partial regression coefficient in the regression equation, we removed the factors that significant level larger than 10 %. Combined with variance inflation factor (VIF) inspection, we weeded out the multicollinearity factors. Finally, there were seven variables selected in the model. After running the regression analysis, the results were as shown in the Table 5.

The regression coefficients for five characteristic variables of the linear model could be obtained by regression analysis. We calculated the housing hedonic price model of Xiamen, as shown in Formula (6):

$$\text{Price} = 7371.611 - 0.114X_1 - 0.182X_2 + 6957.728X_6 + 1538.665X_{11} + 3489.819X_{21} \quad (6)$$

According to the results of regression analysis, we could find the main elements that influence the price of commodity housings in Xiamen: PropertyManagementFee (X_{21}), NextToKeySSchool (X_{11}), SuperMDistHuff (X_2), ShoppingMDistHuff (X_1), LocationDistrict (X_6).

7 Discussion

7.1 Discussion of Influencing Factors

7.1.1 Convenience of Shopping (X_1 and X_2)

Commercial center is a place where business services highly gather. It's one of the most important parts of civil center and has the function of business, service, recreation and so on. Commercial center can provide many kinds of service to meet people's needs. People gather and the land here is developed high-density. According to the effect of differential incomes of land, land price here is higher. Because of the positive external benefits, the residential land around commercial center is expensive. With the advent of the era of consumption and the improvement of people's income level, the convenience of consumption is also an important factor for people chose where to live. Besides, supermarket plays an important role for family's daily purchasing, so the Hedonic model is also reflect influence degree of the convenience from dwelling district to supermarket. It could be seen from the hedonic price model that the two variable coefficients. X_1 is −0.114 and X_2 is

Table 5 Regression coefficient analysis of liner model

Variable	Non-standardized coefficient		Standardized coefficient			
	B	*Std Error*	*Beta*	*t*	*Sig.*	*VIF*
Constant	7,371.611	1,424.913		5.173	.000	
PropertyManagementFee (X_{21})	3,489.819	377.496	.438	9.245	.000	1.619
NextToKeySSchool (X_{11})	1,538.665	673.197	.088	2.286	.024	1.074
SuperMDistHuff (X_2)	−.182	.053	−.144	−3.438	.001	1.264
ShoppingMDistHuff (X_1)	−.114	.045	−.144	−2.545	.012	2.308

−0.182. They suggest that the distance from dwelling districts to shopping centers and supermarkets was inversely proportional to the housing price. With the distance between dwellings and shopping centers expanding, the residential price would fall. More specifically, every 1 km of the increase in distance would cause 114 RMB (Yuan) loss in sale price. Meanwhile, every 1 km of the increase in distance between dwelling and supermarket would cause 182 RMB loss in sale price.

7.1.2 District that Dwelling District Locate (X_6)

Xiamen Island is the administration center, culture center and business center of Xiamen city. It developed early and owns beautiful coastline view. The district outside Xiamen Island fall behind Xiamen Island obviously. For a long time, the willing wish to live on Xiamen Island rather than outside is ineradicable. It could be seen from the hedonic price model that the dwelling district on Xiamen Island is averagely 6957.728 RMB higher than those outside Island. Lack of land, unbalance of supply and demand and urban public facilities' difference are the reason of housing price's variation.

7.1.3 Proximity of the Key Primary Schools (X_{11})

Education was the foundation of the nation, which was related to the quality of people and long-term development of the country. As the high-quality education resources were limited and unevenly distributed, many parents in China rent or even bought houses near key schools in order to offer their children with better education. As can be seen from the above hedonic price model, the key primary schools had a positive effect on housing price and the sale price would decrease 1538.665 RMB per kilometer the distance to provincial-demonstrated primary schools increased.

7.1.4 Property Management Level (X_{21})

Property cost was a symbol of property management level. Decent property management indicated high-quality living conditions. As can be seen from the regression coefficients of characteristic variables, the property management level had a great influence on premises price. Housing prices would increase 3489.819 RMB as long as the property cost increased 1 RMB.

Other variables, such as building area, household number, parking digits, volume rate, greening rate and so on, had little influence on housing price. It meant that the scale of premises can hardly affect people's consuming choice, while the parking digits, volume rate, greening rate, etc. were prescribed in urban controlling detailed planning and could meet people's needs of living.

7.2 Discussion of Causes

7.2.1 Single-Central City Polarization Development

Since the reform and opening up 30 years ago, the construction of Xiamen island had been highly concentrated, tended to saturation, presented as a typical single-central spatial structure [17]. "Dense population, traffic congestion, landscape degradation, overtop property price " had become a common problem on the main island, and urban construction or supporting facilities outside main islands had large lagged behind. The proportion of area between the main island and outside was 1:11, and the proportion of population density was 11:1.65 % of the jobs distribute in the main island, and large cultural,educational and body health facilities, port airports, railway stations and the important traffic facilities were almost located in the main polarization effect of the island was very obvious. As is shown in Fig. 2, most of big size shopping mall distribute on Xiamen Island.

In recent years, some sea-crossing passages (such as JiMei bridge, XiangAn tunnels) have built. The infrastructures, for example, the port and the new railway station, have begun to locate outside the island. The construction of these functional facilities expands the city scale, however, the corresponding public facilities fail to timely follow-up and the city has not formed the strong anti-magnetic function. As a result, the broadening function of the outer island strengthens the single central polarization effect through the "diffusion—echo" effect, and the single-central structure is difficult to break at present. Single-central city structure decides Xiamen housing price on the spatial distribution pattern of a single center Centre.

7.2.2 Unbalanced Allocation of Public Facilities

Some researchers [18] thought that housing price had spatial correlation for two reasons. One was that neighboring areas tended to the same sequence development, resulting in adjacent area had the similar structural characteristics, such as residential size, building age, internal and external design style, etc.. Another reason was the adjoining property of enjoying the same convenience degree of public service facilities. Good accessibility of public service facilities was the important premise ensuring the residents' quality of life. Although single-central polarization would bring a lot of problems, such as traffic congestion, environmental degradation, but high quality public service facilities owned much stronger appealing, which made people unconsciously live around the centre of the city. Consequently, these would lead to high housing price in the core area.

For a long time, Xiamen citizens had a traditional thought —"prefer a bed in the island, not a room outside the island ". In order to change this situation, we must improve the quality of new shopping, transportation, health care, education and other supporting facilities outside the island.

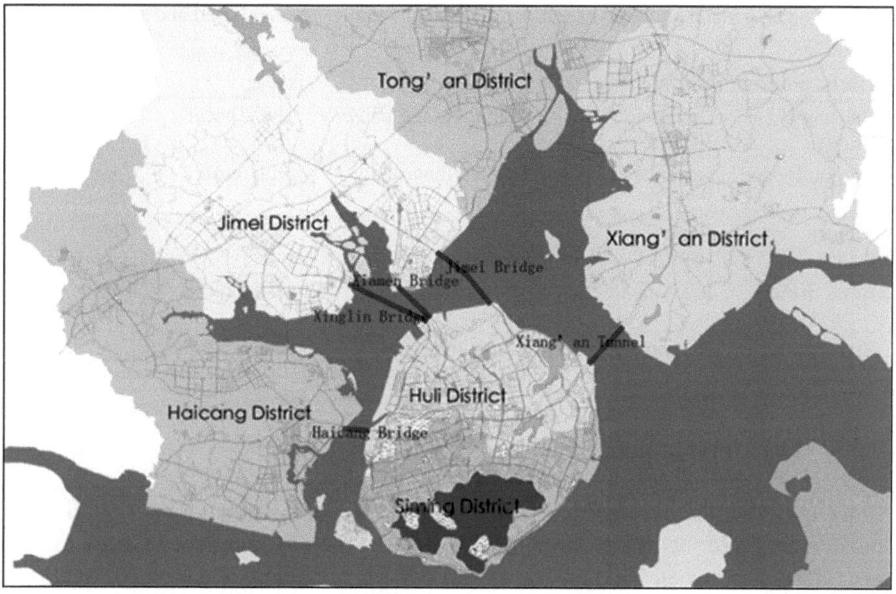

Fig. 4 Passageway across ocean

7.2.3 Geographical Spatial Partition

The gulf made the Xiamen city separated, and caused the urban spatial structure discontinuity. Due to the gulf division, the links between outside areas and the main island are bridges and tunnels. As shown in Fig. 4 and Table 6, the direct link between HaiCang, JiMei and Main Island was earlier (Year 1999 and Year 2008) through Haichang Bridge and Jimei Bridge. The direct link between Xiang'an and the main island is relatively late, and there is no direct link between Tong'an and the main island.

7.2.4 Coastal Landscape Effect

It was obviously that coastline, shoreline resources owned special or extra value because of its beautiful scenery. For Xiamen city, because of its pleasant climate, clean environment, rich historical and cultural background, the value of coastal landscape effects are even greater. A large number of market demand made the real estate price in coastal area higher than other area in Xiamen city. More specifically, in the island, the housing price was decreasing from the coastal line to the inside island, however, in the outside island, the housing price was increased from the inside to the outside coastal area.

Table 6 The general situation of traffic projects connecting Xiamen island and outside

Traffic project	Traffic time	Designed traffic capacity	Connection area	Passing vehicles
Haicang Bridge	December 30, 1999	5 ten thousand pcu/day	Huli-Haicang	Car
Xinglin Bridge	September 1, 2008	3.38 ten thousand pcu/day	Huli-Jimei	Car/train
Xiamen Bridge	December 19, 1991	2.5 ten thousand pcu/day	Huli-Jimei	Car
Jimei Bridge	July 1, 2008	5.5 ten thousand pcu/day	Huli-Jimei	Car
Xiang'an Tunnel	26 April, 2010	5.2 ten thousand pcu/day	Huli-Xiang'an	Car

7.2.5 The Development Strategy

Since establish the development model of bay city, urban development is focus on outside area and the effect is obvious, However, the development strategy of land are different between Xiamen Island and outside area, and development quality is also unlike. Many areas on Xiamen Island are limited to develop to protect coastal landscape. It reduce the alternative of capital investment for land in the development of commercial housing, and the high land price can not amortize to the increased housing. Meanwhile, small plot is hard to configure public facilities. Then it's not difficult to understand the difference of housing price between Xiamen Island and outside area.

8 Conclusion

This paper used Hedonic and Huff model, reveal the spatial distribution of housing price in Xiamen. The research result shows that the convenience of shopping, education resources allocation, dwelling district and the property management are the influencing factors of housing price. Finally, we systematically analyze influence of urban structure, allocation of public service, geographical conditions, coastal landscape effect and the development strategy for housing price.

At the end of the paper, we provide three suggestions. One is to strengthen the rapid transportation construction of the main island and outside, to shorten the space and time distance of the city area, to enhance the different geographical location accessibility, to strengthen the residents travel convenience. The second is to be optimized allocation of public service resources in the city area, especially to strengthen the regional and high quality public service facility configuration outside the island. The last is to weaken the polarizing effect of the island, to relocate parts of the city functions such as foreign transportation, administrative office, entertainment, business and trade of the island.

Acknowledgment The authors acknowledge the financial support of the Chinese Postdoctoral Science Foundation (NO. 2014M551449), and the Project for Special Funding provided by the Fund for Scientific Research in Colleges and Universities (NO. 2012121033).

References

1. Meng B, Zhang J, Wang J, Zhang W, Hao W (2005) Application of spatial analysis to the research of real estate: taking Beijing as a case. Geogr Res 24(6):956–964, 1004
2. Jiang F, Zhu D (2005) A GIS-based study on spatial distribution of land prices-the case of residential land prices in Beijing. Econ Geogr 25(2):199–202
3. Mei Z, Li X (2008) Spatial analysis of houses' price in Dongguan based on ESDA and kriging techniques. Econ Geogr 28(5):862–866
4. Xu X (1997) An analysis on the spatial distribution characters of commercial residential residence price in Shanghai. Econ Geogr 17(1):80–87
5. Luo P, He S, Huang Y, Niu H, Pu X, Du Q (2002) A case study of space-time simulation of Lanzhou's urban housing prices using the SD-GIS model. J Lanzhou Univ 38(4):125–130
6. Montero JM, Larraz B (2009) Estimating housing prices: a proposal with spatially correlated data. Int Adv Econ Res 16(1):39–51
7. Fik TJ, Ling DC, Mulligan GF (2003) Modeling spatial variation in housing prices: a variable interaction approach. Real Estate Econ 31(4):623–646
8. Zhang W, Meng B, Lu X, Liu W (2004) Influence of traffic passages on housing spatial expansion and local residents' selection of housing location-a case study of Beijing. Sci Geogr Sin 24(1):7–13
9. Armstrong RJ, Rodríguez DA (2006) An evaluation of the accessibility benefits of commuter rail in eastern Massachusetts using spatial hedonic price functions. Transportation 33(1):21–43
10. Wang S, Zheng S, Feng J (2007) Spatial accessibility of housing to public services and its impact on housing price: a case study of Beijing's inner city. Prog Geogr 26(6):78–85, 147–148
11. Cohen JP, Coughlin CC (2008) Spatial hedonic models of airport noise, proximity, and housing prices. J Reg Sci 48(5):859–878
12. Sedgley NH, Williams NA, Derrick FW (2008) The effect of educational test scores on house prices in a model with spatial dependence. J Hous Econ 17(2):191–200
13. Visser P, Dam FV, Hooimeijer P (2008) Residential environment and spatial variation in house prices in the Netherlands. Tijdschr Econ Soc Geogr 99(3):348–360
14. Cho S, Kim SG, Lambert DM (2009) Spatially-varying effects of rezoning on housing price. Rev Urban Reg Dev Stud 21(1):72–91
15. Cho S, Kim J, Roberts R, Kim S (2012) Neighborhood spillover effects between rezoning and housing price. Ann Reg Sci 48(1):301–319
16. Liu D, Yang Y, Huang X, Zhang B (2010) Study on the spatial distribution of housing price and its influencing factors in Chengdu based on GIS and hedonic model. J Anhui Agric Sci 38 (29):16519–16523, 16559
17. He Z, Qiu G, Yang Z (2007) Research of Xiamen urban morphology development based on space syntax analysis. Huazhong Archit 25(3):106–108, 121
18. Basu S, Thibodeau TG (1998) Analysis of spatial autocorrelation in house prices. J Real Estate Financ Econ 17(1):61–85

Develop A GIS Based Risk Model to Evaluate the Economic Resilience of Houston Neighborhoods for the Next Oil Bust

Qisheng Pan and Dong Zhai

1 Introduction

Houston is the center of energy in the world and hosts more than 5,000 energy related companies. Oil and gas business has been the major industry in Houston since the early twentieth century. By the end of 1970, Houston had attracted thousands of oil workers because of the oil boom in West Texas. Many young professionals rushed to Houston for high earning jobs in oil and gas companies. However, due to the oil exploration and production in North Sea and the following declined demands, the price of crude oil dropped from 20 dollars to 1 dollar in the early 1980s, which caused serious unemployment problem in Houston. Many people lost jobs and many apartment complexes became empty. The drop of crude oil price caused significant impact on Houston's economical and urban development. With the rising energy demand from developing countries such as China and India from the 1990s, oil prices rose up dramatically in the past decade. Many new technologies have been developed to increase oil production such as Deep Sea and Shale Gas technologies. This change has stimulated the oil and gas industry in Houston and the oil and gas business has been booming again. Many young professionals are rushing to Houston again for oil and gas.

According to some statistics, the population in Houston is increasing at a rate of 2.0 % per year. A very large percentage of jobs are related to oil and gas industry. According to the survey data, 3.4 % of the total jobs in Houston are related to oil

Q. Pan (✉)
Department of Urban Planning and Environmental Policy, Texas Southern University, Houston, TX, USA

School of Urban Design, Wuhan University, Wuhan, China
e-mail: pan_qs@tsu.edu

D. Zhai
Department of Urban Planning and Environmental Policy, Texas Southern University, Houston, TX, USA

X. Chen, Q. Pan (eds.), *Building Resilient Cities in China: The Nexus between Planning and Science*, GeoJournal Library 113, DOI 10.1007/978-3-319-14145-9_18

and gas business. The percentage of oil and gas jobs in Houston is much higher than the average 0.3 % national level. At the same time, many services jobs are also serving the oil and gas workers. The oil and gas industry brings tons of money into the city and the whole city depends on the energy business. However, a possible breakthrough in the research of clean energy technology can cause another bust of oil and gas industry in Houston in the near future.

Many other cities have similar economic structures. For example, Detroit is the capital of auto manufacturing industry and many of the jobs belong to vehicle manufacturing and its related business. San Jose, California heavily relies on the information technology (IT) industry. This kind of single mode of economy has potential risks on future development. With the decline of auto manufacturing, the whole city of Detroit got into economic depression since the late 1970s. Detroit was selling some one-dollar houses in the recent housing market crisis. Many communities became ghost neighborhoods. Relying on a single industry can put cities into high risk if the major industry collapses. A city with multiple industries can be more resilient to this kind of industry collapse.

With the development of green energy, it is possible that oil and gas industry can collapse in one day. How will we respond to such as event? Is Houston ready for changing its traditional oil and gas industry or switching into other industries? In the future several decades, it is possible that a green energy technology can replace the traditional fossil fuel. It is necessary to evaluate the potential risk associated with this event and make us be prepared for future challenges.

2 Urban Resilience

In the literature, urban resilience refers to the capacity of a city to prepare for, respond to, and recover from destruction. Campanella [1] stated that urban resilience is a function of resilient and resourceful of citizens. The recovery of social network with strong citizen participation would be as important as the rebuilding physical infrastructure to ensure the successful revival of a city after major disasters, such as New Orleans after the strike of Hurricane Katrina. Some scholars also made conceptual and empirical discussions on the relevant issues from the aspect of economic vulnerability, such as Cordina [2], Liu et al. [3], and Weir et al. [4].

Urban or regional resilience has been addressed in the literature from different aspects. Lin [5] examines the long term population change across metropolitan areas in the US using regional resilience index that were measured using population data from Census. It is possible to develop an urban resilient index for cities responding to unfavorable events.

High fuel price may have positive effects for the economic sectors related to energy industry but negative effects on many other sectors. It may impose a large gap of social and economic impacts among cities with different combination of industries. Dodson and Sipe [6] developed oil vulnerability index to demonstrate the spatial variability of social and economic impacts of higher fuel prices in Australian cities.

Oil bust has also been addressed in literature. Leigh and Vukovic [7] analyzed the recent change in the supply and demand of oil and investigated the socio-economic and geopolitical consequence of dwindling supply of crude oils. Newman [8] points out most of the U.S. Cities are automobile dependent and thus oil dependent. The progress made by electrical car manufacturers these years can potentially reduce the oil demand significantly. It would be interesting for this study to focus on Houston, the world's energy capital and to develop a resilience index for the city.

3 1980 Oil Bust

In the early 1980s, oil price was about $33 a barrel. The oil and gas industry was booming in Houston. Major oil companies' exploration and production in the US soared and many new oil fields were under development to meet the economic development needs. Houston attracted much attention from job-seekers across America. At the same period, the unemployment rate stood at 17.2 % in December, 1982 because of the huge import of Japanese cars. Many people moved to Houston from Detroit and other parts of the country. Many apartment complexes were built in the Southwest and other parts of Houston to accommodate these migrant workers for oil and gas industry.

However, the oil bubble didn't last for too long. The oil industry bust began in the early 1980s as a result of slowed economic activity in developed countries due to the energy crisis of the 1970s. The energy conservation caused by high fuel prices slowed down the economic development. In the U.S., domestic exploration declined dramatically in the 1980s. Major oil companies put holds on the search for new oilfields. Houston went into recession due to the slowdown of exploration. During these recessions in the early 1980s, Houston's economy collapsed. Many apartments were abandoned and a large percentage of workers were unemployed. Many workers moved out of Houston to seek other opportunities in other cities.

Southwest Houston has a high population density close to approximately 9,000 people per square mile. This area is composed of residential, commercial, and industrial development. There are both single-family residential neighborhoods and numerous low-rise and mid-rise multi-family residential complexes. Southwest Houston has multiple light-industrial parks, as well as a heavy-industrial plant.

Many subdivisions in Southwest Houston were once the victims of the oil-bust recession of the early 1980s. Neighborhoods such as Westwood, and Forum Park, etc. were created over night as large apartment complexes were built to accommodate the oil workers in the 1970s. Oil price dropped to $10 a barrel in January 1986. With 70 % of jobs in the Houston area depending directly or indirectly on the oil industry, the bust caused devastating effects on local economy. Construction stopped overnight and many people lost jobs. In January 1983, unemployment rate in the Houston metropolitan area rose to 9.1 %, the highest among the state's largest metropolitan areas. Many people moved out of Houston and most of these people were renters and many apartments became empty.

Unfortunately, this area went down dramatically with the oil bust in the early 1980s. Almost as fast as these apartment complexes had been built, they were nearly abandoned after the oil bust. A high crime rate was caused by these abandoned apartments. Drug deals and prostitution were made in these abandoned apartments in Southwest Houston. It was similar to what is going on in Detroit today. Many crimes were generated in abandoned homes. The high crime also caused more people to move out of those neighborhoods. The oil bust in the early 1980s put this area into a bad cycle. With the new booming of oil and gas business these days, this area is becoming better than that in 1980s, but there is still some risk related to this area. This area was never fully recovered from this oil bust and remains one of the poor areas in the city.

4 Risk Factors for City and Community Resilience

The risk model assessing city resilience to major economy meltdown needs to cover multiple factors. In this paper we mainly focus on the following factors:

Education attainment
Median income
Homeownership
Availability of public transportation
Percentage of oil and gas employment
Population stability
Employment needs

Education attainment plays an important role in the risk modeling for economic resilience because it determines whether the residents in the neighborhood can switch to other types of jobs if another oil bust happens in Houston. The median income level also plays an important role because higher incomes can allow residents to stay with unemployment benefits. Homeownership is another important factor because people tend to stay longer if they own a house or apartment. Renters tend to move soon after they become unemployed. Availability of public transportation factor is based on the number of bus stops because public transportation can help find jobs during recession especially for the poor people. The percentage of oil and gas employment is the most important factor in this model. The ACS 5 Year estimate provides the percentage of direct oil and gas employment in each census tract. Population stability is measured by the percentage of population that stayed in Harris County for more than 1 year. People who moved from other counties or states tend to move back easily. Employment needs is measured by the percentage of population between ages 18 and 64. All these seven factors have five different rankings based on the corresponding classification criteria. The American Community Survey along with the US Census data provides the best estimation of all these factors at census tract level (Tables 1, 2, 3, 4, 5, 6, 7, and 8) (Figs. 1, 2, 3, 4, 5, 6, 7, and 8).

The overall resilience is a sum of all the ranking scores. It is a good indicator for the future economic resilience in case of another oil bust in the future. It provides an

Table 1 Resilience ranking for education factor

Range (percentage of people with high school diploma or higher)	Resilience score (1–5)
0–13.2	1
13.2–28.1	2
28.1–46.2	3
46.2–65.2	4
65.2 or higher	5

Table 2 Resilience ranking for median income factor

Range (median income in USD)	Resilience score (1–5)
13,257–29,082	1
29,082–43,014	2
43,014–59,539	3
59,539–83,167	4
83,169–133,096	5

Table 3 Resilience ranking for home ownership factor

Range (percentage of homes owned by residents)	Resilience score (1–5)
0–24.99	1
24.99–47.85	2
47.85–64.93	3
64.93–80.09	4
80.09–96.1	5

Table 4 Resilience ranking for public transportation factor

Range (number of bus stops)	Resilience score (1–5)
0–7	1
8–21	2
22–39	3
40–75	4
76–215	5

Table 5 Resilience ranking for oil and gas employment factor

Range (percentage of oil and gas employment)	Resilience score (1–5)
0.084769–0.190149	1
0.051502–0.084768	2
0.028640–0.051501	3
0.011612–0.028639	4
0–0.011611	5

overview of the distribution of resilience indexes in Harris County. In Fig. 8, we can tell that some areas are going to have better resilience to future oil bust than other areas. The risk index gives planners a big picture for the future planning activities.

Table 6 Resilience ranking for population stability factor

Range (percentage of population with more than 1 year residence in Harris County)	Resilience score (1–5)
0.612871–0.798963	1
0.798964–0.884758	2
0.884759–0.936300	3
0.936301–0.970098	4
0.970099–1	5

Table 7 Resilience ranking for employment needs factor

Range (percentage of oil and gas employment)	Resilience score (1–5)
78.3–98.7	1
69.1–78.3	2
63.5–69.1	3
58.5–63.5	4
47.9–58.5	5

Table 8 Resilience ranking for all the factors

Range (summed factor scores)	Category
12–17	1
18–20	2
21–23	3
24–26	4
27–31	5

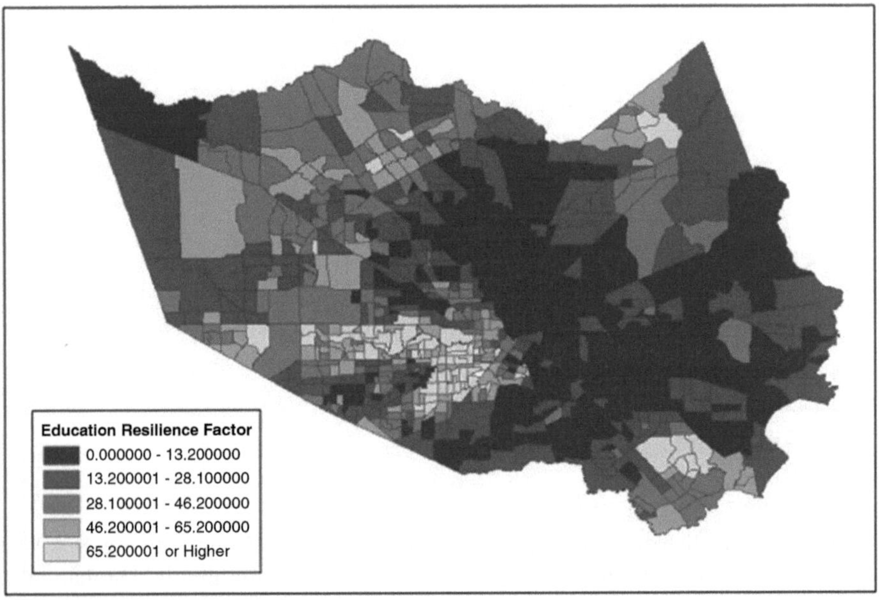

Fig. 1 Education attainment factor in Harris County

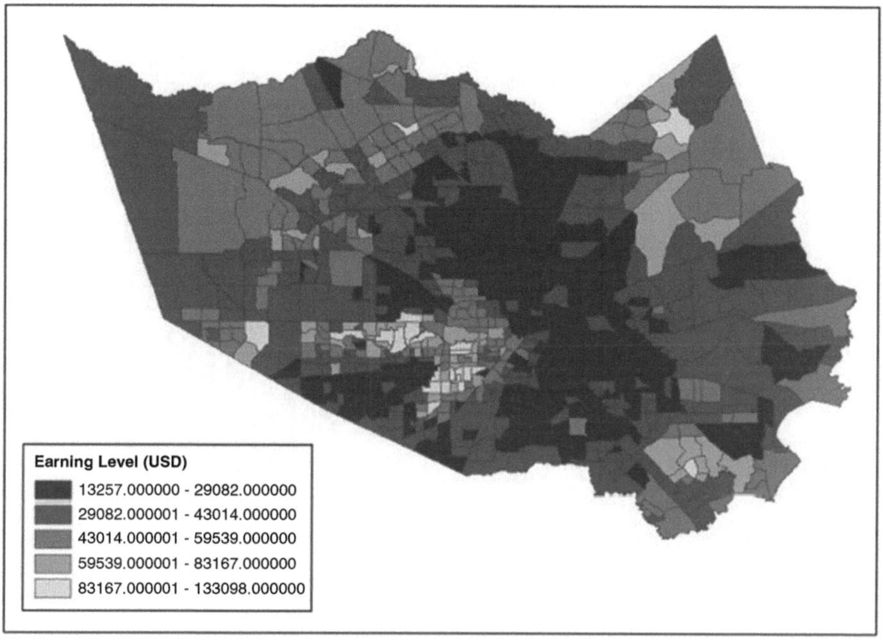

Fig. 2 Earning level risk factor in Harris County

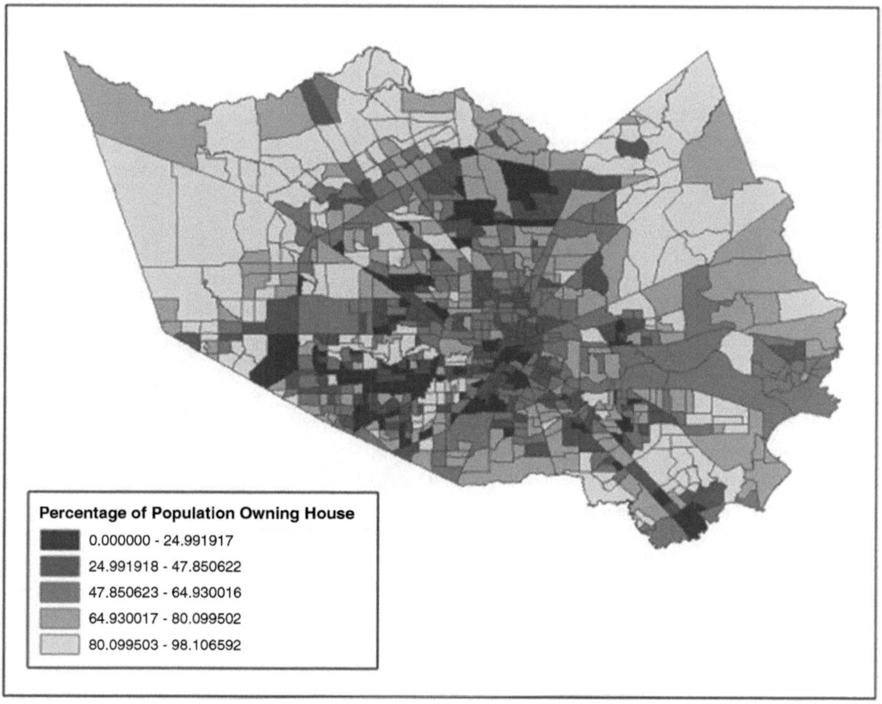

Fig. 3 Home ownership risk factor in Harris County

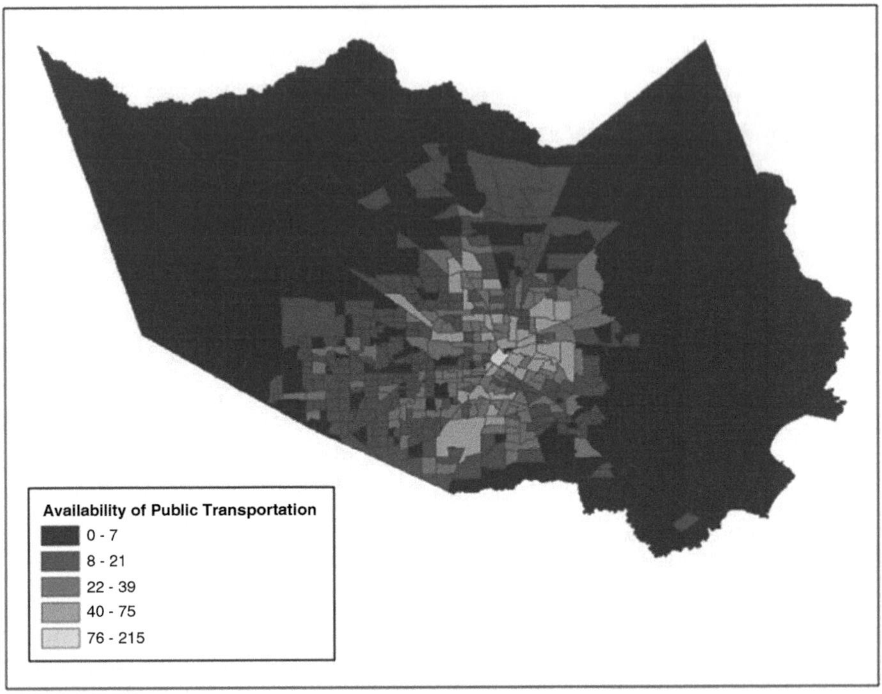

Fig. 4 Public transportation risk factor in Harris County

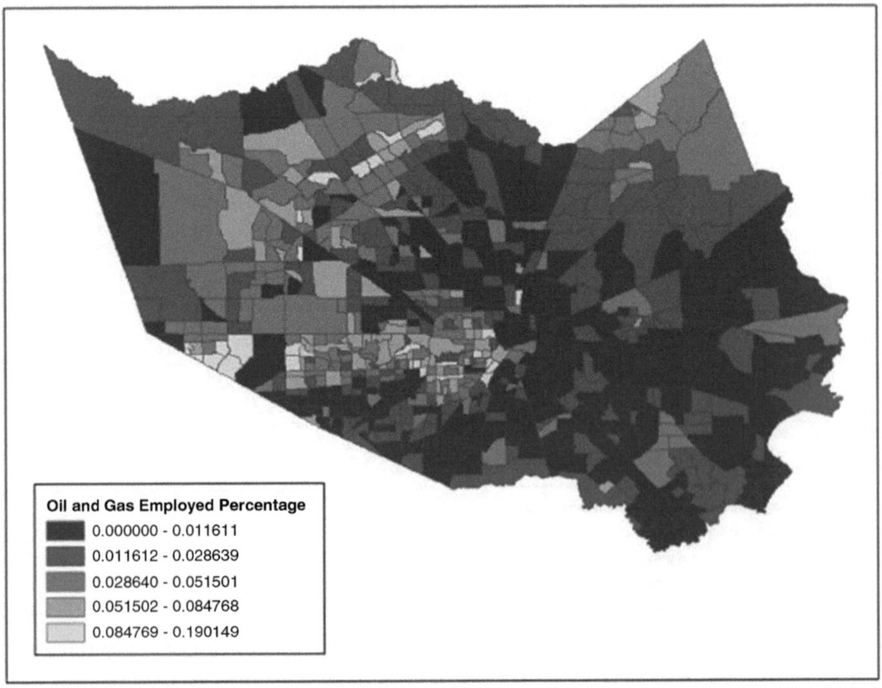

Fig. 5 Oil and gas employment risk factor in Harris County

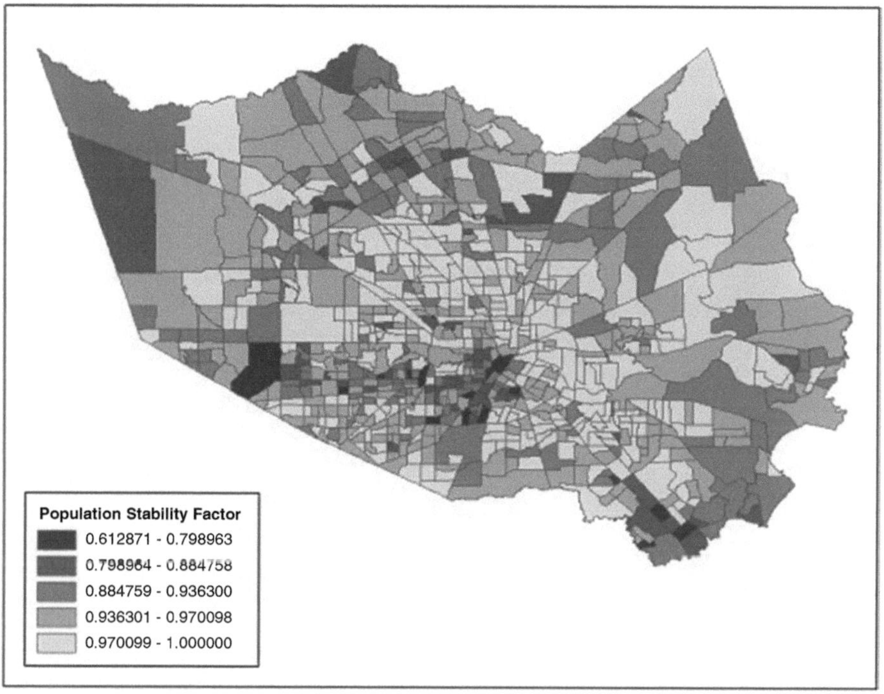

Fig. 6 Population stability risk factor in Harris County

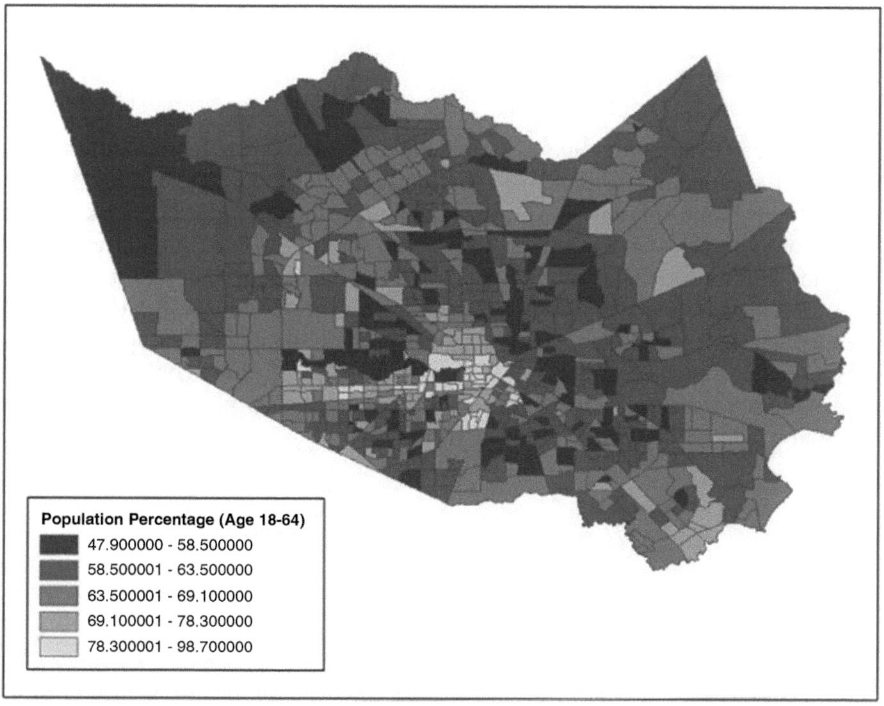

Fig. 7 Employment needs risk factor in Harris County

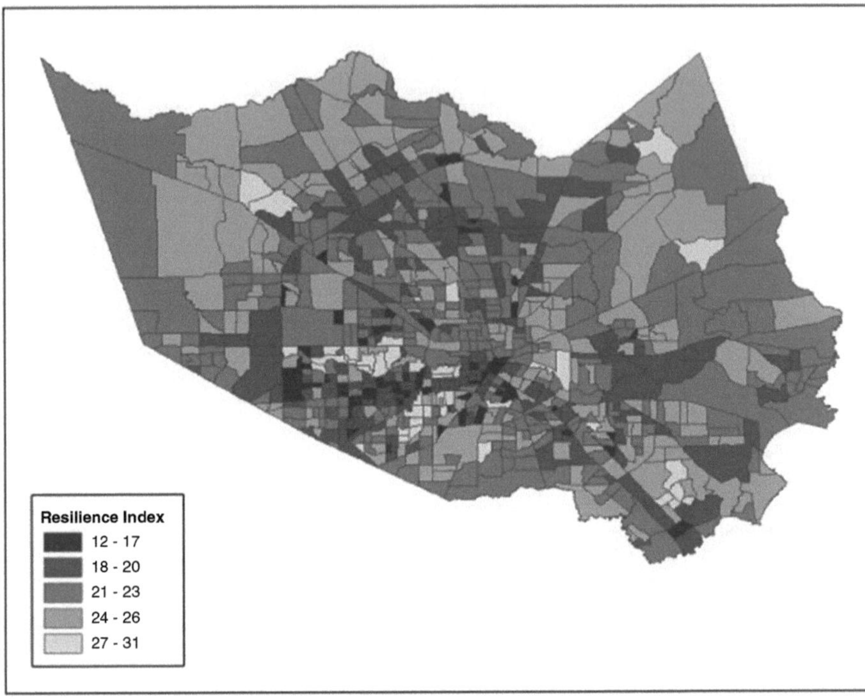

Fig. 8 Overall resilience index in Harris County

5 Conclusion

The American Community Survey (ACS) data sets give us much useful information to model the economic resilience of communities. The model proposed in this paper is a prototype and this model can be expanded to incorporate more risk factors. This prototype gives us a big picture of the economic resilience in Harris County. In order to improve the economic resilience in Houston, we need to improve the education system and prepare more educated work force for the future economic meltdown. At the same time, we also should promote the public transportation and make the jobs more accessible. It is very important for the poor people to get jobs in an economic recession. This model can be very helpful for planners to understand regional economic resilience to the next oil bust.

References

1. Campanella TJ (2006) Urban resilience and recovery of New Orleans. J Am Plan Assoc 72(2):141–146
2. Cordina G (2004) Economic vulnerability and economic growth: some results from a neo-classical growth modelling approach. J Econ Dev 29(2):21–39

3. Liu A, Anglin RV, Mizelle RM Jr, Plyer A (eds) (2011) Resilience and opportunity: lessons from the U.S. Gulf Coast after Katrina and Rita. The Brookings Institution, Washington, DC
4. Weir M, Pindus N, Wial H, Wolman H (eds) (2012) Urban and regional policy and its effects: building resilient regions, vol 4. The Brookings Institution, Washington, DC
5. Lin J (2012) Regional resilience. Working Papers of Research Department of Federal Reserve Bank of Philadelphia
6. Dodson J, Sipe N (2007) Oil vulnerability in the Australian City: assessing socioeconomic risks from higher urban fuel prices. Urban Stud 44(1):37–62
7. Leigh J, Vukovic P (2011) Beyond oil bust: investigating oil economics, society and geopolitics. Atlantic Publishers & Distributors, New Delhi
8. Newman P (1991) Cities and oil dependence. Cities, Aug, pp 170–173

Storm Preparedness: A Case Study of Delaware County, Indiana

Junfeng Jiao, Timothy R. Phelps, and Yuan Li

1 Introduction

Across the United States, a diverse and often destructive mix of weather patterns occurs. One of the most common natural disasters that states across the country face is the tornado [1]. Tornadoes are rotating columns of air that cause varying degrees of intense, violent winds and are often accompanied by hail and thunderstorms. Tornadoes are capable of wind speeds upwards of 250 miles per hour and can leave paths of destruction one mile wide and up to 50 miles long [2]. An average year will be complete with approximately 800 tornadoes resulting in 80 deaths, thousands of injuries, and extensive expenses for home and automotive damages [3].

Although tornadoes are capable of mass destruction, approximately 70 % last fewer than 15 min and cause less than 5 % of the total tornado deaths. The violent tornadoes that cause 70 % of tornado related deaths only occur 2 % of the time, but can exceed an hour on the ground [4]. To combat these diverse storm systems, the US developed a national program that asserts storm preparedness by enabling communities to develop individualized storm management systems; communities that follow the established guidelines are labeled StormReady [5]. In order to be considered StormReady, a community must establish a 24-h warning point and emergency operations center, monitor local weather conditions, provide multiple ways to receive alerts, develop a formal hazardous weather plan, and promote

J. Jiao
School of Architecture, The University of Texas at Austin, Austin, TX 78759, USA

T.R. Phelps
RED LLC Consulting, Noblesville, IN 46060, USA

Y. Li (✉)
Department of Urban Planning, Xiamen University, Xiamen, Fujian 361005, China
e-mail: yako79@gmail.com

X. Chen, Q. Pan (eds.), *Building Resilient Cities in China: The Nexus between Planning and Science*, GeoJournal Library 113, DOI 10.1007/978-3-319-14145-9_19

public readiness through community seminars. These general guidelines are somewhat more relaxed for smaller communities [6].

2 Objective

The objective of this research was to determine if parts of Delaware County, Indiana were lacking storm hazard resources. The County is acknowledged as a StormReady community, and this research sought to seek the effectiveness of the distribution of County resources. This included looking at sirens and emergency shelters to determine smaller scale preparedness compared to the county as a whole [7] (Table 1).

Table 1 The US storm ready guidelines

Guidelines population	<2,500	2,500–14,999	15,000–40,000	>40,000
Guideline 1: communication				
Established 24 h warning point	X	X	X	X
Establish emergency operations center		X	X	X
Guideline 2: NWS information reception				
Number of ways for EOC/WP to receive NWS warning etc.	3	4	4	4
Guideline 3: hydrometeorological monitoring				
Number of ways to monitor hydrometeorological data	1	2	3	4
Guideline 4: local warning dissemination				
Number of ways for EOC/WP to disseminate warnings	1	2	3	4
NWR – same receivers in public facilities	X	X	X	X
Guideline 5: community preparedness				
Number of annual weather safety talks	1	2	3	4
Train spotters and dispatchers biennially	X	X	X	X
Host/co-host annual NWS spotter training				X
Guideline 6: administrative				
Formal hazardous weather operations plan	X	X	X	X
Biennial visits by emergency manager to NWS	X	X	X	X
Annual visits NWS official to community	X	X	X	X

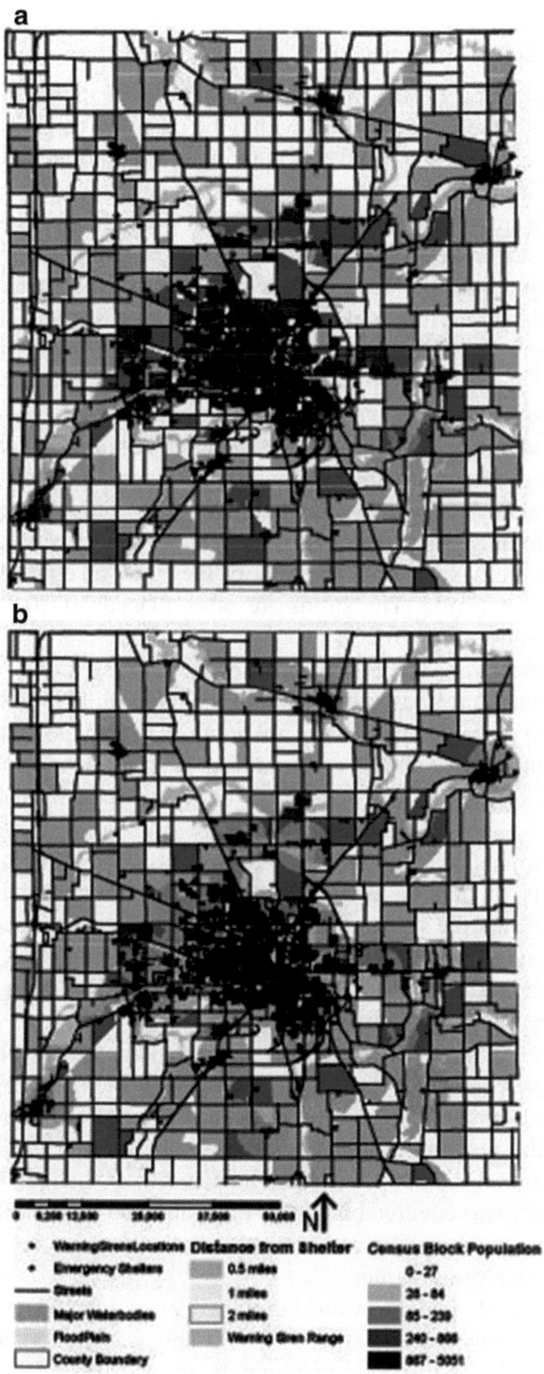

Fig. 1 (**a**) Base map and (**b**) warning siren effectiveness area

3 Methods

The first step in the process was to create a master base map that included all of the warning sirens and emergency shelters within the county. The goal was to be able to compare the county as a whole to the individual, incorporated places within the county. Storm warning sirens and emergency shelters were compared to determine if they were both in relatively the same area. In order for the system to be most effective, it is best if a shelter is in closer range of the siren in that people who can hear the siren can access the shelter. If people are near the shelter but not within sound range of the siren, it would be considered less effective. This was done by using the buffering capabilities of ArcGIS to see how they overlapped in service area (Fig. 1a) [8].

Delaware County operates an integrated public warning system that utilizes tornado sirens and public service announcements over television and radio stations to alert residents of severe weather. Originally known as civil defense sirens, these warning sirens are known by many names and used for various purposes; including use throughout Indiana as tornado sirens. Although the effectiveness of warning sirens severely decreases for residents indoors, they have an effective outdoor range of approximately one mile. The sirens are often placed 30–50 ft high to further increase effectiveness (Fig. 1b).

Once the residents are alerted of severe weather, they are to seek shelter. For residents who do not have basements or safe shelters available within their homes, each incorporated place and the county are required to provide emergency shelters. Emergency shelters are defined as locations, such as schools and community recreation buildings, which are deemed to be safe refuge from natural disasters (Fig. 2a).

For the system as a whole to operate successfully, residents must receive preemptive alerts of and be provided with shelter for protection from severe weather. Additionally, the system must align with the distribution of the population. Initial analysis surveyed the distribution of warning sirens as compared with emergency shelters. The system is less effective if the population is near a shelter but does not hear the tornado siren.

The first level of analysis examined the tornado sirens' effective range to determine what percentage of the population can hear the severe weather warning. Since sound travels in a radial pattern, an airline buffer of one mile was created for each siren and the resulting population overlap analyzed. The next level was to generate network analysis buffers around the emergency shelters to determine how much of the county was covered by a half-, one-, and two-mile drive to each shelter. This assumes that residents are constrained to following the street network when traveling to an emergency shelter.

The two levels of analysis were combined on a map to compare with the preliminary airline buffer results (Fig. 1). The county was broken into census blocks to account for population distribution within the service areas. This helps determine if an area was not serviced because it has a small, distributed population or if there

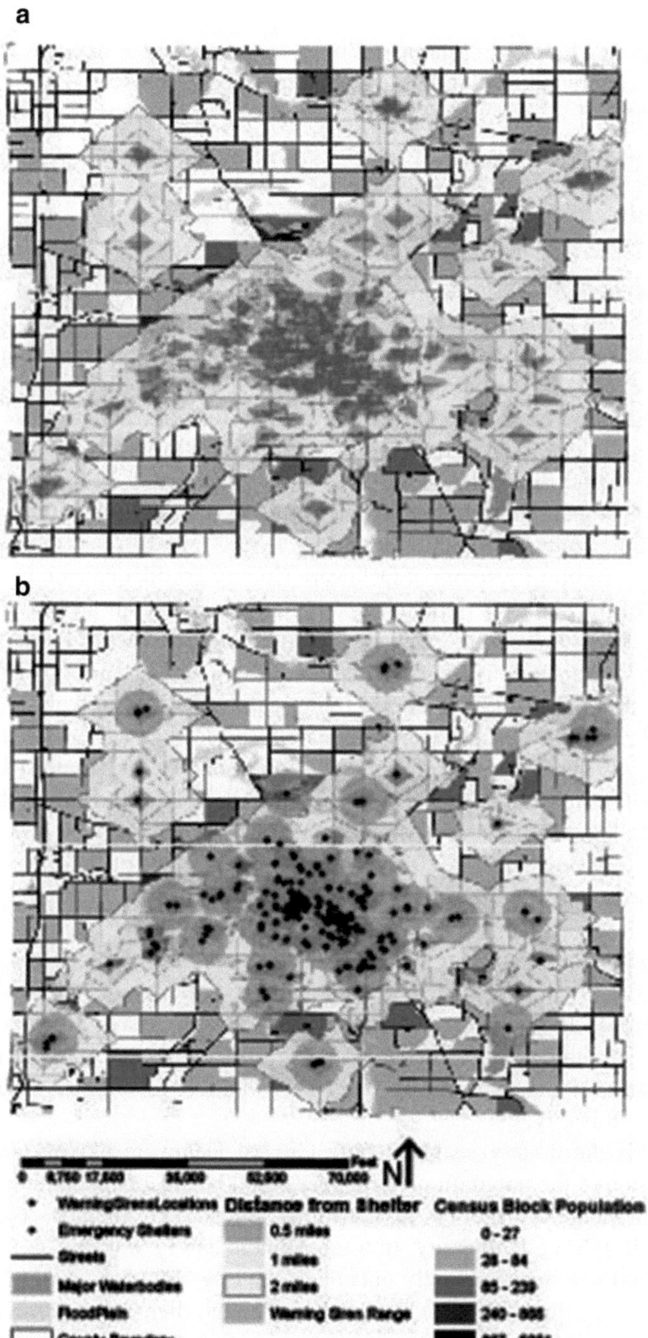

Fig. 2 (**a**) Storm shelter effectiveness area and (**b**) combined storm effectiveness area

is a hole in the service area. The research compared Albany, Daleville, Chesterfield, Eaton, Gaston, Muncie, Selma, and Yorktown as well as Ball State University to determine preparedness on individual, incorporated levels (Fig. 2b).

4 Results

Although there are 33 tornado sirens listed in Delaware County, preliminary research showed that three have been disconnected and that two are proposed but not yet installed. Just over 200 emergency shelters are in the county, 23 of which are not within a mile of a tornado siren. Television and radio public service announcements and mobile alerts can provide warnings to residents in place of tornado sirens, but still leave part of the population unaware. Preliminary analysis showed that nearly 18,000 residents are further than a mile from a warning siren and two miles from an emergency shelter. Much of this area is rural, but combines to make up 15 % of the county population of 117,671.

The storm warning systems of Delaware County are adequate enough to support a majority of the county's population, targeted toward the more heavily populated areas. In Albany, more than 75 % of the population was within a half mile of an emergency shelter. Eaton was easily serviced by a single warning siren and three shelters, with 95 % of the population living within the downtown.

Daleville and Chesterfield were adequately serviced as well, with some shelters overlapping as the cities are adjacent. Gaston is one of the smallest incorporated towns in Delaware County and is easily serviced by a single siren and three shelters (Fig. 3a). Ball State University is home to a large population focus, and is more than adequately serviced with three sirens and a large number of buildings suitable for storms. Selma only requires one siren and one shelter to effectively protect the residents.

Yorktown is the second largest incorporated area in Delaware County (Fig. 3b). The town has three cores for emergency shelters but only has two emergency warning sirens. This city displayed the most service issues. A majority of the population is within a half-mile or one-mile of a shelter but there is a major gap in the service of the warning system. The addition of another siren would be beneficial to this city.

Muncie is the county seat and only City of Delaware County and hosts the largest portion of the population. With over 50,000 residents, it contains the highest volume of emergency shelters and warning sirens. This includes the sirens and shelters of Ball State University. In total, Muncie has 17 warning sirens within the boundary and seven immediately outside of the city boundary. Nearly 70 % of the population is within a half mile of a shelter. Though there are a few holes, the city overall is exceptional in service provided.

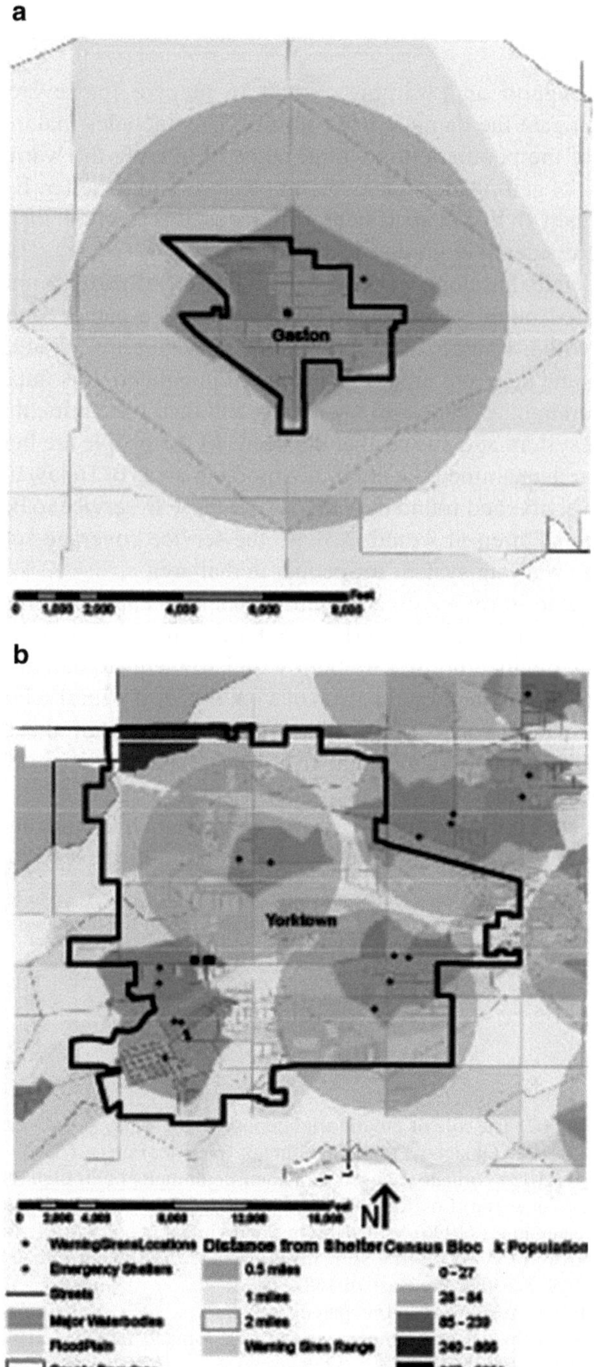

Fig. 3 (**a**) Sufficient preparation – Gaston and (**b**) insufficient preparation – Yorktown

5 Conclusions

Developing a support and warning system to prepare for severe weather is a necessity to mitigate the damage from storms. This includes making sure that the vast majority of the population is within range of hearing the warning sirens, has multiple means of getting alerts, and has ample access to a shelter. Especially given the desires of many to live in apartment complexes, it is essential that there be a safe place for these residents to go during severe weather.

Delaware County has done well and earned its StormReady community status. It has an integrated public warning system for severe weather, and much of the population is within range of the services. As communities change, sirens come on and offline, and different areas become more populated, it is imperative for city officials and community leaders to every now and then reexamine the service areas of the warning system and ensure that the needs of the people are being adequately met. This research examined the eight incorporated areas of Delaware County, plus Ball State University, and found the only severe lapse in service to be in Yorktown. With an additional siren, it would increase the service coverage substantially and provide a better warning system for people in that area.

The downside to storm warning systems is that rural and unincorporated portions of the community will often be overlooked. Further research into the system might include looking at which houses are built with basements to determine if adequate safe areas are available in the rural areas of a county, or if a rural school or similar is nearby that provides a safe place. Though the majority of the county is well serviced, 18,000 is a significant number of people and further research would be worthwhile to determine levels of preparedness in rural areas. Whereas sirens may not be cost effective for rural areas, developing a system and ensuring that everyone has access to safety would be worthwhile for communities everywhere.

Acknowledgment We thank Ms. Jenna Harbin for assistance in the preparation of this paper.

References

1. McCarthy DH (2002) The role of ground-truth reports in the warning decision-making process during the 3 May 1999 Oklahoma tornado outbreak. Weather Forecast 17:647–649
2. Branson-Potts H (2013) Tornado-ravaged Oklahoma community holds high school graduations. In: Los Angeles Times (ed) Los Angeles Times, Los Angeles
3. Monahan J, Steadman HJ (1996) Violent storms and violent people. Am Psychol 51:931–938
4. Brooks HE, Doswell CA III (2001) Normalized damage from major tornadoes in the United States: 1890–1999. Weather Forecast 16:168–176
5. Dengler L (2004) 8. Awareness and preparedness
6. Ashley WS (2007) Spatial and temporal analysis of tornado fatalities in the United States: 1880–2005. Weather Forecast 22:1214–1228
7. Kagawa F, Selby D (2012) Ready for the storm: education for disaster risk reduction and climate change adaptation and mitigation. J Educ Sustain Dev 6:207–217

8. Fox MH et al (2007) Disaster preparedness and response for persons with mobility impairments results from the University of Kansas nobody left behind study. J Disability Policy Stud 17:196–205

Research on the Quality and Level of Urbanization in the Baoshan District, Shanghai

Shan Shuang

1 Introduction

With the regional socio-economic development, the composition of the population, geographical landscape, economic structure and lifestyle, urbanization has become an important phenomenon of socio-economic development in the world today, which is a process of urban characteristics and changes [1]. It has become a basis for the formulation of regional development policies. Recently, the assessment of regional urbanization level and quality has also been a focus of academic research [2]. Baoshan District is one of the suburbs in Shanghai. To an objective, scientific analysis of the development of urbanization in Baoshan, the paper uses location quotient and TOPSIS evaluation method (TOPSIS) to evaluate the quality and level of urbanization in the Baoshan District. Based on the composite score and a comparative analysis, this paper identifies the advantages and disadvantages of the urban development process in Baoshan District, and provides a theoretical basis for policy-making in the future.

2 The Evaluation of the Level of Urbanization in Baoshan

In order to measure the level of development of a regional urbanization, we need to explore the industrial structure, population distribution, employment structure, and many other issues [3]. In assessing the level of urbanization in Baoshan District, using the relevant data of Shanghai census and statistical bulletin in 2010, selecting to reflect the characteristics of the indicators of the urban region, including the

S. Shuang (✉)
School of Resources and Environmental Science, East China Normal University (ECNU), Shanghai 200062, China
e-mail: cissy92990@163.com

X. Chen, Q. Pan (eds.), *Building Resilient Cities in China: The Nexus between Planning and Science*, GeoJournal Library 113, DOI 10.1007/978-3-319-14145-9_20

regional GDP per unit area, population density, the proportion of non-agricultural population, non-agricultural output value the proportion of the four indicators, various regional location quotient calculated as a measure of the relative level of a regional urbanization [4].

In addition to central regions, the weighted average location quotient of the three districts (Baoshan, Pudong, Minhang) are more than 1 in 2010, which shows that the urbanization process of these districts developed rapidly, the level of urbanization is higher than the average level of the city [5].

According to the data, we have selected the population density, per unit of land value-added output (economic density), and the proportion of commonly used land as three indexes, which shows the differences of the process of urbanization of the streets (towns) in Baoshan District.

The results can be found in three different areas: first, high economic density and low population density, such as Miaohang, Gaojing, Wusong, Youyi Road, Songnan, Yuepu, Yanghang; second, high population density and low economic density, such as Zhangmiao; third, low population density and low economic density, such as Gucun, Luojing, Luodian.

Comparing with the average level of the central districts, inter-ring districts and Baoshan district, it can be drawn the relative level of urbanization of different streets (towns) in Baoshan.

3 The Evaluation of the Quality of Urbanization in Baoshan

According to the data in 2009, we have evaluated the development of urbanization in Baoshan District, Minhang District, Jiading District, Songjiang District, Qingpu District, Fengxian District, Jinshan District and Chongming County. Considering all aspects of the quality of urban development, we have selected the evaluation indexes as following (Table 1).

3.1 The Weight Calculation of Indexes

According to the basic principle of the Analytic Hierarchy Process, we have compared the indexes one by one and constructed the judgment matrix of all levels of indexes [6]. Through calculating for each judgment matrix eigenvalues and corresponding eigenvectors and consistency test, we can get the weight coefficients of the indexes at all levels, first level indexes and secondary indexes judgment matrix and weight coefficient [7].

Table 1 List of classification indexes and the index weights

The first level indexes	Weights	The secondary indexes	The total weights
Economy	0.3243	GDP per capita	0.03078
		District-level per capita income	0.02841
		GDP per square kilometer	0.04024
		The proportion of tertiary industry accounted for GDP	0.04261
		The proportion of office building housing area accounted for the total housing area	0.03788
		The proportion of shops area accounted for the total housing area	0.04024
		The proportion of employment of information transmission, computer services and software industry	0.05682
		The proportion of employment of scientific research, technical services and geological prospecting	0.04735
Life	0.2973	Per capita consumption expenditure of urban households (new)	0.06774
		Urban residents' disposable income	0.06021
		The area of residential use per capita	0.03011
		The proportion of housing area (more than eight floors)	0.03763
		The expenditure of urban households per capita(cultural and educational entertainment products and services)	0.05269
		The expenditure of urban households per capita (healthcare)	0.04892
Social development	0.1351	The number of population per square kilometer	0.01002
		The expenditure ratio of technology finance	0.02134
		The area of culture and entertainment per capita (theaters)	0.01552
		The expenditure on education per capita	0.02134
		the number of hospital beds per thousand	0.01746
		The number of doctors per thousand	0.01552
		The expenditure ratio of health care	0.01649
		The expenditure ratio of social security and employment	0.01746
Ecological environment	0.2432	The area of parks per million	0.01737
		The ratio of green coverage	0.03764
		The emissions of SO_2 per square kilometer	0.04344
		The emissions of COD in wastewater	0.04633
		The standard rate of environmental noise	0.04054
		The index of investment in environmental protection	0.05792

3.2 *The Evaluation of the Quality of Urbanization Development*

According to the TOPSIS formula and original data, we can calculate the distance and closeness degree from the first level indexes to the ideal point (Tables 2, 3, 4, and 5), which reflects the rank of the quality of urbanization in these eight districts [8].

Table 2 The distance and closeness degree to the ideal economic quality point

	Minhang	Baoshan	Songjiang	Qingpu	Jiading	Fengxian	Jinshan	Chongming
The distance to positive ideal point	S4+	S1+	S5+	S2+	S8+	S3+	S6+	S7+
	0.30932	0.36299	0.51532	0.55172	0.57915	0.52353	0.51088	0.73162
The distance to negative ideal point	S4−	S1−	S5−	S2−	S8−	S3−	S6−	S7−
	0.62733	0.61708	0.55868	0.54664	0.53104	0.46941	0.41090	0.34615
Closeness degree	**C4**	**C1**	**C5**	**C2**	**C8**	**C3**	**C6**	**C7**
	0.66976	**0.62963**	**0.52018**	**0.49769**	**0.47833**	**0.47275**	**0.44577**	**0.32117**

Table 3 The distance and closeness degree to the ideal life quality point

	Minhang	Baoshan	Jiading	Songjiang	Jinshan	Qingpu	Fengxian	Chongming
The distance to positive ideal point	S4+	S1+	S8+	S5+	S6+	S2+	S3+	S7+
	0.37814	0.42065	0.43750	0.44537	0.50400	0.49917	0.56245	0.62812
The distance to negative ideal point	S4−	S1−	S8−	S5−	S6−	S2−	S3−	S7−
	0.57676	0.55587	0.52946	0.50511	0.46950	0.45835	0.40868	0.39699
Closeness degree	**C4**	**C1**	**C8**	**C5**	**C6**	**C2**	**C3**	**C7**
	0.60400	**0.56924**	**0.54755**	**0.53142**	**0.48228**	**0.47869**	**0.42082**	**0.38726**

Table 4 The distance and closeness degree to the ideal social development point

	Songjiang	Qingpu	Jinshan	Chongming	Fengxian	Baoshan	Minhang	Jiading
The distance to positive ideal point	S5+	S2+	S6+	S7+	S3+	S1+	S4+	S8+
	0.38898	0.42513	0.45526	0.47659	0.47378	0.49356	0.50950	0.62964
The distance to negative ideal point	S5−	S2−	S6−	S7−	S3−	S1−	S4−	S8−
	0.55963	0.52908	0.49420	0.47541	0.45719	0.44505	0.40188	0.30571
Closeness degree	**C5**	**C2**	**C6**	**C7**	**C3**	**C1**	**C4**	**C8**
	0.58995	**0.55447**	**0.52051**	**0.49938**	**0.49109**	**0.47416**	**0.44096**	**0.32684**

Table 5 The distance and closeness degree to the ideal ecological environment point

	Chongming	Qingpu	Fengxian	Songjiang	Jiading	Jinshan	Baoshan	Minhang
The distance to positive ideal point	S7+	S2+	S3+	S5+	S8+	S6+	S1+	S4+
	0.40505	0.50794	0.57791	0.66786	0.68174	0.70337	0.61502	0.26108
The distance to negative ideal point	S7−	S2−	S3−	S5−	S8−	S6−	S1−	S4−
	0.49975	0.40872	0.38435	0.29719	0.29595	0.27559	0.20269	0.07313
Closeness degree	**C7**	**C2**	**C3**	**C5**	**C8**	**C6**	**C1**	**C4**
	0.55233	**0.44588**	**0.39943**	**0.30795**	**0.30270**	**0.28151**	**0.24787**	**0.21881**

In these four indexes, the indexes of the quality of social development and ecological environment in Baoshan have been ranked relatively backward, while others ranked excellent. It shows that we need focus on strengthening the construction of social development and improve the ecological environment in Baoshan. Meanwhile, it is important to enhance the overall quality of urban development in Baoshan.

4 Conclusions

The process of urbanization has developed rapidly in Baoshan District, and the level of urbanization is more than the city average. But there are spatial differences among the streets (towns) in Baoshan. The population density in Zhangmiao is high, but the economic density is low. Both the population density and economic density are low in Gucun, Luojing, Luodian. The economic density is high in other streets (towns), while the population density is low.

Compared with the suburbs, the overall quality of urbanization in Baoshan has at the forefront, while the quality of social development and ecological environment ranked by behind.

Compared with the central districts, the quality and level of urbanization in Baoshan still have a lot of space to develop. That is, first, level of economic development per capita is lower than the central districts', and the proportion of tertiary industry and modern service industry is still lower; second, the level of public service facilities configuration per capita is lower than the central districts'; third, the ecological conditions are better, but the pressure of environmental protection and management is relatively heavy.

References

1. Zhou Yixing (2006) Thoughts on the speed of China's urbanization. City Plann Rev 30:32–40
2. Dadao L (2007) Urbanization process and spatial sprawl in China. Urban Plann Forum 2:47–52
3. George Lin (2007) Chinese urbanism in question: state, society, and the reproduction of urban spaces. Urban Geogr 20:7–29
4. Heikkila EJ (2007) Three questions regarding urbanization in China. J Plann Educ Res 10:65–81
5. Hu Xuwei (2007) Economic globalization and China's urbanization. Urban Plann Forum 2:53–55
6. Chen AM (2002) Urbanization and disparities in China: challenges of growth and development. China Econ Rev 10:407–411
7. Gene HC, Josef CB (2006) The paradox of China's growing under urbanization. Econ Syst 20:24–40
8. The census data and statistical bulletin in Shanghai, 2010

The "Policy-Projects Districts" Model of New Urban Area Development in China and Its Analysis

Wang Jin-Bai

1 Introduction

Western scholars have paid much attention to China's rapid urbanization and its internal processes and mechanisms [1–4], but rarely is research and analysis focused on large-scale new urban area development, which has played an important role in China's rapid urbanization process.

Recently, scholars in China have systematically summarized new urban area development since the Reform and Opening Up [5–7]. However, research mostly references the classic theory of the West (especially the production of space theory) and ignores both the new collaborative relationships between the government and market in the Chinese context, as well as the background history from the planned economy to the Reform and Opening Up.

Currently, *government failure theory* and *market failure theory* has become the international consensus. Eastern and Western countries are both seeking a combination of the two forces, the "the government" and "the market" in their specific environment.

Among China's nearly three decade long, large-scale new urban area development process, from the early sparks of the Special Economic Zones to the "prairie fire" of later development zones and new urban areas, there has been a transition reflecting a significant expansion of space for capitalist markets. Notably though, the degree of centralized control of the government has never been reduced. Indeed, centralized control, which was the legacy of the planned economy, continues to play a role in urban development.

W. Jin-Bai (✉)
Shanghai Tongji Urban Planning and Design Institute (TJUPDI), Shanghai 200092, China

College of Architecture and Urban Planning (CAUP), Tongji University,
Shanghai 200092, China
e-mail: 327865205@qq.com

X. Chen, Q. Pan (eds.), *Building Resilient Cities in China: The Nexus between Planning and Science*, GeoJournal Library 113, DOI 10.1007/978-3-319-14145-9_21

The shifting relationship between centralized government control and free market competition, from the traditional view, presents a new type of collaborative and symbiotic relationship in China. This is beyond the cognitive scope of the traditional theories of market economy and space capitalization, and requires new theoretical studies to explain it.

Especially since the 2008 financial crisis, the traditional Chinese approach of large-scale new urban area development has been increasingly challenged by many factors such as pressure on environmental resources, social stability, and policy adjustments. Many issues became increasingly prominent that indicate that the path may face a major systemic transformation. At this time, reflection on and analysis of this development model is very necessary and urgent.

Looking ahead, China's urbanization level has just crossed the 50 % threshold. As it continues to increase in the future, new urban area development will continue to have a larger share of the country's overall development. Therefore, this research also has important theoretical and practical significance in promoting development for the long-term.

Based on this thinking, the article examines large-scale new urban area development in the Chinese context in order to analyze its internal mechanisms from the perspective of the new collaborative relationships between the government and the market. This research attempts to form a general explanatory model of the logical relationship to explain the evolutionary logic of China's large-scale new urban area development and to forecast its future.

2 The PPD Mode of Large-Scale New Urban Area Development in the Chinese Context

The Chinese government's promotion of large-scale regional development has a long history. In the contemporary era, it can be traced back to the period of the planned economy. In the 1950s, there were 156 large projects built with help from the Soviet Union. In the 1960s, there was the large "Three-Lines" large project construction, and in the 1970s the "complete sets" projects were three of the major undertakings implemented under the government's lead (mostly for industrial development).

Since the Reform and Opening Up began in 1978, China has gone through four rounds of large-scale new development booms: the experiment of Special Economic Zones in the 1980s, the Industrial Development Zones in 1990s, the development of the new urban areas in the early twenty-first century, and the large national strategies since 2008 (Table 1).

Throughout seven total rounds of urban development booms, the central government's control of resources has remained at a high level, but local agents have participated in the developments differently. In the early period of the planned economy, the government implemented large projects directly, and its local agents

Table 1 Seven rounds of large-scale government-led development booms since the founding of New China

	Times	Background	Contents of development	Manner of development	Effects	Typical representative
Planned economy period	1950s	aided by the Soviet Union, early planned economic system built	156 large projects (effectively 150 projects)	The country directly invested an operated	New industrial urbanization	YTO Group Corporation
	1960s	Implementing "Three-Lines Construction" preparing for the International war clouds	approximately 1,100 large projects: military, chemical, machinery manufacturing, etc.	The country directly invested and operated	Locations scattered across the area, development later turned into many industrial cities and towns	Shiyan Motor City
	1970s	The normalization of Sino-US	Introduced complete set projects from western countries, 26 in 1972; 22 in 1978	The country directly invested and operated	Enriched part of industrial urbanization	Shanghai Baosteel Group Corporation
Reform and Opening Up period	1980s	Investment structure reform in initial stage of reform and opening up	Put four special economic zones (SEZ) and 12 development zones into trial	Promotion of local development dynamic through government policy	SEZs develop into new cities, development zones are phased in	Shenzhen SEZ
	1990s	The reform and opening up push greatly, development zone policies promote extensively	Thousands of national, provincial and local development zone	Under national policy, local introduction and operation of projects	Development Zones rapidly develop, industrialization and urbanization speed up	Pudong New Area, Suzhou Industrial Park
	Early twenty-first century	Reform and opening up systemic, new urban areas large expansion	Topics of new urban areas	Localities plan projects for national policy support	Urban and rural areas change, real estate economy inflation, exacerbating societal imbalance	Tianjin Binhai New Area, Zheng Dong New District of Zhengzhou
	After 2008	Economic crisis, expanding domestic demands, development in the middle and west	More than 10 national regional development strategies, several new SEZs, lots of new development zones	Under national policy-driven, local introduction and operation of projects	Development in the middle and west, eastern transformation, urban regionalization	National strategy of Wanjiang City-Chain

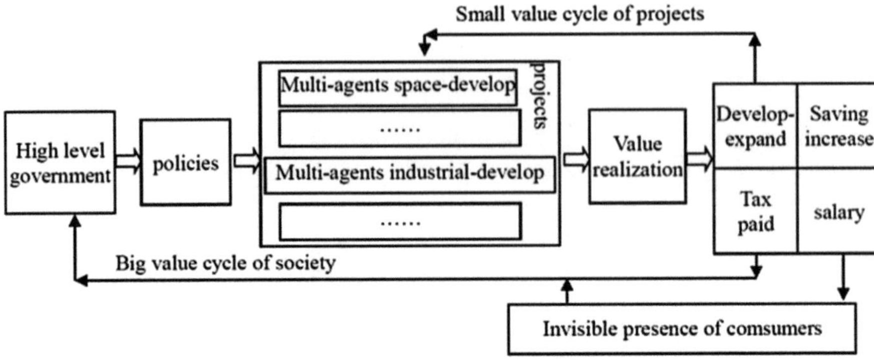

Fig. 1 The general process of the "policy – projects districts" development mode

worked on relatively simple technical tasks. For instance, the factory director was only responsible for production management, and not product marketing or profitability. Since the Reform and Opening Up, practices from Special Zones to Development Zones have formed a new policy-driven regional development pattern. In this case, local agents have transitioned into operating as semi-independent for-profit entities, whose functions include project introduction, implementation, and so on.

Based on such historical trends, we would conclude that throughout the evolution of large-scale new urban development, from large projects dominating local development in the planned economy to policy-driven local development in the Reform and Opening Up period, there is similar development thinking and logic running though different social settings and stages of China. We could generalize it into a general model.

Practically speaking, the model is generally reflected in the circular relationship of policies and projects (Fig. 1). The national or high-level government collectively calls for policies to allocate resources; subsequently, in the planned economy period, it controls the specific projects. Resources and development projects are assigned to local governments and technical enterprises, and then the high-level government recovers and distributes the outputs.

During the period of Reform and Opening Up, there was policy support for the promotion of development in specific geographic areas. In this manner, local governments and entities were incentivized to introduce or plan projects in order to promote local development and obtain profits at the same time, and moreover to further the expansion of the upper policies to achieve those benefits. Output distribution adjusted into a two-level structure: the "small cycles," as in the value of the actual project revenue, and the "big loop" as in revenue of the high-level government. Such a regional development model of the symbiotic "policy" and "project" cycle, hereinafter is called the "Policy – Projects districts" (PPD) model.

This development model is the mainstream model of large-scale regional development in China. Its basic features are the allocation of resources through highly concentrated national (or high-level government) policies (or policy projects) and resource utilization and value-adding through relatively decentralized operation by local agents.

Notably, such a large coherent system has had dramatically different development performance between the planned economy and the Reform and Opening Up time. The system was unsustainable and inefficient in the former period but is thriving to create the "China miracle" in the latter period. It can be inferred that the centralized allocation of resources and the decentralized utilization of resources in different periods of China contains certain inner mechanisms that results in huge performance differences.

In China, the centralized allocation of resources is rooted in the institutional tone of national system. It has not changed fundamentally from the purely centrally-controlled planned economy in the early times, to the policy-driven development during the Reform and Opening Up. To a certain degree, it is considered to be an institutional advantage in achieving rapid economic and social development in China.

Since the Reform and Opening Up period, national policy has been based on a coherent centralized framework. The crucial change is the separation of the control of resources and the utilization of resources, which has promoted diversification and decentralization, and to some extent, created a large number of separate self-interest groups (local government, the restructuring of state-owned enterprises, private enterprises, foreign companies, etc.). It also resulted in a competitive mechanism of resource utilization. In order to survive and obtain development space in a competitive environment, self-interest groups must capture rapidly changing information in the local development process and make timely responses, which results in an overall improvement of the end efficiency of resources use in regional development.

However, this improvement of efficiency in the decentralized form of resources utilization cannot completely substitute for the shortage of local quantity; centralized control entities still can collectively extract large-scale resources through "policies" and "big projects" to promote local development rapidly and powerfully.

Then the combination of "centralized allocation of resources" and "decentralized utilization of resources", cleverly results in a new regional development mechanism with the advantages of "resources beyond the local level" and "end resource utilization efficiency", that may produce effective regional development mechanisms far more than a single advantage, and promote the stunning achievements of China's large-scale new urban development (Fig. 2).

Actually, in the periods of the planned economy, Reform, and Opening up, several huge regional developments were created in China. In this sense, the development of large planned projects and large-scale policy-driven districts have similar initial development efficiency; both achieve development performance far higher than as if done on a local basis, in a short period of time. However, the decentralized utilization of resources in two periods has led to significant differences. During the planned economy period, it is reflected in the technical implementation of the agency, almost without self-benefits and competitiveness, which

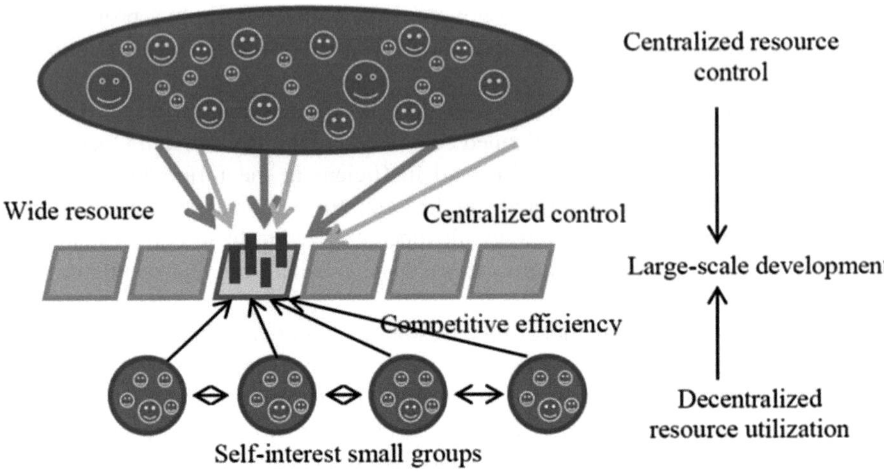

Fig. 2 "Policy – projects districts" pattern's core mechanism schematically

resulted in low efficiency and unsustainability. In the period of Reform and Opening Up, it is reflected in a for-profit business agency, so that operating efficiency improved tremendously, which drove the overall economy to became more dynamic and more sustainable.

3 The PPD Mode's Core Mechanism Model

This section uses the language of the theoretical model to further the analysis of the core mechanism.

A basic assumption is that the final output "policy-projects districts" (PPD) development is a composite function of the centralized allocation of resources and the decentralized utilization of resources, which is expressed in the following terms:

$$\text{P (regional development outputs)} = \int [\text{J (centralized allocation of resources)}, \text{F (decentralized utilization of resources)}]. \quad (1)$$

The centralized allocation of resources in regional development is mainly reflected in the wide resource mobilization and the overall policy control of the development process. A linear and symbiotic relationship is observed: with increasing centralized control and system security, wide resources would be mobilized. On the other hand, the system can also consolidate the centralized control system, and the degree of centralized control used to widen resources' capacity and scale.

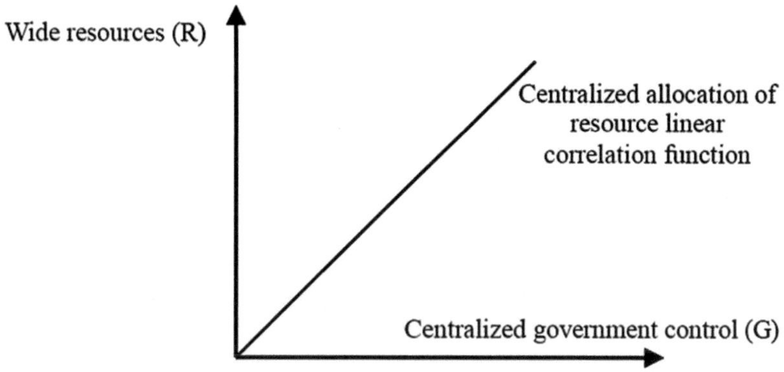

Fig. 3 Linear correlation function of wide resource, central government control in the centralized allocation of resources

For example, in different degrees of centralized control at the city and county level, the provincial- and national-level can mobilize wide resources very differently. The centralized national control of regional development resources even can mobilize nationwide resources to realize development goals. So the factor J in the above formula can be broken down into positive correlation the linear function of sub-elements R (wide resources) and sub-elements of G (centralized government control) (Fig. 3).

The decentralized utilization of resources is reflected in small self-interest groups competitively utilizing development resources. According to mechanism design theory, this method of resource utilization is more effective in informational efficiency and incentive compatibility, which enhances regional development efficiency. Thus, elements F can be further divided into a combined function of sub-elements of S (small self-interest groups) and elements of C (competitive utilization). Moreover, elements of S (small self-interest groups) and C (competitive utilization) are also a pair of symbiotic variables. Only if several active self-interest groups are formed can the competitive utilization be realized. The more active small self-interest groups are, the higher competitive utilization would occur. The relations between the two can be concisely summarized as a positive correlation linear function (Fig. 4).

However, in these elements' two linear functions exist a couple of negative correlation elements. G (centralized government control) and C (competitive utilization) have a negative correlation function relationship – the stronger the centralized control, the lower the extent of competitive utilization, and vice versa. And this negative function of the relationship is not linear, but rather a curve function (Fig. 5), i.e. as centralized control is increased to a certain extent, the extent of competitive utilization will be significantly reduced; the further the improvement in the degree of centralized control, competitive utilization will take advantage of a dramatic reduction to a very low state, such as some countries' completely controlled economy during the World War II. Conversely, when the degree of

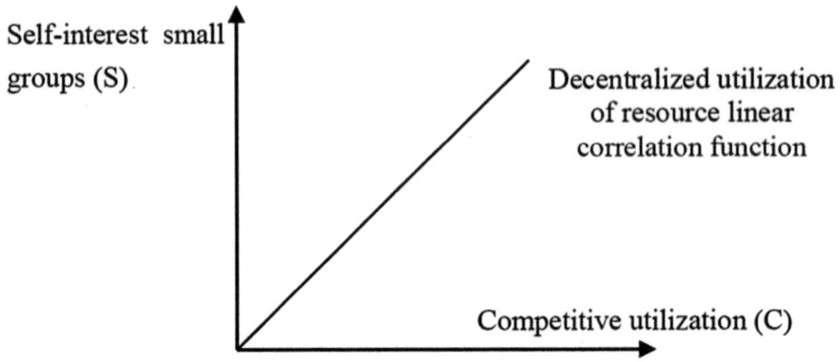

Fig. 4 Linear correlation function of competitive utilization and small self-interest groups in the decentralized utilization of resources

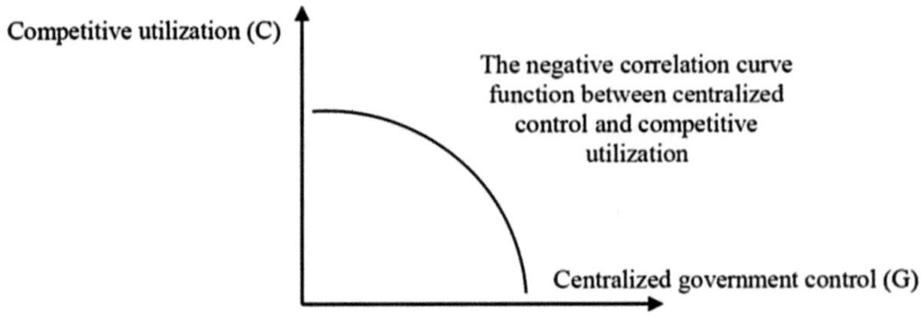

Fig. 5 The negative correlation curve function between centralized control and competitive utilization

centralized control is only modestly reduced, the degree of competitive utilization is significantly improved; if centralized control is reduced to a very low level, the degree of competitive utilization would be increased to an extremely high level, even to an anarchic state of perfect competition.

The combination of the two related linear functions and the negative correlation curve function form the basic function of centralized allocation of resources to decentralized utilization of resources. Its expression is (1):

$$P\text{ (regional development outputs)} = \int [J\,(G\alpha R),\ F\,(S\alpha C)]. \qquad (2)$$

This explanation provides a basic logical relational model of the development mechanisms for the "policy-projects districts" mode. This model's expression is shown in Fig. 6.

The explanation for this model as below:

- Centralized control (G) and small self-interest groups (S) form a corresponding relationship as a horizontal axis; wide resources (R) and competitive utilization (C) form a corresponding relationship as the vertical axis; the vertical axis and horizontal axis constitutes the cross-axis model system of the PPD model;

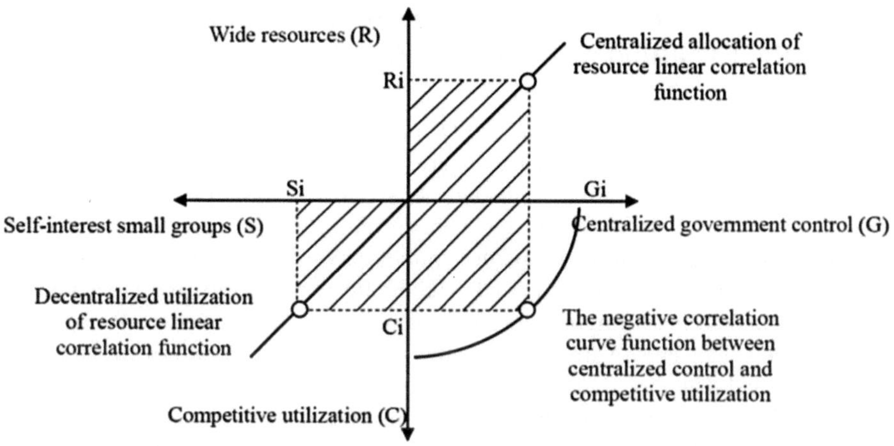

Fig. 6 Theoretic explanation of policy-projects districts development mechanism (general model)

- The upper right quadrant of the cross-axis consists of the linear correlation function of wide resources (R) and centralized government control (G). The vertical axis of wide resources' (R) value Ri stands for its quantity of "policy-projects districts" in the condition of i. The horizontal axis of centralized government control (G)'s value Gi stands for its degree of "policy-projects districts" in the condition of i, the aggregate area formed by Ri and Gi represents centralized allocation of resources' contribution value index in the resources development mechanism of "policy-projects districts" (PPD).
- The lower left quadrant of the cross-axis consists of the linear correlation function of small self-interest groups (S) and competitive utilization (C). The vertical axis of competitive utilization's (C) value Ci stands for its activity level of "policy-projects district" in the condition of i, the horizontal axis of small self-interest groups (S) value Si stands for its activity level of "policy-projects districts" in the condition of i. The aggregate area formed by Ci and Si represents the decentralized utilization of resources' contribution value index in the resources development mechanism of "policy-projects districts" (PPD).
- The lower right quadrant of the cross-axis consists of the negative correlation curve function of centralized control (G) and competitive utilization (C). The value of Gi (horizontal axis) and Ci (vertical axis) stand for its combined output in the condition of i. In other words, it is the correlation coefficient between the centralized allocation of resources and decentralized utilization of resources, which forms another key to the resources development mechanism of "policy-projects districts" (PPD).
- To sum up the overall enclosed area of Ri, Gi, Ci, Si in the upper right, lower left and lower right three quadrants, they constitute the total output of "policy-projects districts" (PPD) development mechanism in the i condition, which characterizes the performance level of the regional development mechanism.

4 The PPD Model's Scenarios and Its Explanation for the Evolution of Historical Practices

This conceptual model is only a basic presentation of an ideal state. It can be further reduced to four specific mechanism scenarios.

The "centralized allocation of resources" scenario plays an absolutely leading role in the regional development mechanism named *Scenario 1* (Fig. 7), which corresponds to the development of the "policy-projects Districts" in the planned economy period. This is evident in early Hubei Shiyan motor city, the Three Gorges Dam, Jianghan Oilfield, Jingmen Petrochemical, Ezhou Iron and Steel, Wuhan Iron large projects, etc. They were basically invested in by the central government, and reached a level that a local origin of resources could not reach, but greatly reduced the space for small self-interest groups and competitive utilization at the local level, which led to inefficiency in the resource's end utilization and lack of long-term sustainability.

In *Scenario 2*, the centralized allocation of resources has contracted but is still on the high side, and decentralized resource utilization has expanded but still ranking the second place (Fig. 8). It corresponds with regional development in the transition from the planned economy to the Reform and Opening Up. For instance, Ningbo Beilun's development mechanism changed from a direct national investment of an ore terminal project to a series of policy-driven developments (development zones, bonded zones, export processing zones, etc.) that were less directly funded by the central government. The initial government investment exceeded the level of local resource availability; the latter policies served to enhance the end efficiency of resource utilization, thereby increasing the overall efficiency and sustainability of Beilun. This is an improvement to *Scenario 1* to a certain extent.

Scenario 3 is found (Fig. 9) when the centralized allocation of resources is further reduced to a subordinate share, and the decentralized utilization of resources

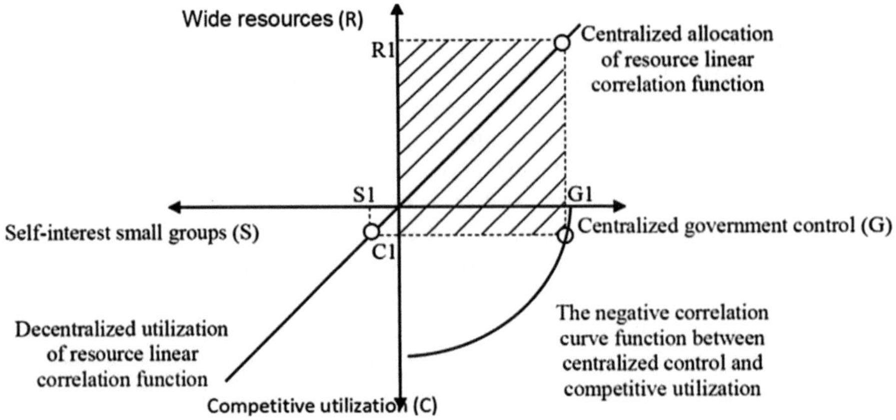

Fig. 7 Scenario 1: the centralized allocation of resources leading the PPD development

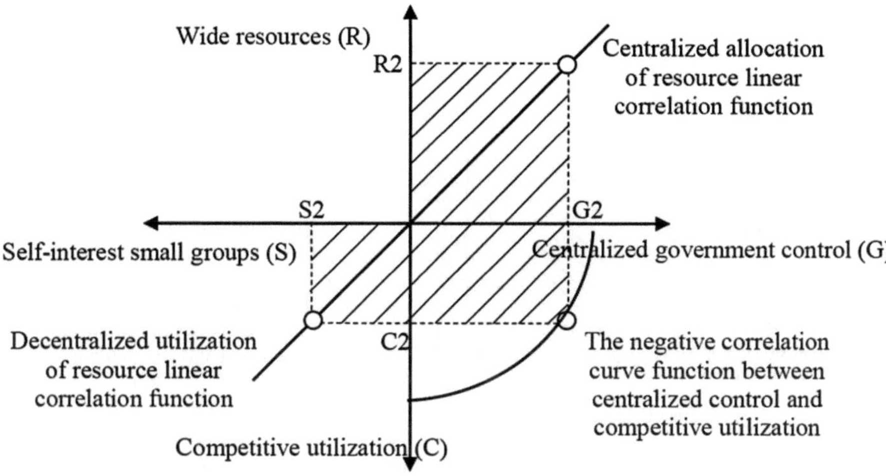

Fig. 8 Scenario 2: centralized allocation of resources on the relatively high side

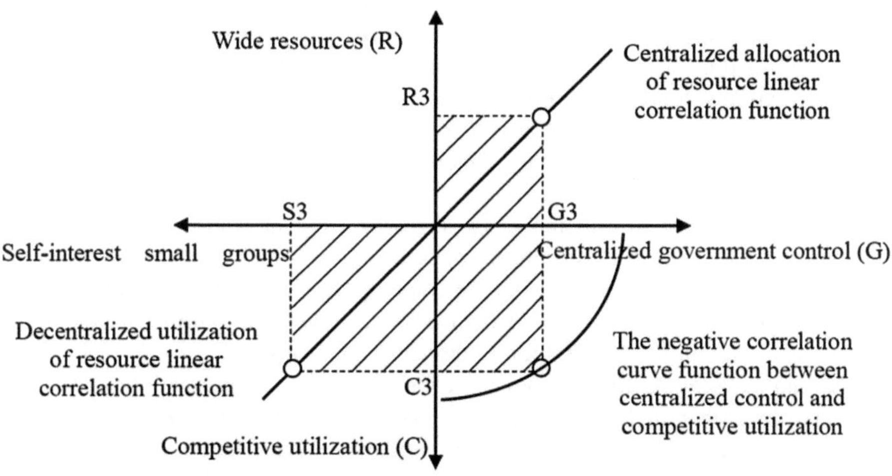

Fig. 9 Scenario 3: decentralization utilization of resources on the relatively high side

is expanded to ranking the major share. This situation corresponds with Pudong New Area development in the new era. National resources input played a large role in Pudong's earlier development, but with the expansion of regional development, the support of national resources decreased gradually, and national policies and local government and competitive enterprise operations increased more and more. That reflects the development of the Pudong New Area trend from *Scenario 2* to *3*. By 2005, after a second startup and 2009 merger with Nanhui district, Pudong New Area's development rarely brought substantial investment from the national government, and local government and private competitive operations has become the main mechanism for Pudong New Area, which notably puts it in the situation of *Scenario 3*.

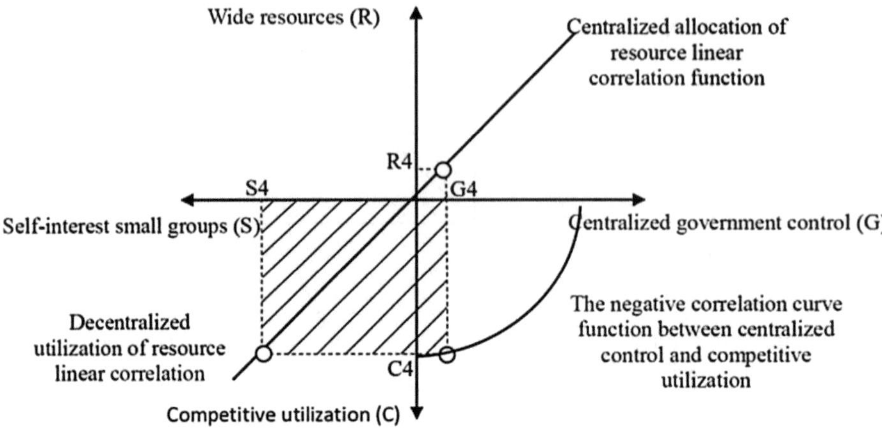

Fig. 10 Scenario 4: decentralized utilization of resources play a leading role

Continuing from *Scenario 1*, *Scenario 2*, and *Scenario 3*, there should be a much "looser" state of *Scenario 4*. That is, the centralized allocation of resources hardly plays a part, and regional development is motivated almost totally through competitive utilization of local resources by small self-interest groups (Fig. 10). This is a highly extreme condition of free competition; it is not the best regional development mechanism, for the "fallacy of composition" and "market failure" goes against their own advantage of the collaboration theory of government and market. Speaking from the practical level, the highly free competition mechanism has played an important role in the early stages of the establishment of the market economy, such as the enclosure movement in English history, the gold rush in the western United States, and so on. In the contemporary context, due to the sparsely populated social interaction and growing social governance, this kind of highly competitive situation is rarely found in modern countries.

As can be seen from the above scenario simulation, the evolution of large regional development from "policy-projects districts" *Scenario 1* to *Scenario 2*, then *3* (Fig. 11) is evident in the 60 years of New China. Hubei Shiyan, Ningbo Beilun, and Shanghai's Pudong New Area represent typical cases at different stages.

In recent years as the government has introduced new strategic local development such as Tianjin Binhai new area, Chongqing Liangjiang new area, and so on, the actual resource inputs and wide resource mobilization is less intensive; under the framework of the "policy-projects districts" mechanism, local government and enterprise in fact dominate development, and resource inputs and outputs during the development are more and more connected to the market environment. This means that China's new urban development mechanism has been basically stabilized close to *Scenarios 2* and *3*.

Based on the international government-market collaborative theory, a reasonable "policy- projects districts" development mechanism still depends on a good

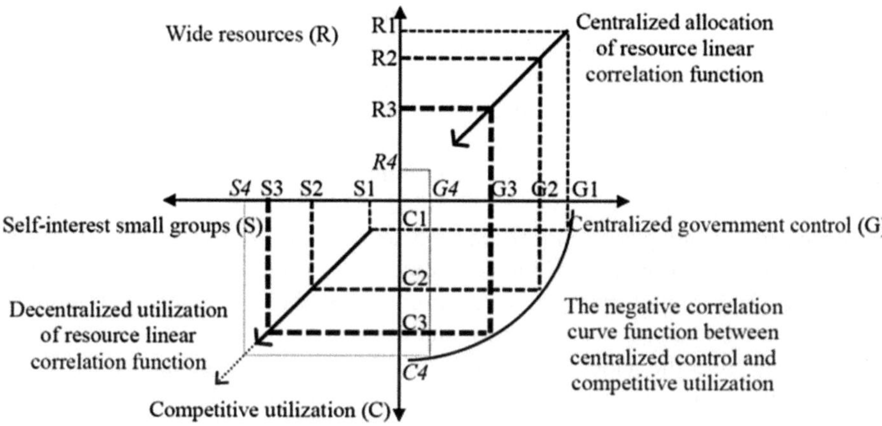

Fig. 11 The evolution of China's large-scale regional development's corresponding "policy-projects districts" mechanism scenario

combination of and joint collaboration between government and market into the future. Even in the current economic slump, some structural transformation trends are appearing: there is opportunity in innovation. That is to say, China's large-scale urban development mechanism, from *Scenario 1*, *Scenario 2* to *Scenario 3*, should look for some kind of combination around *Scenario 2* and *3* rather than to continue linearly to *Scenario 4*.

5 Conclusion

The academic community is still debating the processes and mechanisms driving China's rapid development. On the subject of large-scale new urban area development, upper policy control has been tight, and the operations of decentralized agents has had a huge expansion since China's Reform and Opening Up. This new symbiotic, collaborative "non-shift" relationship of centralization and decentralization is an important feature of China's large-scale development of "policy-projects districts", and this will be helpful in analysis of its internal mechanisms, and will provide the basis for optimization in the next phase of development.

Grasping the core mechanism of the binding consequences of centralized allocation of resources and decentralized utilization of resources, the logical relationship can be further resolved to a linear function of centralized control and wide resources, another linear function of self-interest small groups and competitive utilization, and the third negative correlation curve function of centralized control and competitive utilization, which can construct its logical relational model for the general mechanism's explanation of the mode of "policy-projects districts".

Further reducing the specific context of the general model we can simulate *Scenario 1*'s high concentration of extreme state control and *Scenario 4*'s high degree of free competition, each corresponding at the high level to the planned economic system and free competition system. Both systems have played an important role in history, but they also have all proven to have major defects and are not the advisable choice. On the other hand, *Scenario 2* can be simulated as centralized allocation of resources leading and decentralized utilization of resources in second place, and *Scenario 3* as decentralized resource utilization leading to the subordination of the centralized allocation of resources. Historical precedence and related theory have all proved that these two are the optimized choice of integrated use of government and market.

Deducing from the national context, China's large-scale regional evolution for the past 60 years is actually the evolution of "policy-projects districts" mechanism from *Scenario 1* to *Scenarios 2* and *3*. Recently, China's new national strategic development reflects more policy incentives rather than direct investment of resources. Regional development is actually implemented by local governments and local enterprises. This means China's regional development mechanism is basically stable in a certain state around *Scenarios 2* and *3*. Looking ahead, a reasonable regional development mechanism still depends on multiple forces (like government and market) in order to find an optimal combination between *Scenarios 2* and *3*.

References

1. Ma LJC (2002) Urban transformation in China. 1949–2000: a review and research agenda. Environ Plan A 34(9):1545–1569
2. Zhang L (2004) Chinas limited urbanization: under socialism and beyond. Nova Science, New York
3. Friedmann JR (2005) China's urban transition. University of Minnesota Press, Minneapolis
4. Huang Ya-sheng (2008) Capitalism with Chinese characteristics. Cambridge University Press, New York
5. Wu Tinghai, Yang Baojun, Zhang Chengguo (2011) China's new town (1979–2009) [J]. J Urban Reg Plan 2011(2):19–43
6. Li Xun, Liu Yi (2011) The explanation of the mixture of urbanization patterns of China [J]. Urban Reg Plan Stud 3:70–85
7. Zhu Mengyu, Zhou Chunshan (2013) Research on the management mode and space organization of China's urban new area development [J]. Geogr Trop 2013(1): 56–62

A Study on Transportation Space Improvement and Adaptation in Old Downtown Area of Middle and Small Size Cities

A Case Study for Changxing County, Huzhou

Wen Jie and Huang Jianyun

As the economy grows, Inhabitants are realizing gradually well-off in medium even small city, every household having a motor vehicle (including cars, motorcycles and electric vehicles) is no longer a dream. However, the construction of basic facilities of the old town, especially the transport infrastructure lags far behind their level of development. Obviously, it hinders the development of the urbanization.

In view of this situation, Some cities have segmented, removed, re-planned, re-build the environment. Therefore some big cities have obtained the extremely painful lesson: Thousand cities, one side; A series of civic cultural interface becomes even more precious, such as Guangzhou's arcade, Shanghai's Shikumen and so on. Now, whether should it also continue in small cities to perform?

On the one hand, the old town of traffic space is unable to meet the needs, on the other hand, old town traffic potential isn't utilized fully. We urgently need find a way to improve this situation.

1 Concepts and Contents

1.1 Traffic Space

Traffic is passenger and cargo traffic industry, traffic acts are including static traffic and dynamic traffic. Static traffic mainly refers to the parking of a variety of passenger and freight transport vehicle; Dynamic traffic refers to the flow of people, vehicles and objects. Traffic space is the important physical carrier to realizing two behaviors.

W. Jie (✉) • H. Jianyun
Shanghai Jiaotong University (SJTU), Shanghai 300240, China
e-mail: wenjie096@sjtu.edu.cn; huangjianyun2004@163.com

X. Chen, Q. Pan (eds.), *Building Resilient Cities in China: The Nexus between Planning and Science*, GeoJournal Library 113, DOI 10.1007/978-3-319-14145-9_22

1.2 Adaptation

"Adaptation" is more applied Physics, economics and management science, psychology and so on. In economics and management science, it refers to "as times changing, the enterprise, appraises own strength and the status correctly, and adjust management and operating structure. "Core features of "adaptation" is "fit, tuning, progressive, strain". "Adaptation" can be summarized as: no changing in the main case, Space fine-tune in the structure, function, or the use in order to meet new demands and changes.

To old town traffic space, "adaptation" mainly refers "through the trimming measure of functional substitution, support facilities and policies fine-tuning, old town don't change old fabric of the old town and the road network pattern of cases, enhanced coordination between software and hardware, excavate the traffic potential of space to enhance old town traffic capacity, rehabilitation and old city charm.

2 Issue Analysis

Lacking of transport infrastructure planning in the old town, this is a defect in small city. At that time of the housing shortage, it Considers only houses for men, considers less on driving, parking facilities. With the change of motor and non-motor vehicles, "driving hard, parking difficult" become the problem of the old city.

According to the research, data collection, and reading, The authors believe that the transport space of small cities in the old town there is the following problem:

(A) Lacking of planning is result in inability to adapt to the current requirements of the old town. Because of planning behind, poor plasticity, as the improvement of living standards, urban residents' modes of Transport from the walk, bike into electric cars, motor vehicles, etc. Old hardware (including roads, parking lots, etc) will is on a natural low. A case study of Changxing County: the ratio of the reasonable road network distribution (Trunk roads: Secondary roads : Branch) is 1:1.17:3.33, and Changxing road network distribution ratio: 48.2:21.3:30.5 (see Table 1). The branch ratio is too small. The road's capillary blood supply deficiency. Certainly, its parking lot construction insufficiency; the aging situation of old city's path is serious.

(B) Lacking of management is result in inability to give full play to the advantages of existing transport facilities of the old town. According to the research, traffic potential does not full of release in Changxing's old town, including following some situation:

- The traffic light timing of road intersections in old town is unreasonable, has affected the traffic capacity

Table 1 Changxing County built-up district existing road rank structure data sheet

Road rank	Trunk road	Secondary roads	Branch
Length (km)	153.5	67.78	96.97
Area (km^2)	3.67	0.81	1.10
Length percentage (%)	48.2	21.3	30.5

Table 2 Different traffic ways of traffic quantity proportion (%)

Ways	Bus	Unit auto mobile	Private car	Taxi
Proposition	7.3	4.8	33.5	0.9
Moto	Bike	Walk	Secondary bus	Others
12.8	18.6	10.7	0.2	11.2

- Road markings in old town are not clear, reduce utilization of the transport facilities. The signs are not clear in particular parking;
- The transport structures of small cities are different. The proportion of non-motor vehicles is larger. The geometry condition in old city and the effect of mixed slowed down the traffic capacity significantly.
- As the effect of residents living around, some roads are used for other purposes, as a farmer's market or car park, e.g.: the Jiefang West Road, the Xianqian West Road.
- On the selection of transportation mode, the proportion of the motorcycle, the taxi, the manpower tricycle is high. the public transportation proportion only accounts for 7.3 %. There are 14 lines, 93 km total route, 81 vehicles in The Center of Changxing County (Table 2).

(C) Lacking of ideas is result in brutal destruction of the cultural sector. Nowadays, because of the space limitations of the form and structure, the historical debt, the work of the traffic facilities' reconstruction or consummation in old town has some difficulties. Some cities' approach is the "re-planning" or the "one-size-fits-all", which undoubtedly walk the old route of "one side thousand cities" in the big urban. Even in new planning, Road is for larger rather wider and draws traffic into the old town, which is greatly damaged the city's unique historical context. Apuleya and Lintel's research on three adjacent streets in San Francisco in 1970–1971, shows that when the traffic intensity is in the high level (8,000 veh/day), outdoor activities and neighborhood association will be greatly reduced. No associated, no culture. It undoubtedly is tearing old fabric of the old town, make it fragmented.

(D) Lacking of construction is result in struggling with traffic space of the old town. Deficiencies mainly are in three aspects:

- lacking of money, can't make bricks without straw. As an urban large-scale project, traffic facilities construction is such a long investment cycle, working slowly. The investment which depends on the government be strenuous. If lacking of knowledge on the improvement of the transport infrastructure of the old town, the Government is also easy to keep on intensifying conflicts between transport supply and demand, it restricts the development of old town.

- lacking of advanced technology. The authors see an intersection of the Xianqian Road without traffic lights, which Depends on the manual direction. The backwardness technology has restricted the old city transportation space development.
- lacking of policy support, may promote the cost of traffic construction, weaken the possibilities of market-oriented, reduce motivation of social capital, increases the dependency on Government funding. For example, you need great Government policy support on the implementation of the public transport.

Speaking of this, let's analyze these problems the causes of these problems. The authors think nothing else but below some reasons:

In the process of urbanization, the old town bears a excessive burden of urban functions. The old town of Changxing County (only the core area) is the 13.24 km^2, There is only 29 % of the city center area, but the population is the County's total population 45.9 %. In addition, there are more than 4,300 shops and several schools, public services in more than one place in the region. Land-use development model which is highly concentrated but not intensive is the root of all the traffic problems of the old town.

Yearly's explosive growth in traffic demand overwhelms the transportation network of the old town. The vehicle population statistics only 5 million amount Changxing County statistics stock in 2003. Its stock of motor vehicles has its three times by 2011. According to the traffic department's statistics, Central city motor vehicles number conversion recommended number is 70 %. The vehicle number in the old city is 36,000. In future 2020, this figure will double again. Old town cultural protection is important, but facing this kind of demand, more choice is nothing.

3 Cases Study and Improvement Measures

First lets us understand the object of the research.

Changxing County is China's hundred counties in the northernmost tip of Zhejiang Province, the West side of the Lake, borders with Jiangsu and Anhui Province. Changxing is nearly 110,000 of the resident population of the old town. At 2011 the total length of Changxing County road is 318.25 km. Average per person urban road area is 23.26. Types of land-use are mainly residential land, commercial land and public facilities' land. The old city still is main residential and commercial gathering places of inhabitants in this county (Fig. 1).

At 2011 motor vehicle ownership is 150,000 vehicles, nearly 95,000 for motorcycle (electric vehicles). And 4,000 questionnaires which distributed evenly show

Fig. 1 Changxing of the land-use plans (From Changxing 04 version of the master plan)

that 54 % of respondents prepared to purchases a car in the future 3 years. That residents choosing the car or motorcycle (electric car) is the main way to go out, a total of 73 %.

Old town range: zhicheng old town, located in the heart of the Changxing County, east to the new 104 national road, south to the Xuanhang railway, west to the Xinchang railway in the and North to the Tai lake avenue.

The authors think: solving the small spaces of urban traffic problems, is not only to improve the "hardware" construction (including amenities, features, structural adjustment, land development and building strength, advanced technology, planning controlling, etc), but also to strengthen the transport infrastructure "software" construction (including: concept, the mechanisms of the investment, guiding policies, regulations, and so on).

In view to the traffic problems of the old town, the authors make the following specific improvements.

3.1 The Functional Strategies

As far as the urban layout, main features of the old town are borne residential, administrative, commercial, public services and other functions, it contains several historical attractions (such as Chen Wudi Palace, zhujia Temple). The functions are heterogeneous.

Firstly we should determine the traditional space pattern of the old castle town, the cities shape and the historical culture protection construction in the region, Put an end to reconstruction, expansion or new roads for damage to the old city historical context; Secondly, there will be some administrative and public services (such as education) of the city features stripped out (eg: build administrative centres, relocation of Changxing first Primary school). We can evacuate some residents of poor living facilities group. Old town is Main as a function of commercial, residential and cultural symbols of the town.

3.2 Road System Strategies

We should vigorously promote road infrastructure-building, make up less branch for structural defects, open end roads, form a reasonable level of the road network.

Old Town outlying areas well-built road loop, manifold city traffic; Central Zone encryption support the network system, the traffic and the effective triage and the core areas, the traffic space to be re-assigned to a chronic traffic, public transport, motor vehicle to complement the road network system, and the space for the regulation and of the attempt to increase urban cultural charm.

In the perimeter area of old town we build trunk roads loop, which divert urban traffic; in the centre area we encrypt Branch network system, and which divert the traffic of trunk networks; in core area we formed to slow traffic, public transport for foremost, motor vehicle for secondary road network system, and the adaptation of the space increase the cultural charm of the city.

3.3 Transport Strategies

In small cities, motorcycle and non-motor vehicles are relatively high rate, the conflict points of road intersection are multiplication, it greatly affected the traffic capacity of roads. Motorcycles (electric vehicles) account for about half of the stock of vehicles. In suitable paths of old city, we cross the road to achieve traffic separation, even organize one-way traffic.

3.4 Public Transport and Slow Traffic Strategy

Old town which limited by spatial patterns cost highly for big street. Today, how we ensure the transportation does not restrict the development of the old city in the restriction vehicle development situation?

The authors think we should insist on the development police of public transport priority, reduce the volume of motor vehicle traffic. It requires to improving public traffic service level. According to the survey, Changxing currently opened 14 lines and equipped with 87 buses (Recommended number for 7–8 veh/million, 160–190). The line repetition coefficient is 1.93 (recommendation number is 1.2–1.5), the average bus travel time is 15.9 min, But the time outside the car is 38.7 min, this time is too long. This also is the reason that the share rate of the public transportation system is only 7.5 %. The bus-company should adjust, add lines, purchase vehicles. And according to the amount of traffic we set up public transport hub site (such as the People's Square station), even in the partial area set the public transport system of the bus-only road and a bigger amount transport.

Another auxiliary method of resolving the traffic problem is slow traffic, it which form a slower environment, fit with the old town very high. Changxing slow system can be refined along the angle of point, line and area, such the people's square as nodes to ensure walking space; the old city river system is developed, may establish the slow good landscape belt along the river; in business as a carrier, we can organize stereo slow traffic. By these ways we achieve the Changxing low leisure traffic network of the old town. Of course, the public bicycle rental service is also better able to guide the choice.

This shows that the key to ease the congestion of the old town is devoting major efforts to developing public transport. The key to improve the traffic environment of the old town is slow traffic.

3.5 Parking Systems Strategies

The parking is as an important part of the transport system, the parking problem is one of the contradictions of the traffic problems of the old town. At old town of Changxing the road parking is about 2,000 parking spaces, a shortfall of 167; social and public parking spaces is 1,177, a shortfall of 3,159. Parking space difference is large. Recently we can through following ways to mitigation parking problem:

- add parking spaces, branches as pavement parking or old town of green;
- parking gradually from points and line surface to stereo layout transition (small parking floor and parking tower);
- combine old town reality for parking planning, main: specification road parking, reserve part land to built social public parking;
- Concept should be turned on the public parking, we may implement of parking management system. This system may be taken to the principle of sub-regional

sub-periods, such as peak height peak hours than the flat fee, the road than off-street parking within the parking meters high.
- The introduction of parking guidance system, it can be location, usage of the parking lot, route. it relates the information of road traffic as well, to help drivers to quickly and effectively find the parking lot.

3.6 Intelligent Traffic Management Strategies

As said earlier, the parking guidance system, it's just intelligent transportation systems ITS (Intelligent Transport System) part of it. ITS User Services service areas are divided into: traffic management, electronic toll collection (such as meter: applied to pavement parking), traffic information services, intelligent highway safety assisted driving, operation management, transport, transport infrastructure and ITS data management. It can effectively improve the operation of the traffic of the old town.

3.7 Policy Guidance and Introducing Funding Strategies

We can carries on the synthesis strict management using the legal system method, formulate the stop control laws and regulations. Make the old city traffic planning, construction and management to follow the rules.

The concept that the public utility only invests by the government is changed. The Source of funding comes from loans, bonds, private at home and abroad except from the government. The capital diversification will be helpful to the project operation.

4 Adjustable Utilization Research

In terms of city's development goal, City's refinement is increasingly being advocated (Wenjun Ma). Because of this, people put eyes on these distinctive cities such as Lijiang, Pingyao. The old town of the city's unique imprints and signs of aging well illustrate this point. But now, more cities' imprints completely disappear because of the traffic conflict and so on, let the city become mediocre.

As you know, new roads are only to attract new traffic. Cities like containers, the container is limited and the desire is unlimited. It also declared in advance, the old city even the road be widen in, could not escape the fate of congestion.

I believe that, the town's space is a dynamic process as biological metabolism, not static. The traffic space is also true. It is growing and renewing. Of course, we generally believe that this is a spontaneous evolution. The adjustable utilization of traffic space is to provide an additional possibility. The appropriate trimming will be able to satisfy the old city current even future the transportation change.

However, how to interpret the transit space adjustable utilization of the old town?

4.1 The Adaptation of Ideas

In 1962, the main street of the old town was turned into walking street in Copenhagen. In the Nordic region this was nothing new. At that time, many critics predicted that this Street would be abandoned. But today, the street is still bustling. The old city comes from the walk-bicycle time. It meets the current demand for Non-motor vehicle traffic. But through the interface of the old town of destructive transformation to meet motor vehicle traffic, that is not a wise choice. Old town traffic is from "motor vehicle" back to "people".

The authors think:

- the old town (especially the core area) of traffic should main to slow traffic (walk and bike) as primarily, and public traffic system as auxiliary;
- According to the actual situation, the old city must delineate the scope of protection boundaries, which is considerate from spatial pattern, interface and block streets integrity;
- set up traffic buffer between old town and new town such as public parking and transfer Center;
- To limit the development of motor vehicles in the old town, we can use some ways such as the delineation of the pedestrian area, reducing the number of motor vehicle parking areas, raising parking fees and so on.

4.2 The Adaptation of Functions

The land even if vacated because of the function replacement, the land resources are inadequate in the old town. Old town is still the same, but the residents have had a new demand, the city has a new measure too. The old city had some demands such as the disaster prevention, the afforested and even culture corridors. These requirements do not have enough land to arrangements. That traffic space integrating in these requirements is a very real problem. We have been able to find quite a few examples of references at home and abroad, such as:

Through deepened, Japan 60 % underground car parks are relying on roads without encroaching on architecture, traffic will not be affected; By splitting, in branch-intensive regions by means of function split, the branch is set to slow or walking the road; By integrating, we organize one-way traffic. The rest of the road space as the pavements of the city, such as parking, a green belt and cultural propaganda gallery; by exchanging, we set a section of road to the city commercial walking street or regional activity centre. There are many facilities to set up parks, green spaces, and the underground parking garage for in China. An elementary school in the old city of Kunming for parking tight under the playground sets the underground car park. There is a reference to the traffic space adaptation.

4.3 The Adjustment of Space

The transportation space of the old city more takes "the human" as a criterion. You want to improve the quality of traffic space of the old town. The "people-oriented" start from the "car".

For example: Both sides the slow good path establishes the rest chair; the road by river is set on the landscape and leisure facilities. Through these ways, we may enhance the resident to go on the comfort level of a journey unceasingly, and enjoy the life in the slow traffic.

The traffic flow line of the old town is a flat type. Now that by science and technology it is fully capable of building stereo type of traffic. For example, in the city of Nanping Street Kunming, it set to the commercial walk street, under the street it's the vehicular transportation. Of course, the commercial center of the core area for the old town you can walk though the air corridor to connect (Fig. 2).

4.4 The Adjustment of Developing

We should recognize that space development is growing, even more flexible. The time is developing, the science and technology is also developing. Therefore in the old city development, we should be careful, stick to the scientific concept of development. Through the careful proof, a lot of advanced and mature technologies may utilize in the old city, make it more orderly, efficient operation. For example, through data collection and GIS and statistics, Old city builds the simulation traffic assessment and guides the traffic changes lately. The Auchan supermarket of Changxing establishing brings big traffic pressure in the region.

5 Conclusion

The twenty-first century is the globalization city time, the traffic space of old cities already became the incisive question day by day. Purely from question itself, it can alleviate the traffic pressure temporarily. However, to symptoms has yet to solve. Old town adheres to people as an orient, persists urban and cultural values as a starting point, breaks the pattern, improves the elasticity of traffic space, improves

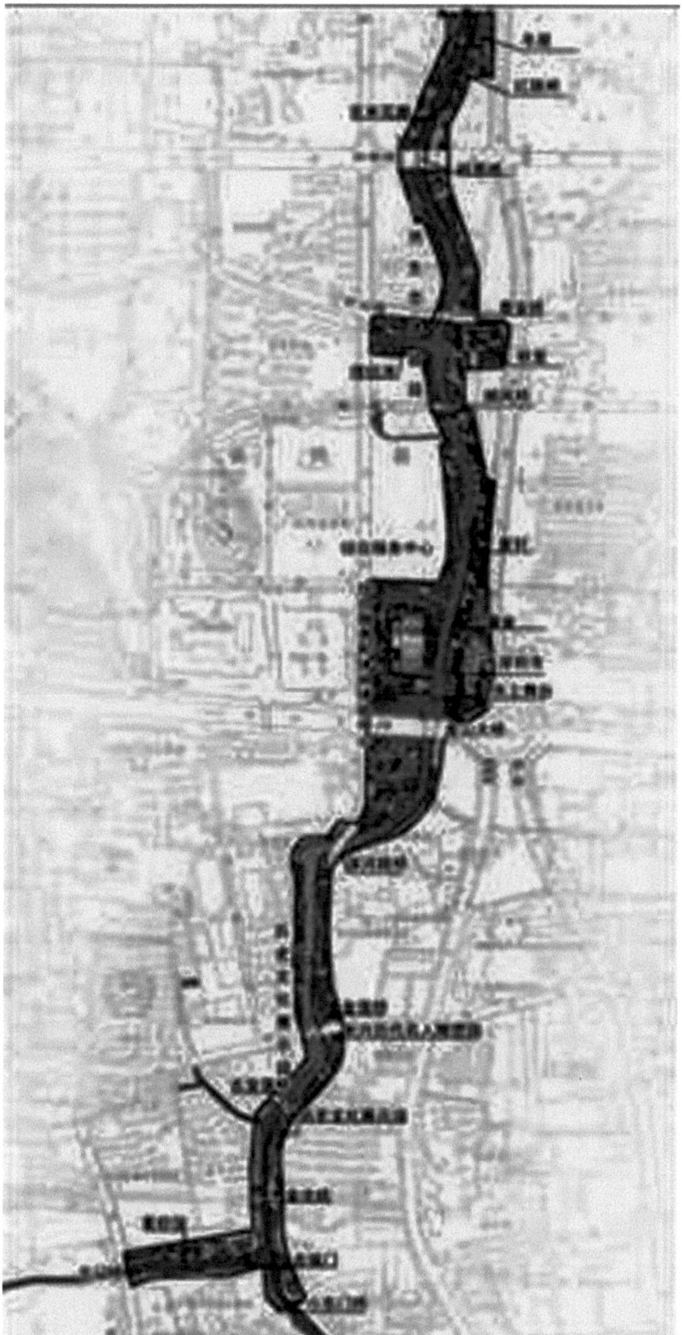

Fig. 2 The long water port moat aquatic tour project plan (Originates in "Changxing County travels plan")

the traffic environment quality. Under the premise of the human and vehicle safe convenient moving, it retains the old city culture value, promotes the old castle town. As a new highlight it's in the development of small town ultimately.

Constructure and Update of Point of Interest (POI) in Urban Open Space

A Case Study of Assessment and Reform of POI in Water Park Tianjin, China

Jun Wu, Naiwen An, Pengbo Li, and Min Zhang

1 Introduction

In recent years, with the social development and economic growth in China, the process of urbanization continues to accelerate; additional urban open space is also constantly emerging. Urban open space known as the city's "living room" and "showcase" [1], brings out vitality and color, diversity and richness of city life. But there are many problems that should not be ignored in urban open space: How satisfied is the public with the urban open space? How the designer and the user thinks about the open space? [2] Where is the gap? How to increase public attraction to the space? To resolve these questions, it requires that the urban open space to have a strong attraction, a consensus must be reached between the designers and the public [3]. And so, the urban open space can attract public interest and strong participation.

So, in reality, there is an urgent task to deconstruct urban space, explore the urban open space elements, understand user behavior and requirements and build point of interest of urban open space to improve the attractiveness and utilization of urban open space [4].

Supported by the Art Science Foundation of Tianjin (No. E10030), the Natural Science Foundation of Tianjin (No. 08JCYBJC05100)

J. Wu • N. An • P. Li (✉)
Department of Architecture, Tianjin Chengjian University, Tianjin, China
e-mail: pengbotj@gmail.com

M. Zhang
Department of Design, Weihai Ente Construction Engineering co., LTD, Shandong Province, China

X. Chen, Q. Pan (eds.), *Building Resilient Cities in China: The Nexus between Planning and Science*, GeoJournal Library 113, DOI 10.1007/978-3-319-14145-9_23

2 Concept of POI of Urban Open Space

2.1 *Interest*

2.1.1 About Interest

Interest is defined as the tendency of people to understand something and engaged in some activity psychologically. It is shown as thinking about one thing [5], selective attitude to some activity and positive emotional reaction of people. Interest is important to people's actions. When people are interested in something or some activity, they will pay more attention to it and it will produce pleasant state of mind, and improve efficiency and quality of the activity [6].

2.1.2 Characteristics of Interest

Orientation of interest: interest always points to some contents. Whether it is material or spiritual, noble or otherwise [7].

Breadth of interest: this refers to the scope of interest. Different people have different scope of interest; one person might have broad interests while others have narrow interests. Generally speaking, people have many interests to gain broad knowledge [8].

Stability of interest: it refers to interest kept on one object or some particular objects for a long time. Once stable, the central interest could be formed and further deepens.

Efficiency of interest: interest always creates efficiency in the activity. Some interests can create significant efficiency in the activity, while others create minor efficiency.

2.2 *POI of Urban Open Space*

2.2.1 Deconstruction of Urban Open Space Environment

In urban open space, activities are happening. It can be described as: Sometime, somebody is doing something somewhere. So, to build the POI, somebody, somewhere and something should be taken into consideration.

Somebody: the user of the place. It is people of different ages, class, occupation hobbies and cultural background. They can choose this opportunity to communicate here freely and equally.

Something: three outdoor activities in urban public space were mentioned in the book "Life Between Buildings Using Public Space", that is, necessary activities [6], spontaneous activities and social activities. The spontaneous activities and social activities are the main activities in POI of urban open space.

Somewhere: the place where something is happening. It is the material environment and should be reasonably designed, and provide suitable space for activities. Cultural factors should be taken into consideration in the design [6].

2.2.2 Concept of POI of Urban Open Space

According to the analysis above, the POI of urban open space is the place that has suitable space and environment; the public can find their desired activities and be willing to engage in those activities.

2.2.3 Characteristics of POI

1. *Clear identification*
   POI should be an identifiable space. People can identify the space according to its special patterns, to know the actual location and the parameter, and to understand the image of the object, characteristics and to find the distinction.
2. *Strong attraction*
   The behavioral environment has certain characteristics; these characteristics can be described as: aesthetic appeal. POI of urban open space has attractive landscape, convenient facilities and appropriate space [7] to attract more residents and visitors.
3. *Significant participation*
   The activities in POI of urban open space can be divided into three categories according to the impact on users: positive activities, passive activities and unpopular activities [9]. The positive activities are both perception and control activities and attract the attention of others to participate in it, such as dancing, singing, practice and other activities.

2.3 Element of POI in Urban Open Space

Importantly, POI is the place that can fit the public activities. But it should take the people and the activity into account. So, the active body, active event and active place are the main element of POI (Fig. 1).

2.3.1 Active Body

It refers to the user of the space. Users of the POI differ in gender, age, culture level, education level, personality, profession and hobbies etc. there are a lot of differences between the people since they have different requirements and time.

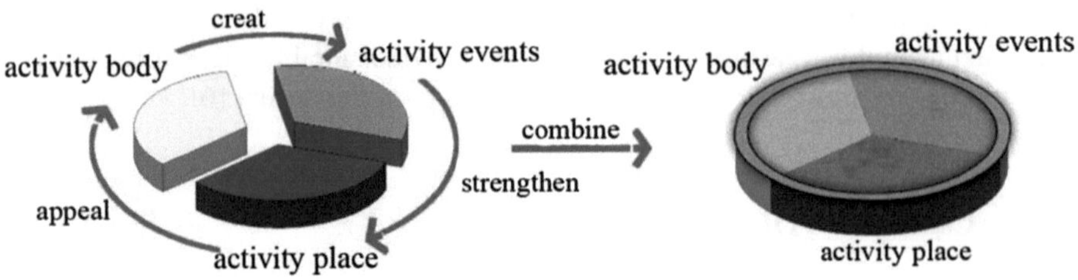

Fig. 1 Relationship of three elements of POI

2.3.2 Active Events

The POI should be built to satisfy the users who take part in these activities with suitable place, appropriate subject and requirements, resulting in more people to take part in the activities.

2.3.3 Active Place

It refers to the place that people can perform activities. It is the material space and satisfies the requirements of activity.

The three element of POI are closely related. Active body creates active event and active event strengthens active place. And the active place provides the material conditions for active body. POI is formed only when the three elements are combined.

3 Construction of POI in Urban Park

Urban park is an urban open space, which can include POIs that can be built to satisfy public activities and attracts people to use the park space. Three elements have distinctive requirements in building the POI in urban park.

3.1 Public: Active Body

The function of modern urban park is to satisfy people's entertainment and fitness requirement. With the methods of humanization design, public's requirements are the most important factors in POI design. The requirements of comfort, safety, personal space, and participation are the focuses of the public [10].

3.1.1 Comfort Requirement

People desires for relaxation and enjoyment when they are active in the park [2]. The POI should be convenient for the public to enter, and through rich activities and landscape appeal to visitor's visual senses, and to provide sufficient facilities to satisfy all kinds of people.

3.1.2 Safety Requirement

Public security refers to the public's psychological needs against external threats of damages to body and spirit. POI design should provide defense for individual field of psychological security, ensuring the person is not affected by surrounding adverse factors and guarantee personal security, at the same time, guarantee the freedom of individual action.

3.1.3 Personal Space Requirement

Ensuring there is no highly visible and open space around individual [2], as it will cause anxiety and uneasiness when there is scope for invasion and interference. There are four kinds interpersonal distance in interpersonal communication (Table 1). POI design should take into account of the personal space requirements.

3.1.4 Participation Requirement

The public wishes to participate in the events and activities, and communicate directly or indirectly with others. According to the models of public participation, there are three layers: active participation, passive participation and on the sidelines [11]. Every layer needs suitable space and conditions to fit the requirement and let people do what they want freely.

Table 1 Interpersonal distance classification [6]

Interpersonal distance	Close range	Far range	Application
Closely distance	<15 cm	15–45 cm	Fighting, affectionate and caress
Personal distance	45–75 cm	75–120 cm	Family members, teachers and students, close friends
Communication distance	1.2–2.1 m		Non-personal things, such as the working relationship with colleagues, social speech etc.
Public distance	3.6–7.6 m	>7.6 m	Lectures, performances and rituals

3.2 *Activities: Soul of POI*

Activities are the link between the spatial environment and body, it can stimulate human potential, form unique passionate feelings, and it is more significant than any forms of physical landscape design alone. It is more exciting and can increase the use-ratio of space.

3.2.1 Content of Activities

People use outdoors to exercise or communicate with other [12]. One person dancing in the park may attract more people to participate in it. This will attract more on lookers and will encourage communication.

3.2.2 Time of Activities

Activities should always occur at a reasonable time [11]. So suitable people can get there and engage in common activities. A space should take time into account and design timetables to fit people's needs.

The duration of the activities in POI will affect the attraction of the POI. It includes direct effect and indirect effect. Direct effect is that the more people active at the same time will increase the use of space; indirect effect is that more people active at the same time in the same place, will increase interpersonal exchange, and stimulate public awareness of potential activities, this will result in a more active and lively atmosphere.

3.3 *Location: Basics of POI*

3.3.1 Accessibility

Accessibility refers to the relative degree of difficulty of people arriving at the space (POI) from an arbitrary starting point. So, Accessibility is a relatively important element in POI [13]. It may decide whether the people go to the landscape area or not. Thus, POI should be an identifiable position, and people can arrive at ease.

3.3.2 Environmental Conditions

The outdoor activities will be subject to environmental conditions. It includes weather condition, sunny, windy and temperature. So, good environmental condition should be created in the POI to satisfy the activities.

3.3.3 Landscape Sequence Planning

POI is the landscape node in the urban space. Planning and designing the POI is not one node but a sequence of landscape [14]. When people have a determined goal, they will go to the place directly. But some people may not have a determined goal. They maybe just strolling and when a place attracts them, they will go and participate. Every node should have its own spatial charm.

3.3.4 Scale of Space

When the crowd begins the activity, they need to occupy certain space, simultaneously hold certain distance with the other activities, this can assure one's own security and privacy [15].

3.3.5 Infrastructure

There are a lot of infrastructures in POI, such as seats, leisure facilities, rubbish bins, landscape lights, lawn lighting, barrier-free access ramp, rain facilities etc. It is the 'furniture' in the space and the public can use them freely and easily. This will ensure a more immersed experience for the public.

4 Case Study of Water Park in Tianjin, China

4.1 Introduction

The area of the Water Park is 125 ha, water surface area is 75 ha, and land area is 50 ha. It is the biggest park of parks in Tianjin (Fig. 2). The Water Park is famous for its waterscape. There are East Lake, West Lake, and South Lake in the park. There are also some islands in the lake. Arch bridges, tortuous bridge and culvert connect them. There are some entertainments in the park, such as flower show, exhibit of lanterns and other entertainment activities. Water Park District Tourism International Village, tower city area, and different architectural styles are in the park. The park also contains the Kobe garden, bonsai garden, water village, Child's Park Village.

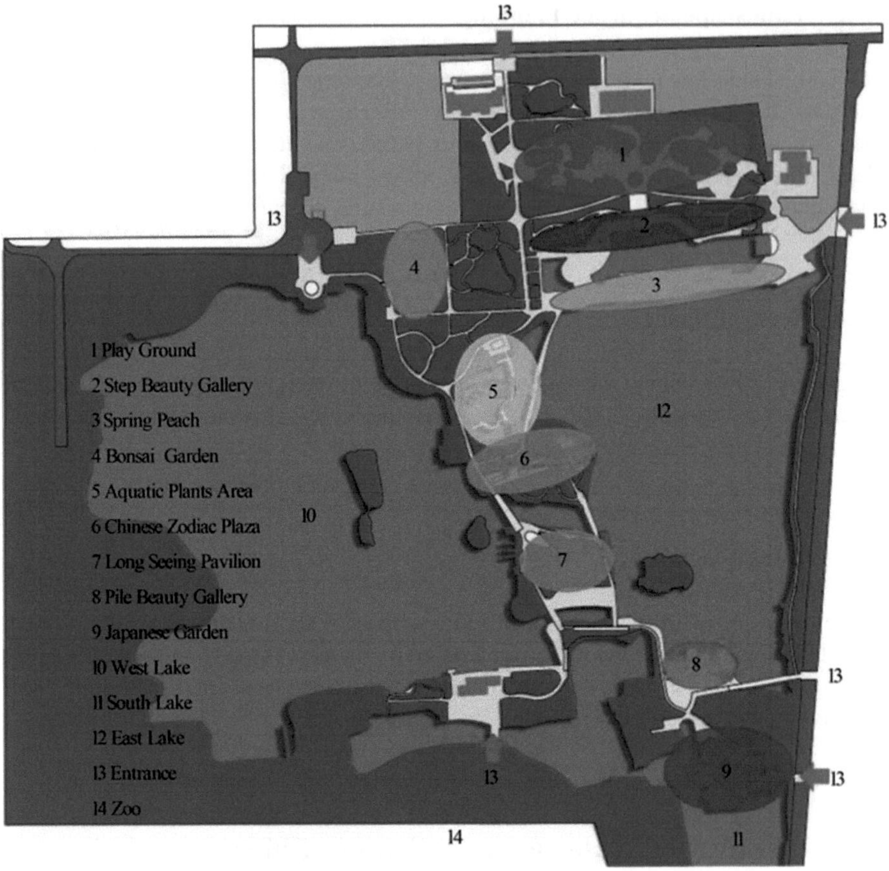

Fig. 2 Main POI in plan of Water Park

4.2 POI in the Park

There are two main routes in the park. One is from North Entrance to the Zoo; another is from East entrance to Five Island Entrance. All of the main landscape nodes are on the visit line.

4.3 Usage Ratio of the Landscape Nodes

Research has been done to understand the usage ratio of the landscape nodes of Water Park. The period of high visitor numbers on work days and vacation days were selected. Using half an hour for statistical time period, three periods were selected for each node. The average numbers of people stayed in each node were

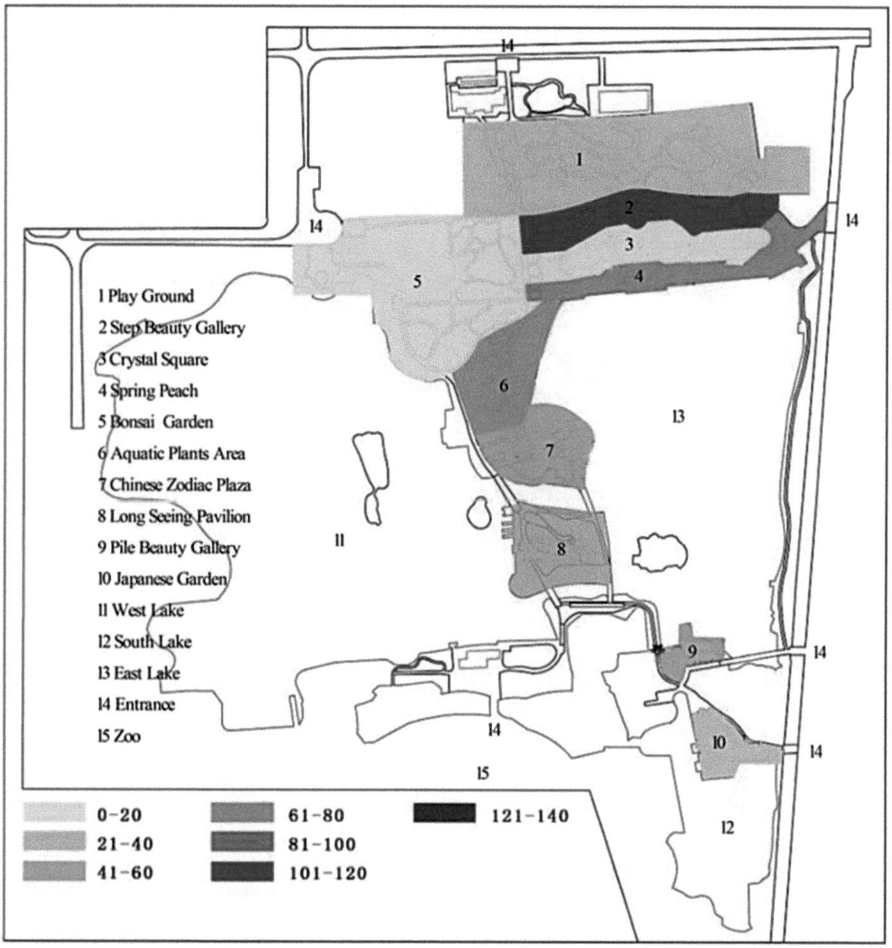

Fig. 3 Number of people in working days

found. The numbers of people stayed are those who were engaging in activities in the node, it included resting, training, entertainment and communication, etc.

According to the result, in working days, the number of visitors in Step Beauty Gallery is the largest; the following is Sping Peach, Aquatic Plants Area, Long Seeing Pavilion, Chinese Zodiac Square, Pile Beauty Gallery, Playground, Waterfront Plaza and Bonsai Garden (Fig. 3). Number of people in vacation days is much the same with the number of people in working day. in the biggest difference are visitors in the Playground. It is almost three times that of working days. Except for Playground, the top five points are Steps Beauty Gallery, Sping Peach, Aquatic Plants Area, Chinese Zodiac Square and Long Seeing Pavilion (Fig. 4).

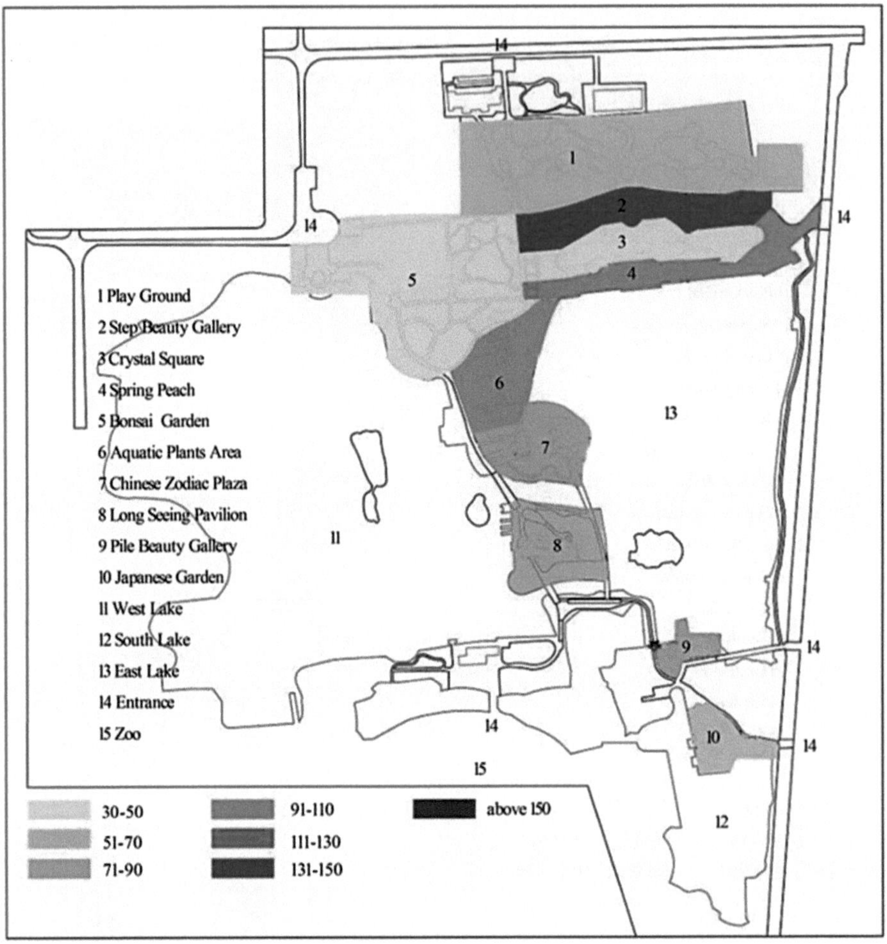

Fig. 4 Number of people in vacation days

4.4 *Analysis of POI in Water Park*

According to the research above and comprehensive analysis of the three factors of POI in Water Park, there are some problems in Water Park nodes. This includes:

4.4.1 People Do Not Care About the Location

When the location is fit to some games, people will have their activities in the place. They did not care if the place is just for the activity. Sometimes, the activities do not last a long time. Some are only done in the morning, for example, plaza dancing and

Fig. 5 Skating people on the Zodiac Square, though it is the road across the square

skating are done in the morning or at night (Fig. 5); others are done at noon, during lunch or teatime, and they would stay in the pavilion or gallery.

4.4.2 Environment Effect the Activities

In Aquatic Plants Area, due to sunshine, people only stay in the area in early morning, in the evening or on cloudy days (Fig. 6). And in the Step Beauty Gallery, because of the crowd, many people cannot stay in the gallery, and the crowd affects the people using the space freely and suitable. So, it is important to build a suitable place for the public to engage in activities on it.

4.4.3 Activities Occupy Other's Space

As some activities do not need dedicated space, the public can play their games everywhere; they even play on the road, this result in them disturbing other people playing games or activities, such as the dancing, skating, etc. These games are important as it lifts the atmosphere around the location. It always attracts many people to visit and participate. So, in the design or landscape rebuilding, the active location design is very important to find the suitable place for the activities.

Fig. 6 People sit on the bridge of Aquatic Plants Area on cloudy days

4.4.4 Less Space for Some Activities

During vacation days, the park is more crowded, and in some places, there is not enough space for the public to sit or stand that seriously affects the suitability and satisfaction of the visitors. Such as in the Playground during weekends, a lot of people go to there, and the place is crowded. People cannot relax leisurely.

4.4.5 Lack of Inside Corner Space

There are some good spaces, but people don't stay there, and no activities are happening. Because there are no shelters or walls, so, the design of inside corner space should be emphasized.

5 Conclusions

The residents living in the city are increasingly expected to enjoy the outdoors. With the continuous expansion of the population and urban construction, urban space is becoming scarcer. Building urban open space to meet rest, recreation, fitness and communication needs in a limited space is the future direction of developing urban opens pace.

5.1 To Meet the Needs of the Public, to Create a User-Friendly Space

People using public space are of different gender, age, work experience, educational background, and have different personalities, hobbies, habits, but they all have equal rights to the urban park space. The different needs of the public should be fully taken into account in positioning the nodes, the type space, and the details to create a user-friendly space.

5.2 Increasing the Experience of Activities to Attract Public Participation

By satisfying the different requirements of activities with suitable space, richness in the type of activities in place, extending the time and the breadth of activities, to continuous enhance public participation, and improve the use of the public space.

5.3 Refined the Open Space, Form the Positive Space

First of all, the choice of the point of interest location should proceed from the overall layout, and reasonable arrangements of spatial sequence. Secondly, for different types of activities, pay attention to the treatment of boundaries of the space. Create positive space through detail treatment to form positive open space and to enhance the use-ratio of the place.

5.4 Enhance the Charm of the Place

The design of the location directly impact user comfort. Design of points of interest should be designed according to the details of the place. By providing a sound infrastructure to meet public use, strengthening cultural expression of the place to attract Public ideological resonance, using terrain, plants, structures and other means to improve the spatial conditions of the open space to guarantee the space is used with comfort.

References

1. Sandstrom UG, Angelstam P, Khakee A (2006) Urban comprehensive planning – identifying barriers for the maintenance of functional habitat networks [J]. Landsc Urban Plan 75(1–2):43–57
2. Matsuoka RH, Kaplan R (2008) People needs in the urban landscape: analysis of landscape and urban planning contributions [J]. Landsc Urban Plan 84(1):7–19
3. Scolozzi R, Geneletti D (2012) A multi-scale qualitative approach to assess the impact of urbanization on natural habitats and their connectivity [J]. Environ Impact Assess Rev 36:9–22
4. Haase G, Richter H (1983) Current trends in landscape research [J]. GeoJournal 7(2):107–119
5. Balej M, Raška P, Anděl J, Chvátalová A (2010) Memory of a landscape – a constituent of regional identity and planning? [J]. Landscape Model: Urban Landscape Perspect 8:107–121
6. Gehl J (1987) Life between buildings using public space [M]. New York: Van Nostrand Reinhold
7. Barter C, Vekasi J, Olmsted Center for Landscape Preservation (U.S.) (2006) Acadia trails treatment plan: cultural landscape report for the historic hiking trail system of Acadia National Park, Maine [M]. Olmsted Center for Landscape Preservation, National Park Service, Boston, p 386
8. Cultural Landscape Program (United States. National Park Service. National Capital Region) (2000) Cultural landscape report: Dumbarton Oaks Park, Rock Creek Park [M], vol 1. The Program, Washington, DC
9. Bohnet I, Smith DM (2007) Planning future landscapes in the Wet Tropics of Australia: a social-ecological framework [J]. Landsc Urban Plan 80(1–2):137–152
10. Yamagishi T (1992) Landscape and the human being [J]. Hum Stud 15(1):95–115
11. Zhang Junjie RN, Wei L (2007) Increasing the attraction of street green land to build a pleasant green space of city [J]. Shenyang Jianzhu Univ 12(1):24–27
12. Shao J (2009) Case study of mountainous landscape planning for metropolitan outskirts [J]. Nongye Gongcheng Xuebao/Trans Chin Soc Agric Eng 25(2):228–234
13. Sun K, Miao Z (2011) Notions of green space system planning in medium and small city. In: International conference on remote sensing, environment and transportation engineering, RSETE 2011, IEEE Computer Society, Nanjing, 24–26 June 2011
14. Jim CY, Chen SS (2003) Comprehensive greenspace planning based on landscape ecology principles in compact Nanjing city, China [J]. Landsc Urban Plan 65(3):95–116
15. Li Z, Huang J (2011) Characteristics and rebuilding of urban landscape of the Xuchang ancient city. In: International conference on consumer electronics, communications and networks, CECNet 2011, IEEE Computer Society, Xianning, 16–18 Apr 2011

Comparison of Two Approaches for Evacuation Plan with Multiple Exits

Jun Xie, Qing Wan, Chi Zhang, and Xiang Li

1 Introduction

Under the situation of rapidly expansion of urbanization and population, it's necessary to make an efficient evacuation plan with the purpose of saving people's lives and property. Approaches have been proposed from different aspects. In the field of network flow, minimum cost maximum flow problem are usually employed to make route and solve the limit capacity when simulates network as undirected or directed graph [1–3]. Heuristics can be applied to find disjoint routes for evacuation. Relaxation-based approach and staged evacuation algorithm can solve the evacuation model heuristically [4, 5].

Based on the "first-in, first out" (FIFO) principle, the shortest path algorithm is improved to adjust the route dynamically when comes to congestion. But all evacuation groups evacuate simultaneously, it may cause congestion when the flow is larger than the road's capacity [6]. Experiment result shows that congestion may obtain higher priority than other factors that affect the evacuation process [7]. And when comparing the effectiveness between simultaneous and staged evacuation strategies in different road network structures, some staged evacuation strategies performs better than simultaneous one under congestion situation [8]. Traffic flow which is assigned to exit can't be predefined but exit and other

J. Xie • X. Li (✉)
Key Laboratory of Geographical Information Science, Ministry of Education, East China Normal University, Shanghai 200241, China
e-mail: xli@geo.ecnu.edu.cn

Q. Wan
Institute of Geographical Sciences and Natural Resources Research, Chinese Academy of Sciences, Beijing 100101, China

C. Zhang
Xinjiang Institute of Ecology and Geography, Chinese Academy of Sciences, Urumqi 830011, China

X. Chen, Q. Pan (eds.), *Building Resilient Cities in China: The Nexus between Planning and Science*, GeoJournal Library 113, DOI 10.1007/978-3-319-14145-9_24

parameters are able to set when modeling evacuation progress. Thus, a super safe zone is added into the network to connect all safe exits. And departure time of evacuees can affects the orderliness and effectiveness of whole evacuation process [9]. However, only departure time of each zone is actually controllable in the staged evacuation modeling. And congestion can be reduced by distributing the demand of the whole evacuation network uniformly [10]. A mixed-integer programming formulation approaches traffic routes without crossing conflicts, and it can be achieved by evacuating groups one by one at different instant with predefined route as well [5, 11]. The staged evacuation algorithm (SEA) can be applied to make a staged evacuation plan which evacuees are organized in different groups and transported to safe exits based on predefined schedule and route without suffering traffic congestions. SEA works well for single exits, but it is not available for multiple exits situation [11].

As a result, two ways, virtual node approach and optimal allocation approach are presented in this paper to compare the efficiency of them. Virtual node, which is usually used in network flow problem, is added into the network to change the topology in the virtual node approach. And the modified network structure can be fed to the original SEA. While optimal allocation approach is intend to treat multiple exit evacuation as multiple single exit evacuation, which is suitable for the original SEA as well. At the same time, a simulation program is developed to imitate the evacuation process.

The following of the paper is organized as follows. Section 2 presents the original SEA. Section 3 describes the principle of two approaches. And Sect. 4 conducts experiments and analyzes the results. The last section concludes the paper.

2 Background

2.1 The Staged Evacuation Algorithm

The typical node-arc model is adopted to describe the road network for this algorithm [12]. Firstly, evacuees gathered in nodes that are on the road network by their geographical position. Thus, evacuees from same node are considered as an evacuation group and enter road network from the node. Assumptions like each evacuation group evacuate with same speed to the predefined exits during the evacuation process and there is no background traffic flow in this organized evacuation plan are made. The total evacuation time, denoted as *TET*, should be the maximum one among all evacuation groups' minimum evacuation time. The minimum evacuation time of an evacuation group calculates from its start time to the instant when it passes through exit completely. Therefore, *TET* can be expressed as

$$TET = MAX\left[\left(t_0^v - t_0\right) + t_{travel}^v + t_{wait}^v + t_{pass}^v\right], \quad \forall v \in [1, \ \delta] \tag{1}$$

Where δ denotes the number of evacuation groups; t_0^v denotes the instant when evacuation group v begins to evacuate; t_0 denotes the instant when evacuation progress starts; t_{travel}^v denotes travel time of evacuation group v along its route regardless of traffic congestions; t_{wait}^v denotes the whole waiting time of evacuation group v which is caused by congestion and t_{pass}^v denotes the travel time of evacuation group v passes through a node completely. Minimizing *TET* and reducing congestion are the goal of an evacuation plan. For a given evacuation group v, the objective can be simplified as minimizing t_{travel}^v, t_{wait}^v and t_0^v. The least travel time can be obtained by employing the classic Dijkstra's method [13] to choose the shortest route to exit because evacuation speed is consistent. Going through exit one by one continuously can help to reduce waiting time and make departure time earlier. Thus, a scheduling algorithm is presented in next section to set departure time of evacuation groups.

2.2 Scheduling Algorithm

The algorithm aims at setting departure time for each evacuation group with the objective of minimizing t_{wait}^v $(t_{wait}^v = 0)$, which is described below. And let $\widetilde{t}_i^v, i \in [0, k-1]$ denotes the instant when evacuation group v starts off at $\widetilde{t}_0^v$ and arrives at arc a_i^v along its route, where $route_v = \{a_0^v, a_1^v, a_2^v, \ldots a_{k-1}^v\}$. k denotes the number of arcs along evacuation group v's route.

1. Let $\widetilde{t}_0^v = t_0$
2. Find all the earliest available time window $\left[t_x, t_x + t_{pass}^v\right]$, $t_x \geq \widetilde{t}_i^v$ on each arc along evacuation group v's route, and $\widetilde{t}_0^v = \widetilde{t}_0^v + \left(t_x - \widetilde{t}_i^v\right)$,
   $\widetilde{t}_{i+1}^v = \widetilde{t}_i^v + \left(t_x - \widetilde{t}_i^v\right) + l/speed$ where l denotes the length of arc a_i^v.
3. $\widetilde{t}_0^v$ is the earliest tentative starting time for evacuation group v.

The above algorithm repeats until the departure time of all evacuation groups are set. In this algorithm, earlier departure time is determined with the consideration of minimizing waiting time and intersection conflict.

3 Methodology

3.1 Virtual Node Approach

Virtual node approach is conduct by modifying the topology of the road network, which means a virtual node is added into the road network to form single destination situation. Thus, it can be easy to solve the multiple exits situation. As shown in Fig. 1, it transforms *m* origins and *n* destinations into *m* origins to one destination [14]. Detailed steps are as follows: First, add a virtual node (i.e., *D**) into the road network as the single virtual exit and its capacity is supposed to be infinite for accommodating evacuees who come from all exits. Second, link each exit (i.e., *D1*, *D2*) to the virtual node with virtual arcs whose impedances are zero, i.e., no travel time. Third, let the capacities of real exits be zero. As a result, the modified road network is able to organize evacuees from their starting node to final single virtual exit, and that can be fit to the original SEA.

3.2 Optimal Allocation Approach

The principle of allocation approach is to treat each exit as single exit problem respectively. Evacuation groups go to their nearest exit and form exit groups first, and then population among exits is reallocated to balance the traffic flow. And *TET* of the entire network is the maximum one among the total evacuation time of exit groups in this approach. Different exit groups have different *TET*, and they may even have a large time gap due to the unknown network and population distribution.

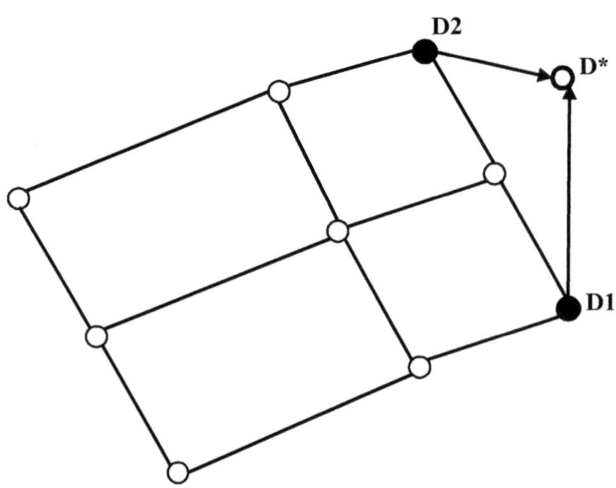

Fig. 1 Illustration of modifying road network topology

Thus, an adjustment is made to reallocate population among exit groups formed initially. The adjustment can be described as follows. Let *r* denote the distance ratio between the farthest exit and the nearest one for each node. It represents the possibility of being reallocated. Then, find two exit groups, *a* and *b*, which have the maximum *TET* and the minimum one respectively. Reallocate the node with smallest *r* in exit group *a* to *b*. The above steps continue until all nodes that have the smallest *r* and in the exit group with maximum *TET* are reallocated. The *TET* of the entire network can be obtained during the process.

This reallocation may lead to the crisscross of evacuation groups because two readjusted exit groups are not adjoining. Hence, every direction should be examined to find time window in scheduling algorithm. And let the longest one be the time window for an evacuation group.

4 Experiment and Results

The two approaches are implemented by C# in Microsoft Visual Studio 2010. The road network of Jing' an district, Shanghai, with 187 nodes and 310 arcs is taken as an experimental data which is shown in Fig. 2. Population and exits generate randomly and we suppose every node in this network has evacuation group to be evacuated. Then, the number of evacuation groups is 184. The average normal adult's walking speed 1.5 m/s is taken as the speed of evacuation group. Four

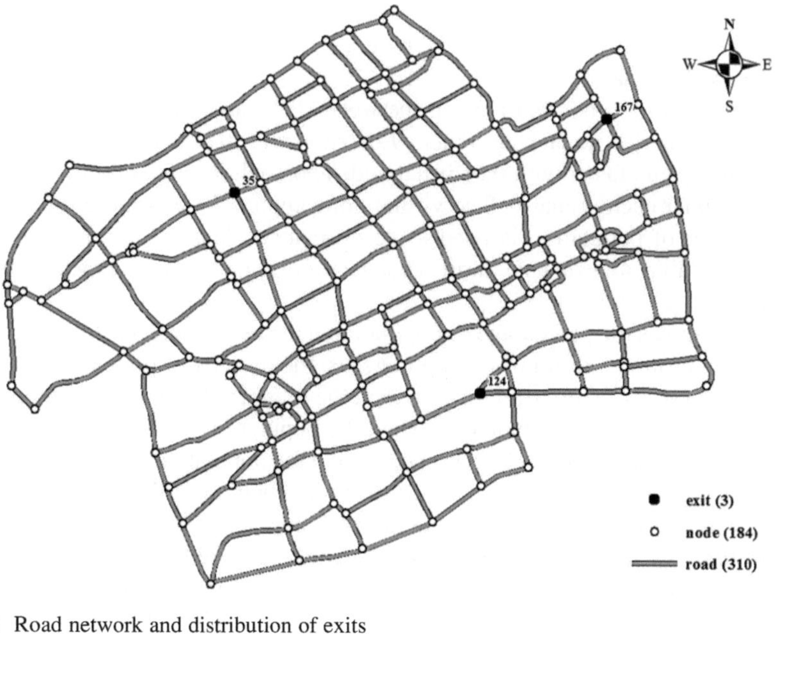

Fig. 2 Road network and distribution of exits

Table 1 Definition of experimental scenarios

	Exit ID	Population
Scenario 1	35,124,167	42,321
Scenario 2	35,124,167	80,076
Scenario 3	35,124,167	158,510
Scenario 4	35,124,167	210,425

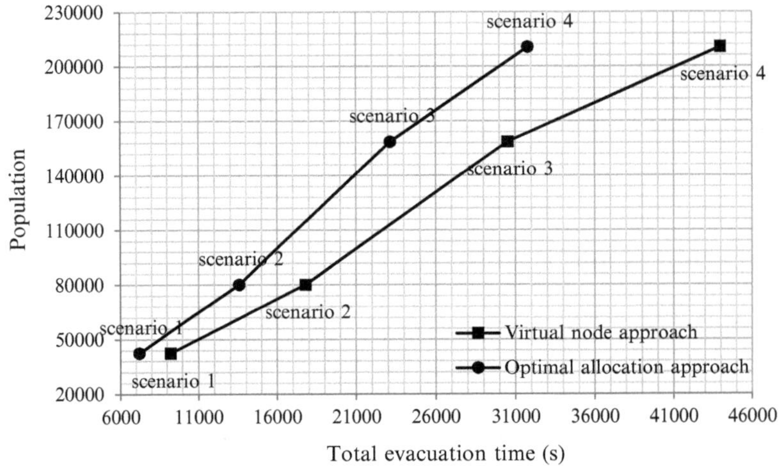

Fig. 3 Relationship between total evacuation time (TET) and population under different scenarios in two approaches

scenarios are defined in Table 1. Three exits are predefined and population increases with scenario.

Figure 3 shows the relationship between *TET* and population under four predefined scenarios in each approach. *TET* rises with the increase of the population, but it goes up faster with the virtual node approach than the optimal allocation approach under each scenario. The virtual node approach costs more time when a large number of population is need to be evacuated. And the optimal allocation approach is able to evacuate more evacuees during the same evacuation time period.

Figure 4 shows the percentage of population that pass through each exit with two approaches. From the result of two approaches, the number of population that pass through exit various with different exit. And the number of population among each exit group can almost keep balanced when the population of the whole network is large. However, the optimal allocation can make the population gap smaller in each scenario when compares to the virtual node approach. And the gap can be even smaller with the increase of the whole network population. This gap is caused by the way that evacuees choose their exit. Generally, people would like to go to the nearest exit, as the exit assignment in the virtual node approach. But it will lead to the consequence that more people go to one exit while others are unoccupied, and

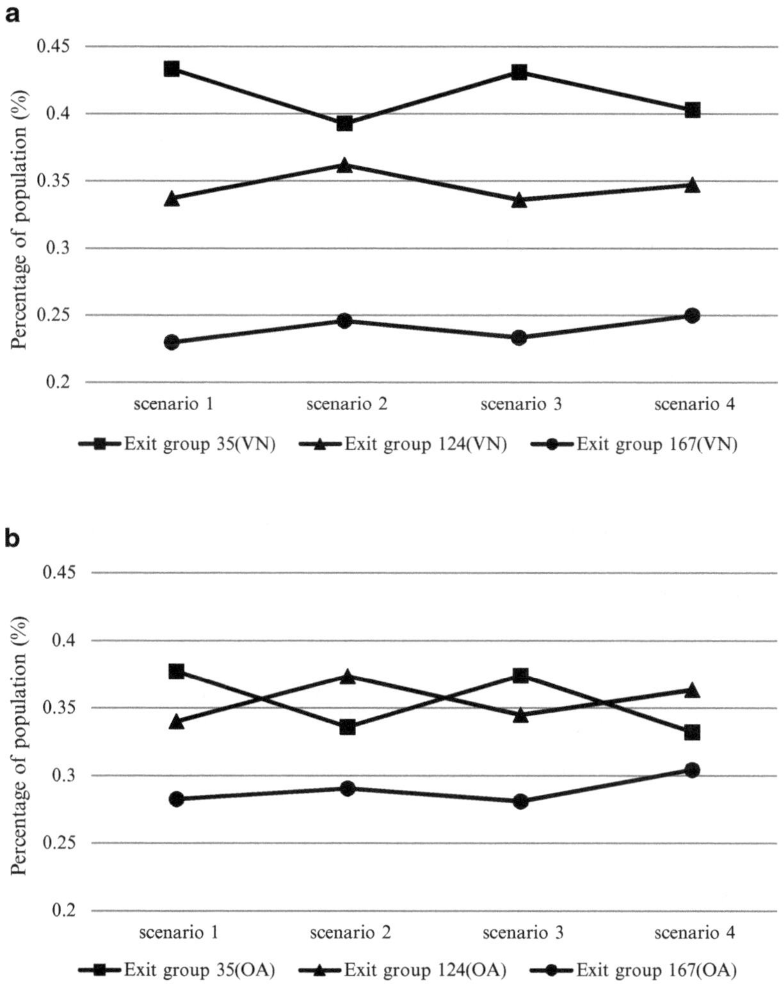

Fig. 4 Percentage of population through each exit under different scenarios in two approaches. (**a**) Virtual node approach (VN). (**b**) Optimal allocation approach (OA)

the simulation program verifies it. Thus, the process of evacuation is prolonged. The optimal allocation approach makes some changes to the choice of exit in order to balance the population among exits and improve *TET*. It can be demonstrated from Fig. 4, population in exit group 35 has been allocated to exit group 167 by optimal allocation approach.

TET is an important indicator to evaluate an approach. And more evacuees are evacuated within a short time period is also an objective of evacuation. In summary, *TET* is relevant to the route length which is due to the complexity of road network and the distribution of population has high impact on it as well. The less gap of population among exits, the less total evacuation time will cost.

5 Conclusion

This paper introduces two approaches to expand the staged evacuation algorithm (SEA) to an evacuation with multiple exits. And the efficiency of them, i.e., virtual node approach and optimal allocation approach is compared under different scenarios with the road network of Jing' an district, Shanghai. The original SEA can organize evacuees into different evacuation groups and transport them to a safe exit based on predefined schedule and route without suffering traffic congestions. In order to make it applicable to multiple exits, a virtual node is added into the network to change the topology in the virtual node approach. And the multiple exit evacuation can be treated as multiple single exit evacuation in optimal allocation approach. A simulation program is also realized to visualize the evacuation process. Results illustrate that the distribution of population has high impact on the *TET* and the optimal allocation approach can allow more people to evacuate in the same time period.

Acknowledgement This paper is sponsored by the National High Technology Research and Development Program of China (No. 2013AA122302), Innovation Research Grant of East China Normal University (No. 239201278220056), Shanghai Natural Science Foundation (No. 11ZR1410100), and National Natural Science Foundation of China (No. 41271441).

References

1. Yamada T (1996) A network flow approach to a city emergency evacuation planning. Int J Syst Sci 27(10):931–936
2. Kamiyama N, Katoh N, Takizawa A (2006) An efficient algorithm for evacuation problems in dynamic network flows with uniform arc capacity. In: Algorithmic aspects in information and management. Springer, Berlin, pp 231–242
3. Lim GJ, Zangeneh S, Reza Baharnemati M, Assavapokee T (2012) A capacitated network flow optimization approach for short notice evacuation planning. Eur J Oper Res 223(1):234–245
4. Campos V, Bandeira R, Bandeira A (2012) A method for evacuation route planning in disaster situations. Procedia Soc Behav Sci 54:503–512
5. Bretschneider S, Kimms A (2011) A basic mathematical model for evacuation problems in urban areas. Transp Res A Policy Pract 45(6):523–539
6. Zheng YF, Li X (2009) Approach to planning for an organized evacuation. J East China Norm Univ (Nat Sci) 2009(5):45–52 (in Chinese)
7. Li Q, Fang Z, Li Q, Zong X (2010) Multiobjective evacuation route assignment model based on genetic algorithm. In: Geoinformatics, 2010 18th international conference on 2010. Beijing, China, pp 1–5
8. Chen X (2003) Agent-based simulation of evacuation strategies under different road network structures. University Consortium of Geographic Information Science. http://www.

researchgate.net/publication/237510346_Agent-Based_Simulation_of_Evacuation_Strategies_under_Different_Road_Network_Structures
9. Chiu YC (2004) Traffic scheduling simulation and assignment for area-wide evacuation. In: Proceedings of the 7th international IEEE conference on intelligent transportation systems. Sophia Antipolis, France, pp 537–542
10. Liu Y, Lai X, Chang G (2006) Cell-based network optimization model for staged evacuation planning under emergencies. Transp Res Rec J Transp Res Board 1964(1):127–135
11. Li X, Huang B, Liu Z, Zhang X, Sun J (2012) A novel method for planning a staged evacuation. J Syst Sci Complex 25(6):1093–1107
12. Sheffi Y (1985) Urban transportation networks: equilibrium analysis with mathematical programming methods. Prentice-Hall, Englewood Cliffs
13. Dijkstra EW (1959) A note on two problems in connexion with graphs. Numer Math 1:269–271
14. Han LD, Yuan F, Chin S-M, Hwang H (2006) Global optimization of emergency evacuation assignments. Interfaces 36(6):502–513

Study of Urban Patterns Optimization Employing CFD Method: A Case Study of Chenjiazhen Experimental Ecological Community, Chongming, Shanghai

Chao Liu, Peng Xu, Weizhen Chen, Liang Zhang, and Weilin Li

1 Urban Patterns and Wind Environment Introduction

Urban micro-environment mainly includes wind environment, light environment, thermal environment, sound environment, environmental pollutants and so on. In this series of micro-environment, the wind environment is greatest influenced by urban planning and design, which is closely related to the city building layout, physical characteristics, spatial relationships, building envelope production, the choice of related technology, personnel comfort, and energy use [1].

Urban patterns defined in this paper have two layers of meaning: how buildings with different function distribute, and how buildings with different height distribute. Urban patterns will affect the climate situation within the city blocks, and form the so-called "micro-climate" – building patterns will change the flow of the wind within the city blocks; building shield phenomenon existed affects the distribution of solar radiation in the community.

Wind environmental impact on the city's energy consumption is significant: in summer, the small outdoor wind speed is not conducive to a natural cooling and will increase the probability use of air conditioning and refrigeration, and the large outdoor wind in winter will result in the increase of building exterior surface heat

C. Liu
Shanghai Tongji Urban Planning & Design Institute (TJUPDI), Shanghai 200092, China

College of Design, Construction and Planning, University of Florida, Gainesville, FL 32611-5701, USA
e-mail: liuchao1020.pku@gmail.com

P. Xu (✉) • L. Zhang • W. Li
College of Mechanical Engineering and Energy, Tongji University, Shanghai 200092, China
e-mail: xupeng@tongji.edu.cn

W. Chen
College of Architecture and Urban Planning, Tongji University, Shanghai 200092, China

X. Chen, Q. Pan (eds.), *Building Resilient Cities in China: The Nexus between Planning and Science*, GeoJournal Library 113, DOI 10.1007/978-3-319-14145-9_25

dissipation and heat penetration, which will lead to an increase in heating load. Therefore, urban patterns affects the wind environment, thereby affects residents' life quality, energy consumption and greenhouse gas emissions. How to assess and optimize the urban patterns from the perspective of low-carbon and energy saving under the premise of ensure or even improve the wind environment comfort is the problem concerned and will be solved in this paper.

2 Urban Wind Environment Research and Application Methods Introduction

2.1 *Urban Wind Environment Evaluation Methods*

Currently, there are three main methods for the evaluation and optimization of wind environment: simulation test, field test and numerical simulation methods [2]. Simulation test is based on the principle of relativity, and the model is fixed on the artificial environment ground, and stream and other external conditions is man-made which used to simulate real process of various phenomena; Field testing is testing and researching of field parameters by distributing limited measuring points or flow measuring points in the field; Numerical simulation is using computer as a means to achieve study purposes of engineering problems and physical problems and various issues of nature by numerical calculation and image display methods. Advantages and disadvantages of these methods are shown in Table 1.

In the beginning of urban planning and design, carrying out short cycle, low price and controllable numerical simulation is significant for improving the city wind environment and rational planning for buildings.

2.2 *International and Domestic Urban Wind Environment Overview*

The research of single building wind environment is focused on surrounding air flow and wind pressure distribution, while the study on the wind environment of city and community is complexity and particularity, besides the content of single building. Ventilation will change for the morphological characteristics of building community like blocking, channels and etc. If urban planning and design can fully and reasonably make use of these characteristics, through the reasonable layout of buildings especially in high-rise buildings, the discomfort caused by high wind speed in some areas will reduce. Making use of natural ventilation could reach energy-saving purpose without add extra money or technology.

Natural ventilation can improve energy saving of the whole community. By the mechanism of the natural ventilation formation, under the premise that achieve the

Table 1 Method's of ventilation environment analysis and their pros & cons

Method	Advantages	Disadvantages
Simulation test	Relatively accurate, reproducible and can be considered to set the boundary conditions	High cost, long cycle, the test required to satisfy the principle of relativity
Field testing	Accurate, reliable	High cost, long cycle, laborious
Numerical simulation	Low cost, short cycle, controllable	Accuracy is relatively poor, need to verify

same cooling purpose, compared with the conventional mechanical ventilation in air conditioning, natural ventilation is a free technology, without any energy consumption. The research results of literature show that compared to the building use the mechanical ventilation in air conditioning the building use natural ventilation can save cooling energy about 14–41 kWh every square each year, and cost saving about \$1.3–\$3.6 [3]. IEA 2000 summary report indicated that in European countries, the offices using natural ventilation almost have 50 % reduction in building energy consumption [4]. Thus, the natural ventilation is very energy-efficient.

There are a lot of numerical simulation softwares, mainly for the single building, such as PHOENICS, ANYSYS-CFX, FLUENT, STAR-CD, ICEM. Recently, some scholars have been studied urban wind environment, especially the wind environment system by using numerical modeling method. 1996, ZHANG Y Q, Arya S P and Snyder W H used the k-epsilon turbulence model for modeling flow and dispersion of a single cube model, they used the Froude number indicates stratification (turbulence) levels of the flow [5]. 1998, HASSAN A A and Crowther J M used the k-epsilon turbulence model to simulate a two-dimensional street canyon flow and diffusion. Simulation results showed significant vortex flow in street canyon and the highest concentration of pollutants was in the east wall lower edge of upwind buildings, the lowest concentration of pollutants was in the west wall upper edge of downwind buildings [6]. In 2001, Shaviv E, Yezioro A and Capeluto Isaac G. used software SustArc to simulated urban light environment and used FLUENT to simulated urban wind environment, and obtained the maximum available space to ensure good lighting and adaptive scheme to good ventilation condition [7]. Zhou Li, Xi Guang simulated High-rise building wind farms and wind pressure and obtained flow field and wind pressure distribution of irregularly shaped single building under different wind directions, and simulated three in-line high-rise buildings, the results were compared with the monomer building [8]. Cheng-Hsin Chang and Robert N. Meroney compared Fluent simulation results with wind tunnel experimental results and found that Fluent simulation results and observations of street canyons wind and pollutant concentration field simulated by wind tunnel experiment agreed well [9]. 2003, Capeluto L G, Yezioro A and Shaviv E used Fluent to evaluate the rationality of a proposed downtown design and have achieved good results [10]. 2004, Jae-Jin Kim and Jong-Jin Baik used the RNG k-epsilon turbulence model to simulate air flow and diffusion phenomena of the inner city blocks underlying surface represented by Cube matrix group, RNG k-epsilon scheme is developed on the standard k-epsilon program, which make up some deficiencies of the standard k-epsilon closure plan [11]. Lu Ping, Yuan Jiuyi and Zhang Wenyu used the numerical model to study the impact of street canyon aspect ratio and symmetry of the building on both sides on vehicle emissions and dispersal of pollutants in the street canyon [12].2005, Wang Fei and Xiao Yongquan used commercial software PHOENICS for some actual complex plot wind field simulation, the result showed PHOENICS can realize the simulation of building wind environment, and it has a guiding significance of the optimization design of building wind environment [13]. 2006, Cocea l O, Thomas T G, Castro I P and Belcher S E carried out direct numerical simulation (DNS) study of flow within the urban surface which representation with matrix group [14]. Tseng Y H, Meneveau C and Parlange M B carried out large-eddy simulation of flow and pollutant dispersion within the real urban surface. They believed bluff-body distributed 6–8 grid points and the sub-grid associated Lagrangian dynamic model is the required minimum mesh size and the best sub-grid model of large eddy simulation for flow and pollutant dispersion within real urban surface [15]. Zhang Ning and Jiang Weime used large eddy simulation to simulate airflow field and pollutant dispersion around on a single building [16]. 2007, Piotr K. Smolarkiewicz, Robert Sharman, Jeffrey Weil, Steven G. Perry, David Heist et al. used large eddy simulations to simulate the underlying surface of a pentagonal complex models. Draw a conclusion that immersed boundary method is more effective of, but fitted coordinate transformation method is more accurate than immersed boundary method, and both methods are not limited to flat land surface, but also for complex underlying surface [17]. Ma Jian, Chen Guobiao and Mao Yalang used FLUENT software and numerical methods based on the Reynolds averaged NS equations and the RNG k-epsilon turbulence model to simulate the wind environment surrounding blocks of buildings consisting of square cross-section single-block building and rectangular cross-section multi-building [18]. 2008, Xie Z T, Coceal O and Castro I P. carried out large eddy simulation of flow within urban surface which represented with randomly distributed matrix. They compared the average velocity distribution within the underlying surface, Reynolds stress distribution, turbulent kinetic energy distribution of the results with uniform distribution of matrix groups results studied by previous large-eddy simulation and direct numerical simulation and found that although there is a certain similarity in the statistical amount of average space, but exist significant differences in a turbulent statistic [19]. 2009, Zheng-Tong Xie, and Ian P. Castro carried out large eddy simulation of flow and point sources diffusion within underlying surface in a London central region. And believed that the use of 1 m small-scale spatial grid and 1 s time step can make accurate simulation of the real urban surface [20]. Shi Ruifeng, Cui Guixiang, Wang Zhishi, Xu Chunxiao and Zhang Zhaoshun used Large-eddy simulation method to study Macao Netherlands Park District's wind field and the distribution of traffic pollutant concentration. Using a method of combining distributed resistance element with immersed boundary, and using a Lagrangian dynamic sub-grid model to closed equations [21]. In 2010, Bao Yi and Luo Kun of Zhejiang university used Renault average method to simulate the flow wind environment in a residential area in Hangzhou [22]. Xie Yi and Ge Wenlan simulated air flow field of building internal and external environment quickly and easily through combination of BIM (Building Information Modeling) and CFD techniques, and analysis and evaluation of the fluid environment of the building internal and external environment of air flow [23]. 2011, Xie Yi took BIM three-dimensional model for the platform, used ECOTECT software light environment and the FDS developed by American Standards and Technology for CFD simulations [24]. 2012, Li Meihua, Xia Haishan and Li Xiaobei introduced the BIM technology applications of wind environment simulation in urban planning from the designers perspective [25]. 2013, Shang Tao and Qian Yi used the RNG model of Airpak software simulated and evaluated Winter and summer wind environment of Wuhan University tea port district on the basis of Wuhan area's natural climatic conditions. They used this simulation to guide local residential district planning and program decisions from a well-constructed wind environmental perspective [26].

In summary, the mainly wind environment simulation models for city are k-epsilon model, RNG k-epsilon model, direct numerical simulation (DNS) and large eddy simulation. Although there are many scholars attempt to simulate the wind environment of cities, but the use of CFD in urban planning is still in its infancy, and data of wind speed, temperature and the surface pressure of building surface under the large area fine model simulation is completely feasible. Researches generally use the FLUENT software, and its disadvantage is that using ordinary PC computer needs a long time, and working implementing procedures aren't integrated and normative, however, for the non-research project, using ECOTECT to simulate light environment and Airpak or PHOENICS to simulate wind environment can meet the requirements.

3 Case Study

3.1 Chenjiazhen Experimental Ecological Community Introduction

Chenjiazhen Experimental Ecological Community is located in Shanghai's Chongming Island, close to the East China Sea, has a subtropical monsoon climate, shown in Figs. 1 and 2. Chongming Island with the goal of building a "world-class eco-island" and low-carbon model city, which is a strong support background of low carbon development of Chenjiazhen Experimental Ecological Community. The Master Plan of Chenjiazhen Experimental Ecological Community was developed in 2004, detailed planning enacted in 2011 and now the very young community is in full construction period. Chenjiazhen Experimental Ecological Community with planning area of 406.43 ha, planned population of three million people is expected to be completed the goal of new town construction by 2020. It is hot in summer and

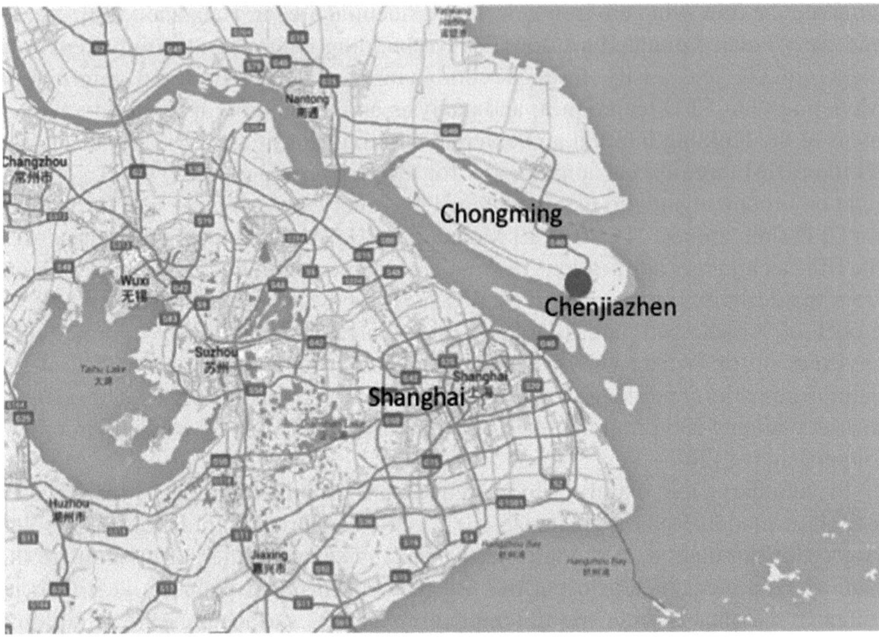

Fig. 1 Location of Chenjiazhen Experimental Ecological Community

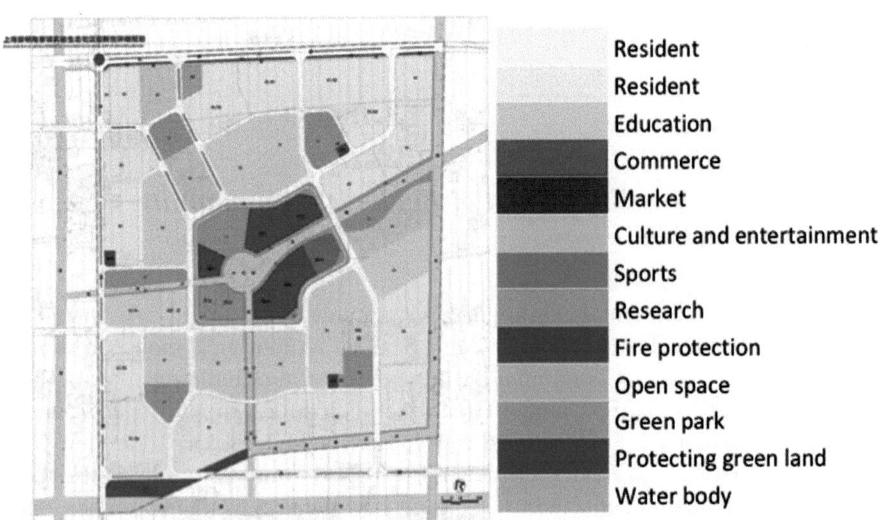

Fig. 2 Land use map of Chenjiazhen Experimental Ecological Community detailed regulatory plan

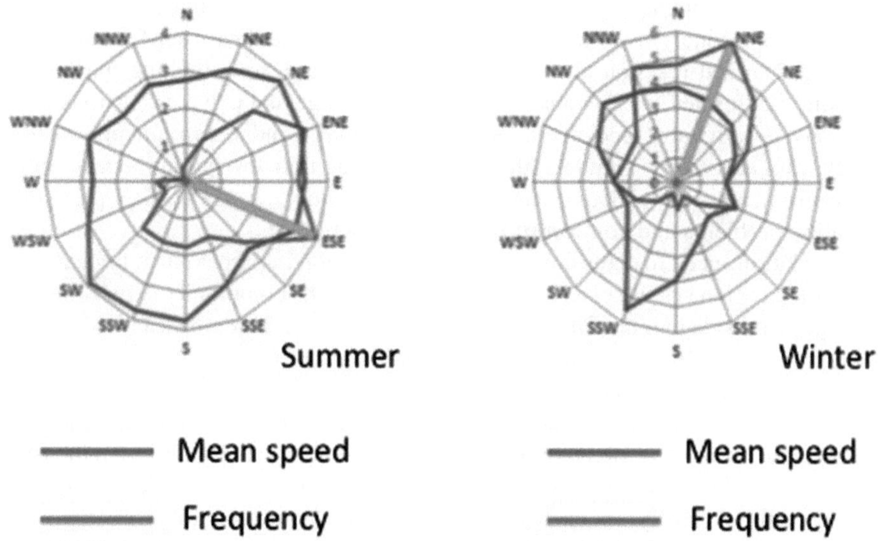

Fig. 3 Wind rose diagram in Shanghai area: summer and winter

cold in winter, the average temperature is between 13 °C and 20 °C, the average annual precipitation is generally 800 mm ~ 1,600 mm.

The planning and construction site of Chenjiazhen Experimental Ecological Community is 372.77 ha, the total construction area is 1,616,000 square meters, and construction land per capita is 124.26 m^2/person. In the planning and construction land, residential land is 221.59 ha, accounting for 59.44 %, land for public facilities is 25.36 ha, accounting for 6.8 %, roads and squares land is 58.70 ha, accounting for 15.75 %, municipal utilities accounting for 0.13 % is 0.47 ha, and green-land is 66.65 ha, accounting for 17.88 %. In addition, the water area is 33.66 ha and detailed planning of land use plan is shown in Fig. 3.

The living spaces of Chenjiazhen Experimental Ecological Community include two types of residential land: type one class residential land and type two. The residential density was arranged divided into two-circle in accordance with "inside high outside low", volume rate were controlled at 0.3–0.6 and 0.7–1.0 range.

3.2 Chenjiazhen Experimental Ecological Community Wind Environment Simulation Analysis Method

This study considered CFD as the method to analysis wind field of community and made recommendations for optimizing community morphology according to the analysis results. Use Airpak3.0 and FLUENT 12.0 to simulate the communities built environment. FLUENT is a computational fluid dynamics (CFD) commercial software to simulate the complex flow from incompressible to highly compressible. FLUENT are widely used in conversion and turbulence, heat transfer and phase change, chemical reaction and combustion, multiphase flow, rotating machinery, dynamic / deformed mesh, noise, material processing, fuel cells, etc. Airpak3.0 is an analysis software specialized oriented artificial wind environment systems, which can accurately simulate the air flow, heat transfer, pollution and other physical phenomena, and can accurately simulate air flow, air quality, pass heat, pollution problems and comfort of the ventilation system.

Simulations need to determine the simulation region in Airpak3.0 at first, and the simulation region size is set as follows: the length of the community is four times as the windward side length of the community, the width is three times as the width of the community, and the height is 2.5 times as height of the community. Secondly, import geometric models of buildings and trees within district into the analog area with CFD. Tree simplified rectangular geometry combinations. Use the grid generator in Airpak3.0 to generate grid, and the grid number was 16 million in the form of unstructured meshing hexahedral meshes. Mesh quality meet the Airpak internal quality testing. Thirdly, set each grid boundaries of the analog area.

The construction of the border is set to Block, the ground is set to Wall, the windward side of simulation area is set to air boundary, atmospheric boundary set to the default parameters, wind speed in winter and summer is 3.5 m/s and 3.4 m/s, respectively, and wind direction is North to east 15° and east to South 15° respectively. The boundary of the wind condition is set to the pressure outlet, and the remaining three surfaces are set to Symmetry. Using RNG equations for solution, and considering air density and gravity factor, we set the default settings for processing of solution. After 541 iterations, the equation convergence and get the velocity field of entire simulation space. Intercept the velocity distribution in 2 m for comparison.

The wind farm direction of outside wind according to the meteorological files (CSWD file) northeast wind is identified as the prevailing wind in winter, and the dominant wind is southwest wind in summer. Summer and winter wind roses shown in Fig. 3.

3.3 Chenjiazhen Experimental Ecological Community Wind Environment Simulation Results

This community winter environment issue is relatively clear, as shown in Figs. 4 and 5. The main problem is from the community landscape corridor is the same direction with Northwest winter wind, no external building block resulting that the inlet winter wind velocity of region community A is large, more than 3 m/s. District B is a neighborhood park, so a larger wind is unsuitable for residents rest. The reason is the winter winds return here after blow the around building and form vortex and turbulence here. Area C and Area A problem are similar which belong

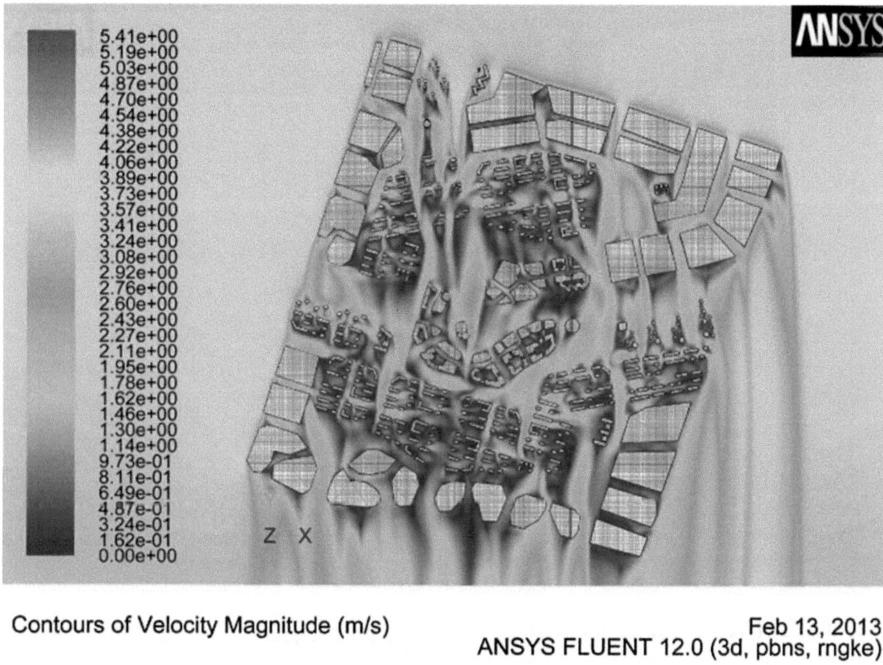

Fig. 4 Community ventilation speed distribution (two meters high) in winter (northeast monsoon)

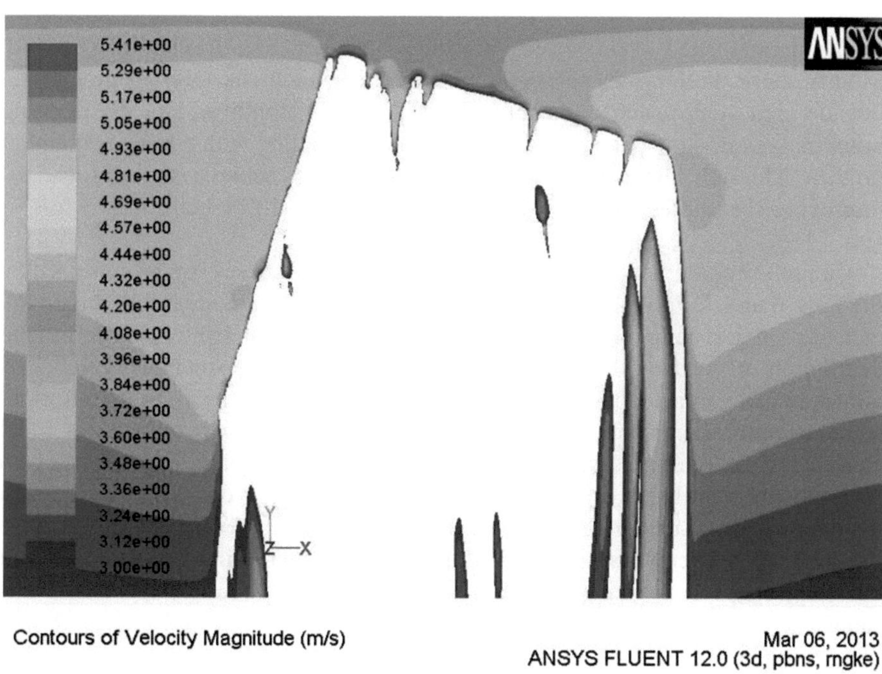

Fig. 5 Community ventilation speed greater than 3 m/s distribution (two meters high) in winter (northeast monsoon)

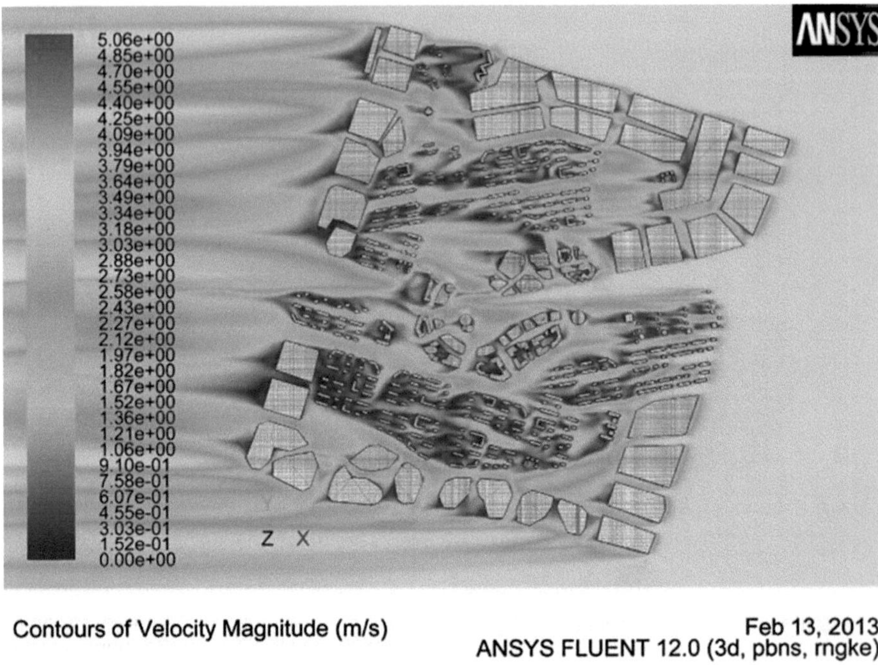

Fig. 6 Community ventilation speed distribution (two meters high) in summer (southeast monsoon)

to the unobstructed open space as winter air inlet, and formed large wind. Modify these three issues respectively: planting windbreak trees in green area outside area A, locating compact low-rise villas in south of area B, planting arbor in Area B, and open space of area C function calling with nearby residential groups. Through these three forms fine-tuning to achieve the purpose of improving the winter winds environment, the speed will be control at 3 m/s or less.

Community summer wind environmental problems' causes can be analyzed by the Figs. 6 and 7. Public buildings ventilation conditions are ideal, but due to the lack of southeast summer wind gallery and the residential spacing is too dense, the ventilation of residential areas is poor. In particular, the residential region E is almost in calm area and residents have a poor sense of summer wind. In addition, summer ventilation conditions of public buildings in district D are not ideal caused by the buildings shelter in east and south. Therefore, the strategy to improve the ventilation in summer, on the one hand to open up and expand community airy corridor in southeast direction, on the other hand is to increase the distance between residential building which is increasing the height of the building to increase spacing in the premise of the residential group land area is same.

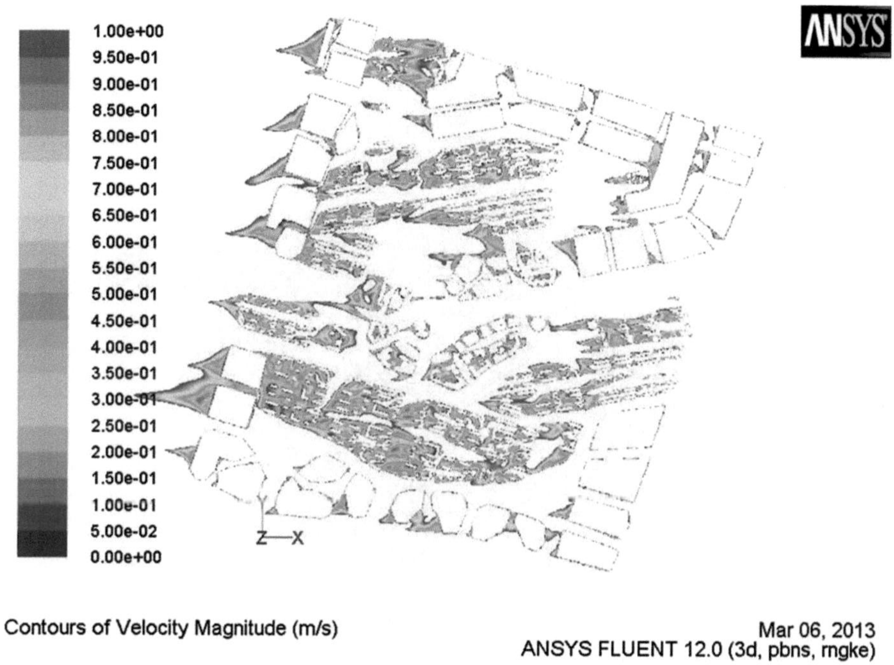

Fig. 7 Community ventilation speed distribution less than 3 m/s distribution (two meters high) in summer (southeast monsoon)

3.4 Optimize Community Patterns Based on Wind Environmental Analysis

According to the wind and the light environment simulation results we can obtain the following shape optimization principles of community, and optimization results shown in Figs. 8 and 9.

1. The main channel of community consistent with summer dominant wind direction as far as possible, to ensure that all areas exist air flow area which is parallel or nearly parallel to the dominant wind direction, and avoid channels parallel with winter dominant wind direction as far as possible.
2. Setting up space in the community summer dominant frontal boundary, in order to facilitate the outside air come into the community; avoiding the existing of the channel which is parallel to the prevailing wind in winter.
3. Not to block each other in the height set of the building. The general recommendation is tall buildings within the community and relatively little building is located at the border.

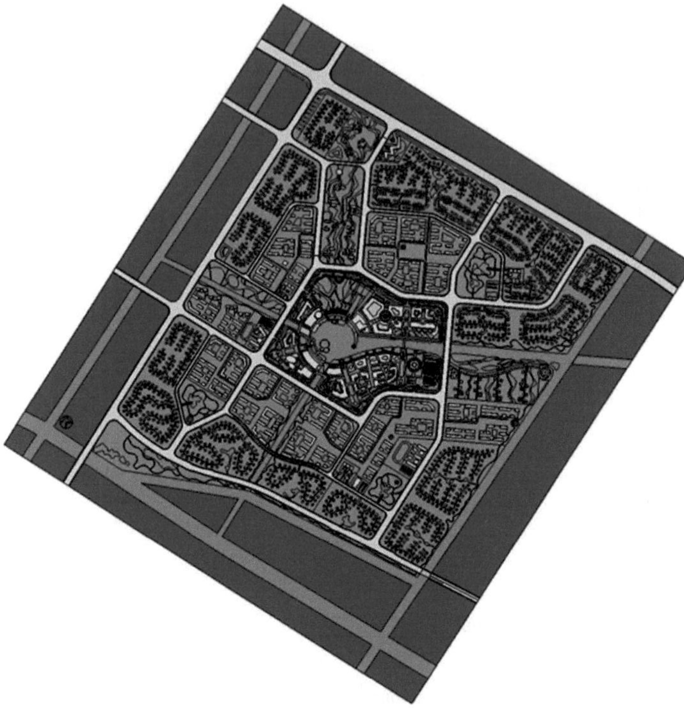

Fig. 8 The original community special pattern plan

In order to visual display the optimizing region and methods in the above images, Fig. 10 is a space detail optimization effect diagram of the community

3.5 Energy Savings Analysis of Morphological Optimization

The morphological optimization in this paper will achieve the purpose of cooling energy savings by improving the natural ventilation. More and more engineering practice proved that the use of natural ventilation can improve building tenants comfort while reducing cooling energy consumption. Winter and summer speed before and after optimization in two meters section shown in Table 2.

Energy savings benchmark for the district is considered as the buildings all utilize electric refrigeration for cooling in summer (cooling season considered as June 1 to September 30). For the optimization model, the CFD simulation results can be seen that the optimized dominant wind direction and the wind speed in the

Fig. 9 The optimized community special pattern plan

summer around construction can meet the requirements (greater than 2 m/s). To simplify the calculations, it is assumed that when the outdoor wind speed greater than or equal to the prevailing wind (3.5 m/s), the wind speed around the building can meet the demand of natural ventilation (2 m/s).

Natural ventilation saving rate calculation can be obtained as follows:

$$\Delta = \frac{T_2}{T_1} \tag{1}$$

Where, T1 is number of hours for the cooling season, T2 is the number of hours to meet natural ventilation cooling season speed, temperature and humidity.

According to the Shanghai weather file (CSWD) to complete the T1 and T2 statistics. The result is T1 = 2,928, T2 = 148 and the final energy saving rate is 5 %, the main reason of energy efficiency is not impressive is that outdoor humidity of Shanghai is excessive in summer.

Community northwest perimeter planting windbreaks

Community development space in northern planting high-forest

Fig. 10 Detailed community pattern optimization schematic diagrams

4 Conclusion

Through the above analysis, we obtain the following conclusions:

1. Wind environment simulation on community-scale can provide an important reference for the optimization of urban patterns.
2. To consider natural ventilation in the design phase of urban patterns can achieve passive energy conservation.

However, we can notice that:

1. In the simulation, using certain assumptions (such as the approximate simulation of trees) to simplify the calculation, which may have some impacts on the

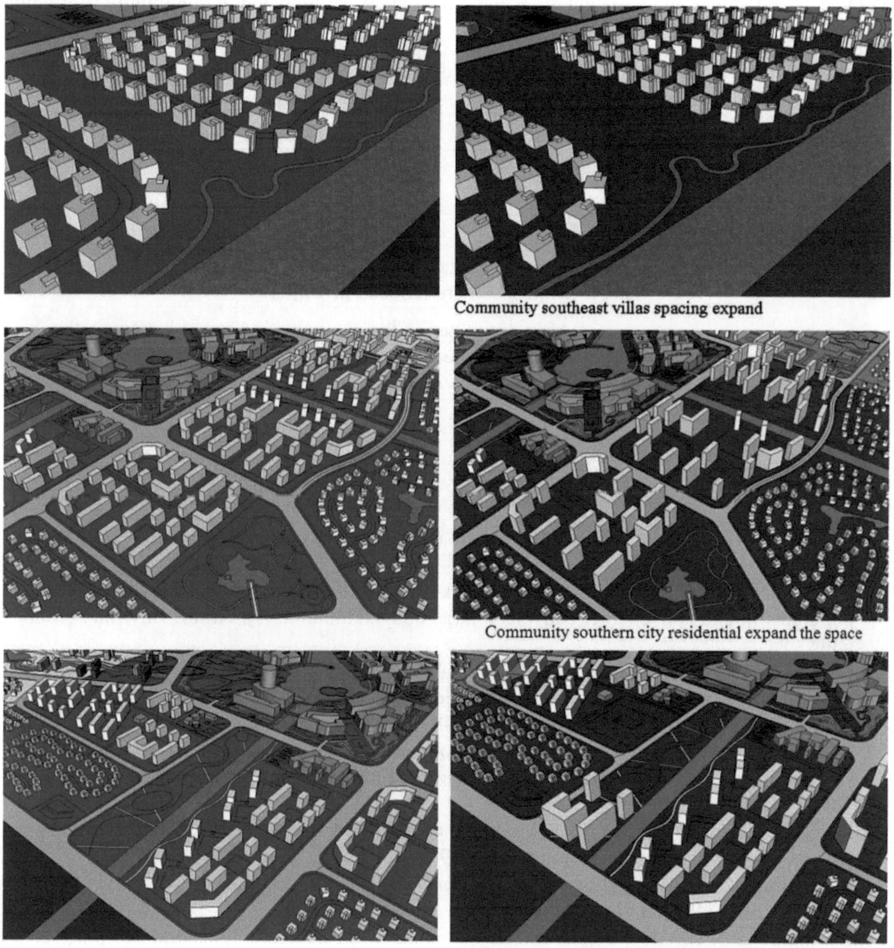

Community southeast villas spacing expand

Community southern city residential expand the space

Exchange west entrance to the park and the internal community residential location, and increased floor space

Fig. 10 (continued)

accuracy of the simulation results. So it needs to consider the relationship between the simplified calculation and accuracy of simulation in future studies.

2. Other environmental factors (outdoor temperature and humidity, especially humidity) also have influence on the energy-saving effect of natural ventilation, therefore it is required to comprehensive consideration when discuss the applicability of natural ventilation.

In the context of low-carbon and energy saving, this paper has practical significance in attempts of passive design on urban planning areas. From an academic point, this paper should be regarded as an initiate. Wish to more and more researchers can actively explore the potential applications of this field, and struggle for the ecological and energy efficient city goals.

Table 2 Facet values comparison before and after pattern optimization of community ventilation environment in winter and summer

Tangent plane at two meters	Area weighted average speed	Maximum speed	Minimum speed
Winter before optimization	1.786	5.622	0.005
Winter after optimization	1.606	4.701	0.011
Summer before optimization	2.057	7.235	0.011
Summer after optimization	2.094	11.244	0.014
Comparison results	After optimization, the average wind speed and maximum wind speed are reduced in winter		
	After optimization the average wind speed increases, the maximum speed increases, the minimum speed increases in summer		

References

1. Yi B (2010) The numerical simulation of urban district microenvironment [D]. Zhejiang University, Hangzhou
2. Baker J, Walker HL, Xiaoming Cai (2004) A study of the dispersion and transport of reactive pollutants in and above street canyons-a large eddy simulation [J]. Atmos Environ 38:6883–6892
3. Shaolin P (1998) Global change and sustainable development [J]. J Ecol 17(2):32–37
4. Howard L (1833) Climate of London deduced from metrological observations [M], vol 1, 3rd edn. Harvey and Dorton Press, London, p 348
5. Zhang YQ, Arya SP, Snyder WH (1996) A comparison of numerical and physical modeling of stable atmospheric flow and dispersion around a cubical building [J]. Atmos Environ 30:1327–1345
6. Hassan AA, Crowther JM (1998) Modeling of fluid flow and pollutant dispersion in a street canyon [J]. Environ Monit Assess 52:281–297
7. Shaviv E, Yezioro A, Capeluto IG (2001) Sun and winds in a new business district in Tel Aviv [J]. Build Simul 2001:151–158
8. Li Z, Guang X (2001) Numerical analysis of the wind field on high buildings [J]. J Xi'an Jiaotong Univ 5:471–474
9. Chang C-H, Meroney RN (2001) Numerical and physical modeling of bluff body flow and dispersion in urban street canyons. J Wind Eng Ind Aerodyn 89(14215):1325–1334
10. Capeluto Isaac G, Yezioro A, Shaviv E (2003) Climatic aspects in urban design – a case study. Build Environ 38:827–835
11. Kim J-J, Baik J-J (2004) A numerical study of the effects of ambient wind direction on flow and dispersion in urban street canyons using the RNG k-εturbulence model [J]. Atmos Environ 38:3039–3048
12. Ping L, Jiuyi Y, Wenyu Z (2004) Numerical simulation of the effect of street canyon structure on wind and pollutant concentration fields inside street canyon [J]. Plateau Meteorol 04:534–539
13. Fei W, Yong-quan X (2005) Simulation and evaluation of building cluster's wind environment with PHOENICS software [J]. J Shandong Univ Archit Eng 05:39–42
14. Coceal O, Thomas TG, Castro IP, Belcher SE (2006) Mean flow and turbulence statistics over groups of urban-like cubical obstacles [J]. Bound Lay Meteorol 2006(121):491–519
15. Tseng YH, Meneveau C, Parlange MB (2006) Modeling flow around bluff bodies and predict in grub and dispersion using large eddy simulation [J]. Environ Sci Technol 40:2653–2662

16. Zhang Ning, Jiang Wei-me (2006) A large eddy simulation on the effect of building on atmospheric pollutant dispersion [J]. Chin J Atmos Sci 2:212–220
17. Smolarkiewicz PK, Sharman R, Weil J, Perry SG, Heist D, Bowker G (2007) Building resolving large-eddy simulation and comparison with wind tunnel experiment [J]. J Comp Phys 227:633–653
18. Ma Jian, Chen Guo-biao, Mao Ya-lang (2007) Study of wind environment around building complex based on CFD [J]. J Zhejiang Univ Technol 03:351–354
19. Xie ZT, Coceal O, Castro IP (2008) Large-eddy simulation of flow sover random urban-like obstacles [J]. Bound-Lay Meteorol 129:1–23
20. Zheng-Tong Xie, Castro IP (2009) Large-eddy simulation for flow and dispersion in urban streets [J]. Atmos Environ 43:2174–2185
21. Shi Ruifeng, Cui Guixiang, Wang Zhishi, Xu Chunxiao, Zhang Zhaoshun (2009) The large eddy simulation of air flow and traffic pollution in urban residential [J]. Prog Nat Sci 02:212–221
22. Yi B, Kun L (2010) Numerical simulation of micro-environmental flow around city resident sub-districts [J]. Energy Eng 5:47–53
23. Xie Yi, Ge Wenlan (2010) BIM-based micro-environmental simulation and evaluation in urban planning [J]. J Inf Technol Civil Eng Archit 3:51–57
24. Xie Yi (2011) BIM based micro-environment simulation in Beijing road urban planning [J]. J Inf Technol Civil Eng Archit 2:63–68
25. Li Meihua, Xia Haishan, Li Xiaobei (2012) The application of BIM technology in urban planning microenvironment simulation [A]. Urban Planning Society of China. Diversity and inclusiveness – 2012 China urban planning annual meeting proceedings (14. Urban planning techniques and methods) [C]. Urban Planning Society of China, p 10
26. Shang Tao, Qian Yi (2013) Wind environment simulation and evaluation of the residential district in Wuhan: a case study of Chagang district in Wuhan University [J]. Huazhong Archit 01:48–51

Assessing Public-Sector Transportation Finance and Planning in Urban China

Jiawen Yang and Chuanglin Fang

1 Introduction

As happened in many other developing countries, transportation in China's mega-cities and economic advanced regions is badly congested. How to supply the much needed mobility and accessibility through transportation investment and planning is a significant challenge. It is worth noting that China's urban transportation problems cannot be simply attributed to under-investment. By the end of 2007, China had 54,000 km limited access highway, ranked No. 2 in the world, only after the United States [1]. During the recent global economic crisis, China's central government budgeted four trillion Chinese Yuan (one USA dollar equals about 6.5 Chinese Yuan) to stimulate growth, over one third of which is for transportation infrastructure.

Literature has consistently suggested that the growth of travel demand in the fast growing economy could lead to significant mobility and accessibility barriers if transportation institutions and planning process cannot be reformed timely [2]. The need to reform China's public-sector transportation is particularly obvious as the market-oriented reform shifts the pattern of urban growth, increases motorized transportation and leads to demand for transportation infrastructure and services that cannot be timely supplied in the pre-reform institutional context [3].

China, however, is indeed making significant changes. In 2008, China announced two significant decisions regarding surface transportation finance and

J. Yang (✉)
School of Urban Planning and Design, Shenzhen Graduate School, Peking University, Shenzhen 518055, China
e-mail: yangjw@pkusz.edu.cn

C. Fang
Institute of Geography and Natural Resources, China Academy of Science, Beijing 100000, China
e-mail: fangcl@igsnrr.ac.cn

X. Chen, Q. Pan (eds.), *Building Resilient Cities in China: The Nexus between Planning and Science*, GeoJournal Library 113, DOI 10.1007/978-3-319-14145-9_26

planning. First, responsibilities of planning and management for urban transportation would be transferred to the Administration of Communication from the Administration of Housing and Urban-rural Development. Second, a national fuel tax was announced in December 2008 and became effective on January 1st, 2009. These two inter-related reforms signalize a significant deviation from China's pre-existing path of transportation finance and planning.

China's fiscal reforms have been systematically reported, particularly how its municipal finance can be better structured in order to sustain urban development [4, 25]. However, how China's institutions and finance for urban transportation planning and investment have evolved has not been systematically studied. This article aims to fill this literature gap by providing an overview of public-sector transportation finance and planning reform that is relevant to today's urban China. Planning institutions are examined together with the associated finance mechanism for a better understanding of transportation investment outcomes in the public sector. The article particularly explains how economic development and urbanization has shaped the demand for road-based transportation mobility and thus driven the reform of transportation finance and planning institutions. As urbanization and motorization are common challenges faced by many developing countries, China's experience assessed in this article can shed light for other countries.

2 Transportation Institutions in the Planning Economy

Public-sector transportation in today's urban China should be examined in the large context of national institutions for transportation, particularly those inherited from the planning economy. A brief overview of transportation finance and planning in the period of the planning economy can help understand what lead to today's reforms, particularly the regional transportation problems after fiscal decentralization.

Before China decided to open its door to the outside world around 1980, it had maintained a highly planned economy. China's fiscal system, correspondingly, was quite centralized. Local governments collected various revenues and submit them to the higher level governments. They also report a regulated annual budget to the higher level government and received funds for local projects of economic and social development. Very few surpluses could be retained on the provincial or lower level to reward good government performance. Local governments acted like a local executive agent of higher level governments.

Responsibilities of transportation development and planning under this fiscal system featured a top-down structure plus a city-region division. Figure 1 shows the overall administration hierarchy from the national level to the city level. National agencies that oversee transportation development include the Ministry of Communication (MOCM), the Ministry of Railway (MOR), and the Civil Aviation Administration (CAA), and the Ministry of Construction (MOC). These four agencies have different statutory responsibilities, with MOR overseeing national railway,

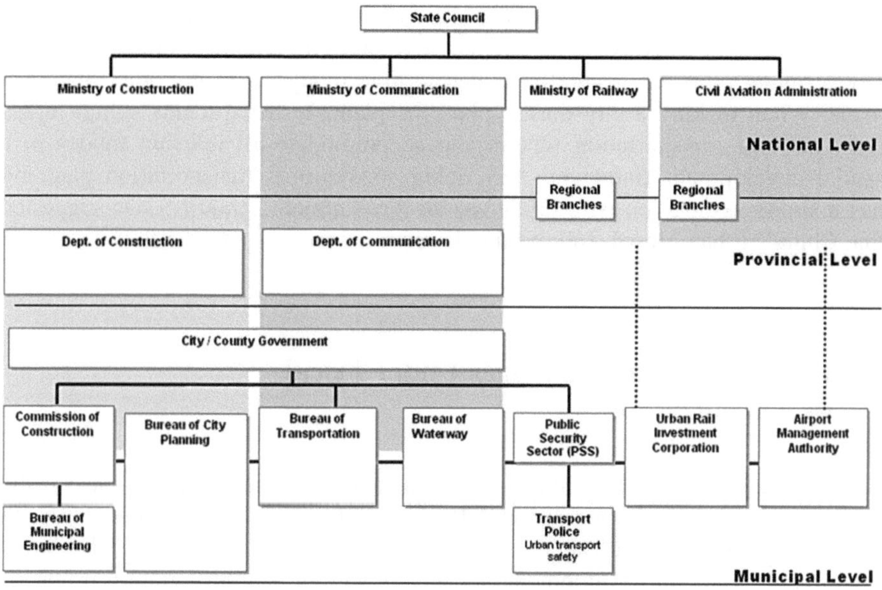

Fig. 1 Administrative structure of public-sector transportation institutions

MOCM for national inter-city road and water transportation, CAA for national air transportation and MOC for urban transportation.

All four central administrations implement their transportation investment decisions and policies through local branches. Their relationship with the comprehensive government at the locality, however, varies. The regional branches of CAA and MOR are not part of local governments (province or city governments). They report directly to the national administrations and they receive funds from and implement system development decisions by their corresponding central administration. This structure was selected to emphasize the national integration of an intercity and interregional network [5].

Local offices of MOCM and MOC have been closely aligned with the local comprehensive governments, at both province and city levels. Responsibilities of rural, inter-city, and water transportation are assigned to the Bureaus of Transportation (local offices of MOCM). Responsibilities of urban transportation are assigned to the Commission of Construction (local offices of MOC). These local offices are part of the local comprehensive governments. They develop their own plans and invest in local transportation systems by following the guidance from the higher level governments and by serving the economic-social development strategies at the locality. In the pre-reform period, funding for those local offices came mainly from the higher level offices, rather than the local governments [6].

Urban areas are the natural connecting points of different transportation modes. City master plans, which are managed by the city commission of construction, are assumed to coordinate all these transportation investment activities within urban areas. The city master plan has a spatial framework that not only specifies land allocation and major streets, but also the location where urban transport can be linked to regional railway, waterway, air, and inter-city road network. In doing so, city planning units are generally invited to give inputs when the other plans are made. When making a city master plan, the planning departments solicit inputs from relevant transportation sectors and accommodate all relevant modes in a spatial development framework. By doing so, regional transportation planning has a strong association with urban spatial development. Zhou [7] has suggested that China's urban growth follows the direction of inter-city traffic flow.

3 Transportation Challenges Under Fiscal Decentralization

The above mentioned sector-oriented planning and investment decision-making did not change during the period of fiscal decentralization introduced in the 1980s. As local governments take advantage of the fiscal reforms and speed up infrastructure investment for urban development, several categories of transportation problems have emerged. Most of them are relevant to the integration of systems developed by different city governments, integration of different modes developed by different government units in a single city, and the integration of urban and rural passenger transportation in a single economic region, they are also relevant to the decentralized fiscal system itself by varied extent.

3.1 Toll Collection and Public Trust

China's fiscal decentralization arrived at the period of the national ideology shift toward economic development. Local governments have tried to impress the higher level government leaders with quantitative measures such as GDP growth, government revenue, and the inflow of foreign direct investment. They have also strived to generate more tangible achievement of urban physical development, such as modern infrastructure, and mega-projects of art, sports and high-rise buildings. Economic development policies cost resources. They include not only relevant tax packages to attract mobile investment, but also efficient public administration, timely infrastructure supply, and coordination between public and private investment [8]. Particularly, transportation infrastructure is important to land development and regional economic development. How to generate enough revenue to finance infrastructure development became a fiscal challenge for China's local leaders, similar to the situation in Latin America [9]. In order to close the gap between public revenue and public expenditure, China's local governments were awarded flexibility to collect extra-budgetary revenues. How this decentralized fiscal institution helps develop the overall urban infrastructure in China has been well explained by Wu [10] with a case study of Shanghai.

This fiscal flexibility has surely helped local governments meet the mobility demand in an era of fast economic and urban development. City governments have primarily relied on land lease revenue, either directly or indirectly. The expected land value appreciation following urban infrastructure investment has been purposefully used by the city government to leverage bank loans [11]. For regional road construction, maintenance and management, various fees have been charged to transportation service providers and private automobile owners. Those items include annual road maintenance fee, annual vehicle registration fee, annual transportation service management fee, etc. Typically, those fees are not pooled into the local government general revenue. For example, the road maintenance fee was collected by the bureau of transportation and reserved for inter-city and rural road construction and maintenance. Land lease revenue was used by the city government for urban infrastructure development, and seldom used for inter-city transportation projects.

Enterprise operation has been widely employed for capital intensive projects such as limited access highway, tunnels and bridges, and intercity roads if access control can be relative easily set up. Since 1984, the State Council has encouraged local governments to finance road infrastructure with bank loans and toll collection. Decision-making on toll road construction and toll collection was decentralized to the province government. By the end of 2007, toll roads were 94 % of the limited access highway and 60 % of first class express ways.

While the above user fee-based approach is generally viewed as a more efficient approach to transportation finance, public criticism on highway toll collection, however, became intensified in the past decade. Toll collection has been widely criticized as a cause for traffic delay and mis-use of public trust by the city government. According to the Public Road Act and the Toll Road Management Regulation by the State Council [12], county, prefecture and province governments have the authority to construct toll roads based on local needs. The toll revenue by regulation is only allowed to pay back the bank loan, assuming that annual maintenance fee paid by vehicle owners will be apportioned for toll road maintenance. Once the loan is paid off, the toll should be removed.

Installment of a toll booth is subject to the approval from the province-level government. The central government sets policy standard for the local government to follow. For example, the regulation by the Ministry of Communication specifies 40 km as the minimum spacing of adjacent booths. If the road is financed with a bank loan, the maximum length of toll collection is 15 years, and it could increase to 30 years in the relative underdeveloped middle and western part of China. If the road is financed with private sector money, the maximum length of toll collection should not go beyond 25 years and it could be extended up to 30 years in the middle and western part of china.

This regulation, however, appears to be overlooked by the local government. With the increase of traffic flow, toll revenue was so attractive that the local government and the bank would love to keep it there forever. Cash flow of toll collection was never transparent to the public. The most controversial example probably is Beijing's Airport Expressway, which is one of the earliest limited access highway constructed in China.

Case 1: Airport Expressway in Beijing

The airport expressway was built to shorten the travel time from central Beijing to the airport. The government financed the roads with a bank loan in 1993. Several years' after its operation, the public began to inquire the rationale to continue the toll collection. They doubted that either a high percentage of the toll revenue is not used to pay back the loan, or the loan had been paid off.

Despite complaints and enquiry from the public, the local government did not stop toll collection. Instead, it sold the toll collection right to another public corporation – the Capital Development Corporation – and then lengthened the toll collection to 30 years. The story was never clear until the central government's general accounting office investigated this issue. It found out that over the 14 year period up to 2005, the accumulated toll revenue is over three billion, which is much higher than the original investment amount (1.2 billion). Under the pressure from the public and the higher level government, the city government lowered the toll from 15 Yuan to 10. While toll collection can surely be continued for other reasons such as congestion mitigation, the lack of transparency in toll revenue disposal and the use of the highway as a cash cow surely damages public trust in urban transportation finance.

3.2 *Challenges for Multi-City System Integration*

Roads in China are divided into different groups according to who have the major responsibilities for road finance and maintenance. The central government manages national backbone roads, which serves a relative high percentage of cross-province traffic. The province government manages the provincial backbone road network, which serves a relative high percentage of cross-city traffic. The other roads, which mainly serve the local traffic, are managed by the prefecture/county/city government. In the planning economy, public transportation fund came mainly from above. Roads belonging to the national backbone network was much better constructed and maintained, compared to the local roads.

Fiscal decentralization has dramatically shifted access to road finance. Urbanization and motorization enabled local governments to collect ample infrastructure fund in the format of land lease revenue, road maintenance fee and highway toll. City governments can raise more resources for road finance than the provincial and central government, particularly in the relatively developed Pearl River Delta. While doing so, city governments are more interested in serving the intra-city traffic than the inter-city traffic. The relative under-maintained segments of national and provincial roads could become bottlenecks for cross-city or cross-province traffic in relative developed regions. This connectivity issues has been caused not only by pavement quality, but also by toll collection, as illustrated by the case below.

Case 2: Road Connectivity in Guangzhou and Foshan

Foshan and Guangzhou are two neighboring metropolitan cities in the core areas of the Pearl River Delta, one of the most advanced economic regions in China. Roads within each city are well paved and designed. The connection between them, however, tends to be overlooked. For a period of time, new roads seldom extended beyond city boundary. Existing roads, when upgraded, seldom extended to the border. For the several upgraded roads that indeed connect these two cities, toll collection booths were set up at the city border.

The toll collection was selective, based on the home city of vehicle registration. Guangzhou's toll booth charged vehicles driving into Guangzhou but registered in Foshan. Vehicle registered in Guangzhou would not be charged. Similarly, Foshan's toll collection booth only charged vehicles driving into Foshan but not registered in Foshan.

This practice has its root in road finance. Local government's road maintenance expenditure is assumed to be covered by the annual road maintenance fee paid by vehicle owners. The fee collection is based on vehicle registration location. Guangzhou's vehicle owners pay the fee to Guangzhou City government when they register their vehicles annually. The fee, however, was not shared with bordering Foshan city. As a result, Foshan charged those vehicles whenever they are driven on its roads.

This system connectivity problem did not bother those who live and work in the same city. However, it was a big hassle if one wanted to drive beyond city boundaries. As urban expansion increased volumes of cross-border traffic, waiting lines in front of the toll booth could easily become congested. In 2008, after Guangzhou and Foshan realized that economic benefit of free roads is bigger than the revenue they could collect from the cross-border traffic, they reached a cost-sharing agreement on road construction and maintenance. The toll booths were eventually removed.

3.3 Challenges in Urban-Rural Integration

In a typical city region in China, urban transportation and rural transportation are planned, developed and managed by separate government units. Rural roads and intercity passenger transportation are managed by the bureau of transportation, but urban roads and urban passenger transportation are managed by the city commission of construction. As urban areas expand outward into the pre-existing rural areas, confusion often arises in terms of which government unit should do what. Since urban expansion typically suggests a transfer of service supply responsibility among government units for the newly added urban communities, under-supply or delayed supply of transportation services is not unusual.

Case 3: Under-Supplied Transit Services in Panyu
Panyu, once a suburban county administered by the Guangzhou City government, has been a place of affordable housing for many who work in the central part of the Guangzhou City. In the early 1980s, the Pearl River separated the then Panyu County in the south from the urban blocks of the Guangzhou City, which was limited to the northern side of the river. In 1988, the Luoxi Bridge was constructed, with generous donation from Yingdong Huo, the rich businessman in Hong Kong. This tolled bridge links Guangzhou to Panyu. Several new residential communities were developed, including Luoxi new town, Lijiang Garden, Guangzhou Olympic Garden. In 1999, Panyu Bridge and Huanan express way was completed. Eight huge residential communities were established. The biggest one has a land area of 7,500 mu (a Chinese area unit. One mu equals to 666.66 m^2). All these housing units were built to attract those who work in Guangzhou city, with its lower housing price and better access to natural scenery.

Transportation services for those communities, however, have always been a problem. Bus service from Guangzhou to Panyu was classified as inter-city service, which was planed and managed by the bureau of transportation. Urban transit within Guangzhou City was planned and operated by the municipal engineering company. Those two categories of services are operated under different service standard. A transfer between the urban and inter-city systems could cause a lot of pain for time constraint commuters.

What makes the problem worse was an administrative transition. In 1997, Panyu city (a county-level city) was scheduled to become Panyu District of the Guangzhou City. This change implies that Panyu will lose its control of land planning, land leasing and project permitting, which further implies that Panyu's local government would lose control of the land lease revenue (an extra-budgetary revenue item) once the annexation is completed. Before the actual change happened, Panyu's city government was eager to lease as much land as possible and also permit as many projects as possible.

Uncertain about its own fiscal ability to expand municipal services to those newly planned urban communities, the land lease and project development permitting came with a clause that the developers are responsible to provide a broad spectrum of urban facilities and services, including schools, libraries, parks and hospitals. The developers were willing to do so primarily because they negotiated a very cheap land price.

The fast growth of rural Panyu is surely not well budgeted by the Guangzhou city, who became in charge of Panyu's municipal and transportation services after the annexation. Particularly, Panyu's passenger transportation service was not updated to the urban standard. Transportation connection to

(continued)

the central Guangzhou City, however, is the precondition to attract housing buyers. Under the pressure to market the housing units, the developers promised commuters buses that connect the community with major transfer points in the Guangzhou City. Those services are more expensive to operate because of the lack of scale economy. The service is also less attractive compared to those offered by the city government.

3.4 Challenges for Multi-mode System Integration

As urban areas expand, urban transportation modes tend to become diversified. The task to maintain seamless connection across different modes becomes more challenging, particularly when China's fast rate of urban growth and the capacity of planning institutions are considered.

First, comprehensive urban transportation planning is not required by law in China. It is up to the local municipality to decide whether a separate transportation plan is needed. The problem, as identified by Gakenheimer [13], is that transportation policy-making in Chinese cities is divided among different government sectors, without a comprehensive policy center to synthesize them together. The plans from separate government branches may conflict with each other. Lack of integration has been reported. For example, in Chongqing, many stations of urban rail are distanced from nearby bus stops, making it inconvenient for passenger transfer.

The mis-integration happens more frequently when the transportation facilities are planned and developed under the guidance of different governments. In cities with railway, how to connect rail terminal with urban transportation network has always been an issue. One particular example is the Beijing Western Railway Station. The station is only several blocks from the nearest subway station. However, no underground tunnel is provided. Railway passengers rely on on-ground bus services or taxi, which causes severe congestion around the station.

While sector-oriented transportation development existed in the pre-reform period, the modal integration was not so bad. During that period, the relative slow pace of urban growth gave relevant government units enough time to adjust its own subsystem for integration purpose. In addition, with the higher-level government working as the supervising agency and local government as the implementation agency, the local units will not be resistant to ideas of better integration as long as the higher-level government is willing to burden the additional cost. With decentralized finance, however, relevant government units are under the pressure to finance and develop its own system. Any additional cost associated with better integration could discourage the agency from doing so. Occasionally, this mis-integration can result from purposeful show-off of modern transportation modes. One interesting example can be observed in Beijing. Station entrance along several lines is sited away from sidewalks. Pedestrians have to walk across bike lanes and even several motor lanes before entering the station.

Urban planning is assumed to integrate all those transportation components. However, urban transportation system in the post-reform period has become larger and more diversified, which requires improved knowledge of transportation planning and system integration. With urban transportation planning as an optional component in urban planning, the human resource for transportation planning is generally under-equipped in the municipal commission of construction. Only a very few big cities have fully staffed independent professional transportation planning institutions. Some inconsistency within a single plan or conflict between different plans may be realized in the project permitting process. But no planning process exists to correct these detected problems. No reward mechanism has been institutionalized to encourage transportation professionals to walk across existing organizational lines.

All above challenges suggest that reform is needed not only for transportation planning, but also the overall system of transportation finance. The problem of planning and finance are interconnected. Under fiscal decentralization, the toll- or fee-based local transportation finance has reinforced the sector-oriented transportation planning, and led to various challenges in transportation system integration. As new economic environment drives the demand for seamless transportation across multi-mode, multi-city and urban-rural boundaries, reforms are needed to restructure existing institutions for transportation finance and planning.

4 Transportation Supply Through Institutional Innovation

Signs have shown that institutional reform is on the way. A significant change came in April, 2008, when Chinese central government decided to move the functionality of urban road transportation planning and management from the Ministry of Housing and Urban-rural Development to the new Ministry of Communication. Part of the reason was to consolidate the planning and management of different modes of urban transportation such as road transportation, subway, buses, and private motorized transportation. Following this announcement by the central government, the National Academy of Transportation Research was contracted to study the institutional reform possibilities for local governments [14]. The report recommends a municipal comprehensive transportation policy-making commission above current mode specific institutional structure [14].

The above leadership by the central government is understandable given its involvement in metropolitan transportation investment in key megacities. The central ministry has budgeted big grants to create eight railway hubs, and eight highway hubs and eight comprehensive transportation hubs. Those hubs are selected to be within the big cities such as Beijing, Shanghai, Guangzhou, Wuhang, Xi'an, and Chengdu. A planning capacity that matches the investment tasks is essential for an effective use of public resource.

This national leadership also reflects the demand for reforms at the locality. A group of Chinese megacities have taken a lead in creating its own metropolitan transportation policy centers [14, 23], partially to better address the mobility and accessibility challenges at the locality. Eight major Chinese cities, including Shenyang, Haerbing, Wulumuqi, and Xining, have already consolidated its urban transportation and rural transportation management functionalities. A group of other cities, including Beijing, Guangzhou, Chongqing, Chengdu, Shenzhen, Wuhan, has innovated in multi-mode integration.

Case 4: The Shenzhen Experience

The earliest and also the most comprehensive metropolitan transportation reform has been observed in Shenzhen. Shenzhen Municipality had a similar institutional setup for urban transport as most other mid-size and large Chinese cities, with a multitude of bureaus and departments overseeing the planning, construction, operations, and maintenance of different transportation modes. The main problem associated with this setup, as discussed in a previous section, was weak coordination.

In 1996, the Shenzhen municipal government established an urban transport management headquarter, which is a coordination agency. Its main functions are to oversee and coordinate urban transportation planning, financing and implementation process, and ensure urban transport development follows policies, the city master plan, and budget. In 2001, successful experience moved the city government to create a comprehensive urban transportation commission. Its responsibility covers almost all transportation modes and services, including roads, urban transit, waterway, seaports, railway, urban rail, airport, freight logistics, and postal services.

Reforms of planning institutions, like that in Shenzhen, are timely. In most Chinese cities, however, this transition has not happened yet. Transportation system expansion is still implemented with ad hoc project-based investment decision-making. It does not encourage comprehensive and coordinated planning in which trade-offs are made and priorities set on the basis of relative benefits, costs and impacts. Even in cities that have tried alternative urban transportation institutions, few cities have a comprehensive multimodal urban transport plan. If a plan exists, it is most often a collection of different and independently prepared projects brought together in one document. Stronger national leadership is expected to help reform existing planning institutions and also to encourage integrated planning. The fuel tax reform, which came in January 2009, provides a new opportunity.

5 Innovating Transportation Finance

It is worth noting that China's fuel tax reform was not oriented toward public transportation finance at the very beginning. In 1994, the fuel tax was firstly introduced as part of a new wave of comprehensive fiscal reform. This fiscal innovation was stimulated by a desire to strengthen the central government's ability to affect the macro economy. Under the fiscal responsibility system, the extra-budgetary revenue was sufficiently enough for the local government to pursue their own development objectives, regardless of the macro policy set by the central government. The central government's solution was to restructure the local fees into taxes managed by the central government. After 1996, more than one fourth of fees were completely abolished. Most of the rest were transformed from irregular charges that were usually subject to the discretion of local governments into regular taxes that must fall under the surveillance of budgetary accounts. The new system is called tax assignment system as it specifies portions of tax revenue that will go to the central and local governments respectively.

Under this background, relevant national government units discussed the possibility of a national fuel tax as a replacement for various transportation fees collected by the local governments. In 1994, the national government selected Hainan Province as the experimental region to test the idea of a fuel tax.

Case 5: The Hainan Experience

Hainan Province is an island in southern part of China. It was part of the Guangdong Province. It became a separate province in 1988 when China's national government decided to use this island to experiment some brave reform ideas. Hainan started its fuel policy in January 1994. It replaced all preexisting road maintenance fees, tolls, and transportation management fees with a fuel surcharge. As a result of this finance institution, Hainan has no toll collection booth.

The surcharge was selected to be 30 % of the fuel price. As price increases over time, the amount of surcharge per liter has increased for several times to maintain the same percentage. In 2006, when the fuel surcharge increased by 0.32 Yuan per liter, the government received many complaints, particularly from the taxi drivers. In order to reduce fuel cost, about 90 % of taxi fleet in Hainan has been converted to natural gas power.

Fuel surcharge has been viewed as a stable revenue stream for highway finance. According to the local officials, banks are more willing to finance highway projects when the payoff ability of the project is not tied to the specific facility, but the surcharge revenue throughout the province. In 1994, Hainan had only 60 km limited access freeway. In 2008, the total length reached 660 km.

The Hainan experience was used to support the movement toward a national fuel tax. In 1997, the National's People's Congress (NPC) passed the Public Road Act, which for the first time mentioned fuel surcharge as a replacement for the existing road maintenance fee. The plan then was to implement this finance reform as early as Jan 1st, 1998. In 1998, the idea changed from fuel surcharge to tax. The State Council proposed amendment

(continued)

to the Public Road Act, which was firstly rejected and then approved in October, 1999. This amendment provides the legal support for a national fuel tax. In Dec 2008, after observing half year's price drop in the international oil market, four national governments units (National Development and Reform Commission, the Ministry of Finance, the Ministry of Communication, and the National Bureau of Taxation) jointly published a draft of the Fuel Sale Tax Reform Plan and solicited public inputs. In 1 week, 48,643 pieces of comments were received from online forums, by fax or by mail. (Unfortunately, these comments are not circulated for public review.) On Dec 18th 2008, the State Council circulated the final plan, which became effective on Jan 1st, 2009.

The final tax scheme reflects a compromise between the central and local governments. At the early stage of policy design, a replication of the Hainan experience was proposed. That was to eliminate not only maintenance fees, but also all tolls. This early proposal was strongly opposed by the central transportation administration and by the local governments, who have relied on this revenue stream for road construction and other purposes. As a result, toll collection on the limited access highways and first class expressways are left untouched. This tax revenue is used to replace six categories of fees: public road maintenance fee, waterway maintenance fee, public road transportation management fee, public road transportation surcharge, waterway transport management fee, and waterway transportation surcharge.

This reform appears to suggest an efficiency improvement in the finance system itself. When the government raises fees or taxes to finance public services, the management cost is the societal cost for service delivery. The fuel tax reform does not create any new type of tax, but increases the tax rate of an existing tax. The gasoline sales tax increases from pre-existing 0.2 Chinese Yuan per liter to one Chinese Yuan. The diesel tax increases from 0.1 Chinese Yuan per liter to 0.8 Chinese Yuan [15]. There will be no significant cost increase for the tax collection system. The 300,000 employees released from the pre-existing fee collection system can now work on other activities. Automobile owners no longer need to take a separate trip to pay the annual fee. The fuel tax is paid when fuel is purchased.

In addition, one might expect that this tax could affect individual consumption behavior to some extent. The pre-existing annual road maintenance fee is a fixed cost for car ownership, regardless the annual mileage. The reform replaced the fixed annual charge with a fuel tax, which varies according to mileage. It decreases the cost of annual car ownership and increases the cost of driving. It has the potential of encouraging car ownership and transit riding at the same time. Which impact will dominate is an interesting question. This fuel tax could also help compact car owners save annual driving cost by 10–20 % [16], thereby encouraging the ownership of more fuel-efficient vehicles. It sends a positive message to the environmentalist.

6 Envisioning a New Era of Transportation Investment and Planning

It is still too early to predict how this tax reform could be used to leverage changes in transportation investment and planning in metropolitan cities. However, it indeed creates a fiscal capacity for the central government to start a strong national leadership for urban transportation. According to the current regulation, all tax revenue goes to the national government and is reserved for is reserved for transportation capital projects and maintenance. The tax replaces pre-existing fees charged by the local governments. In order to recover the revenue loss of the local government, the fuel tax revenue is transferred to localities based on a certain formula.

An immediate step the central government has already taken is to eliminate toll booth on secondary highways. At the early period of toll road investment, local governments under-estimate traffic growth. They selected relative low-cost secondary road. Later, as fast traffic growth justifies high-volume first class expressway and even limited access highway, traffic was diverted away. The toll booth on the secondary roads in many localities can hardly collect enough revenue for the interest payment of relevant bank loans. The central government has decided to use a portion of tax revenue to eliminate toll collection booth on secondary roads. The plan is to use fuel tax revenue to cover 50 % of unpaid bank loan, and the local government covers the remaining 50 %. (In the pre-existing situation, local governments are responsible for 100 % of the loan.)

More significantly, the national fuel tax could be used to leverage institutional changes at the local level. The national fuel tax provides a revenue source that is beyond a single mode, beyond urban-rural division, and beyond city boundaries. The monetary transfer from the central government to the local government can surely be used to promote changes in planning institutions, processes and methods for integrated urban transportation system in China, as has happened in the USA [17]. The creation of metropolitan planning organizations and the associated transportation planning processes in USA metropolitan areas have been greatly facilitated by the regulated access to federal gasoline tax revenue: Transportation plans by metropolitan planning organizations, not component city or county government, are needed in order to finance metropolitan area transportation projects with federal transportation grant. The national fuel tax in China could play a similar function if used appropriately.

This new tax could also be used to promote investment in public transit. In the past two decades, investment in China's urban transportation has been dominated by road transportation, despite that China's high density urban regions are compatible with mass transit rather than automobile driving. This investment bias has its root in local finance. Local government investment in roads can be more likely rewarded with land lease revenue [18]. After decades' silence on urban transportation issues, in 2007, China's Ministry of Transportation and the Ministry of Construction jointly announced the transit-priority strategy for megacities, which is part of the broad national strategy to build an energy-saving and environment-friendly society. However, no complementary policies have been created to assure that transit projects can be prioritized in gaining right of way, capital investment, and operational subsidy. No project evaluation procedure and method exist to make sure this priority can be implemented. The availability of this national fuel tax revenue could be used as a financial tool to leverage local governments' efforts for public transit.

7 Conclusions

China's discourse of public-sector transportation reform has been mainly driven by the need for economic development in relative advanced urban regions [19], which are the national or regional centers of economic and social activities. The pace of economic development and the changing demand for transportation infrastructure and services challenge the capacity of pre-existing institutions to supply mobility across multiple modes, multiple cities and urban-rural boundaries [20, 24]. Concentration of transportation infrastructure, such as airports, railway terminals, river/sea ports, and high capacity roads in dense urban areas also naturally raise challenges for transportation planning.

While fiscal decentralization has fast-tracked economic development and urbanization by providing much-needed transportation infrastructure and mobility in past decades, the evolution of urban and regional economy is demanding a new format of mobility that emphasizes multi-modal integration, urban-rural integration, and seamless connection from one city to the other. Problems of system integration observed in Chinese cities have their connection with the pre-existing planning process and institutions, and also have their root in public transportation finance. The decentralized and sector-oriented transportation finance has reinforced the transportation planning and development practice by mode, by urban-rural division and by city boundary.

Planning and development of integrated urban transportation systems requires efforts at both local and higher level governments. Local initiative of reforming metropolitan transportation planning institutions in cities like Shenzhen has clearly demonstrated that the demand for alternative transportation planning process exists. The power of comprehensive urban transportation policy centers, however, has been rather limited because of lack of fiscal resource to reward innovative planning practice.

The opportunity is emerging. The newly introduced national fuel tax surely gives the central government a fiscal capacity to build a relative strong national leadership in surface transportation. The national fuel tax provides revenue that is beyond a single mode, urban-rural division, and city boundaries. The monetary transfer from the central government to the local government can surely be used to promote changes in planning institutions, processes and methods for an integrated urban transportation system.

Distance of course exists between the above reform potential and observed institutional reality. It is still unclear how determinant China's central government will use the tax reform to leverage changes at the locality. It is even unclear how this new fiscal environment might help to create a coherence package for urban and regional planning decisions, including land use, energy and environment regulations. Particularly, no planning framework exists to make sure conflict between land development plan and transportation investment plan will be identified and solved as early as possible. All above issues are important. In China, strong evidence suggest that the land development has strong connections with urban transportation [18, 21, 22]. Better integrated urban-rural mobility will expand housing options in China's megacities. Without well coordinated land development strategies, an enhanced transportation supply can lead to sprawling development in the suburban areas. This is a challenge not only for developing nations like China, but also developed nations like the United States, where the coordination of regional transportation and regional land development has long over-due.

Acknowledgment This research was funded by China's National Natural Science Foundation Grant 412271177.

References

1. The National Geographic, China, Special issues, May 2008
2. Gakenheimer R (1999) Urban mobility in the developing world. Transp Res A 33:671–689
3. Pendakur VS (1992) Urban transportation in China: trends and issues. Transportation research record no. 1372, Nonmotorized Transportation
4. Bird RM (2004) Getting it right: financing urban development in China. International tax program paper 0413, Institute for International Business, University of Toronto, Nov 2004
5. Zhang Y (2009) Plan integration for coordinated urban growth in China, Chap. 9. In: Smart growth in China. Lincoln Institute of Land Policy Press, Cambridge
6. Lin GCS (1999) Transportation and metropolitan development in China's Pearl River Delta: the experience of Panyu. Habitat Int 23(2):249–270
7. Zhou Y (1998) Urban growth and its major external transportation flow. Urban Plan (in Chinese) 2:22–25
8. Cheshire P (2009) Choosing areas for spatial policy interventions: principles to guide integrated spatial plans, Chap. 13. In: Smart growth in China. Lincoln Institute of Land Policy Press, Cambridge
9. Zegras C (2003) Financing transport infrastructure in developing country cities: evaluation of and lessons from nascent use of impact fees in Santiago de Chile. Transp Res Rec J Transp Res Board 1839:81–88
10. Wu W (1999) Reforming China's institutional environment for urban infrastructure provision. Urban Stud 36(13):2263–2282
11. Yang J, Gakenheimer R (2007) Assessing the mobility and accessibility consequences of land use transformation in urban China. Habitat Int 31:345–353
12. China State Council (2004) Regulations for toll roads. National State Council, Beijing China, 1 Nov 2004
13. Gakenheimer R (1996) Roles of urban transport planning in China. In: Conference proceedings transports urbains dans les Pays en development conference proceedings CODATU, New Delhi, Feb 1996

14. China's Academy of Transportation Science (2008) Continuing transportation management reform in major cities. Research report, Academy of Transportation Science, Beijing China
15. Zhou Y (2009) An expert view of the national fuel tax reform. In: China business, 6 Jan 2009
16. Xiao Q (2009) Xi'an: the quiet national fuel tax reform. Retrieved 8 June 2009 from http://news.sohu.com/20090107/n261612697.shtml
17. Weiner E (1992) Urban transportation planning in the United States: an historical overview, report at USA federal Department of Transportation
18. Yang J, Feng C, Cao G (2007) Land and transportation development in China: an economic analysis of government behavior. J Transp Res Board 2038:78–83
19. Huang, X (2007) Speech at the 24th meeting for transportation reform in China's major cities, 29 Oct 2007. Retrieved 5 June 2009 from http://www.moc.gov.cn/2006/06buzhangwy/huangxianyf/buzhangjh/200710/t20071030_443197.html
20. Fang Z (2009) Calculating expenditure one month after the national fuel tax. Retrieved 4 Feb 2009, from Metropolitan Times. http://auto.sina.com.cn/news/2009-02-04/0836454520.shtml
21. Pan H, Shen Q, Zhang M (2009) Influence of urban form on travel behaviour in four neighbourhoods of Shanghai. Urban Stud 46(2):275–294
22. Yang J (2010) The spatial and social characteristics of urban transportation in Beijing. Transp Res Rec 2193:59–67
23. China's Academy of Transportation Science (2008) Reforming for the greater transportation institute in local governments. Research report, Academy of Transportation Science, Beijing China
24. Fang C (2007) Theories of regional planning and spatial governance. Business Press, Beijing
25. Wong CPW, Bird R (2005) China's fiscal system: a work in progress. Working paper 05–20 in Georgia State University, Andrew Young School of Public Policy

Rational Traffic Dispersion

Take the Example of No Left Turn in the Rush Hours Along Renmin Road in Suzhou City

Yao Yangyang

1 Introduction

In order to deal with citizens' pressing concern about the traffic jams in the central urban area, and meanwhile maintain the style and features of the ancient city, the traffic administrative department of Suzhou City implements the rational traffic dispersion scheme to ease traffic jams at Renmin Road since March 15th, 2010 (Fig. 1), that is, no left turn in the rush hour of public utilities. This scheme drives motor vehicle "right in and right out" and carries out the measure of "No Left Turn in the Rush Hour" at eligible entrance and exit on Renmin Road in the ancient city Suzhou. For the purpose of alleviating traffic jams in rush hour on Renmin Road, the department scientifically selects locations for vehicles to turn around, adopts physical isolation in some key parts, and allocates the saved phase time of signal to non-motor vehicle and vehicle going straight ahead to ensure the continuity of traffic flow on the artery.

But after the implementation of "No Left Turn in the Rush Hour of Public Utilities", while it eases the traffic jams on Renmin Road and even central urban area, is traffic pressure increased on Renmin Road and its surrounding road network so that it causes inconvenience for citizens driving cars? Different groups response and comment differently. Hence, beginning from the angle of urban planology, this article will carry out in-depth investigation and analysis of the implementations of the scheme from overall traffic benefit, effects on people's life and many other aspects, trying to find out existing problems and conflicts, put forward with suggestions for improvement and provide new reference for other roads faced with similar problems in this ancient city.

Y. Yangyang (✉)
School of Architecture and Urban Planning, Huazhong University of Science and Technology (HUST), Wuhan 430074, China
e-mail: 15314648@qq.com

X. Chen, Q. Pan (eds.), *Building Resilient Cities in China: The Nexus between Planning and Science*, GeoJournal Library 113, DOI 10.1007/978-3-319-14145-9_27

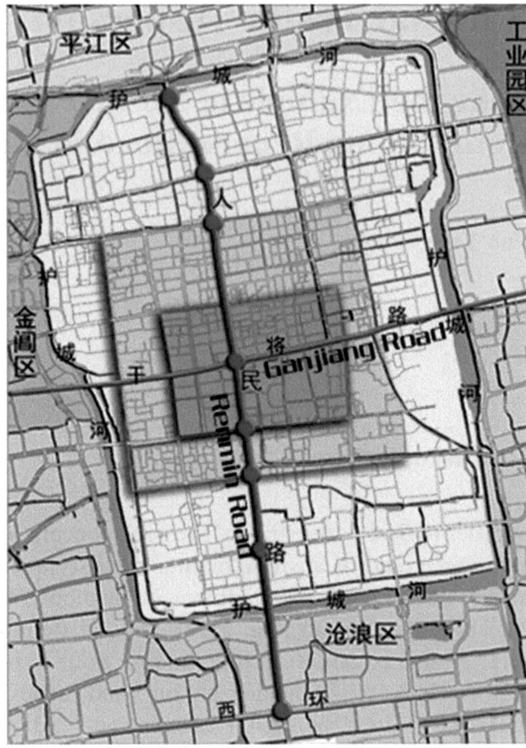

Fig. 1 Location map of Renmin Road in Suzhou (Source: author complete)

2 Scheme

2.1 Background of the Scheme

The implement object of this scheme is Renmin Road, which connects southern and northern Suzhou, the ancient city, goes to Suzhou Railway Station in the north, leads to Wuzhong District in the south, and is the only artery running north and south inside this ancient city. It undertakes such traffic functions as gather and distribute traffic along the line, connect the inside and outside of the city, public traffic corridor and transit passage of vehicle. But due to the special surroundings of the ancient city Suzhou, road engineering construction is under restrictions. Besides, with the urban development and expansion of Suzhou, traffic demand is increasing day by day and the following severe problems occur to the traffic functions undertaken by Renmin Road.

2.1.1 The Gather and Distribution of Units Get In and Out Along This Line Affects Overall Traffic Benefit

Lots of traffic attractions gather on both sides of Renmin Road, including two business districts Guanqian Street and South Gate in Suzhou, various commercial buildings, cultural tourism attractions, large-scale education facilities which are distributed along the street and 37.5 % of enterprises and public institutions within the scope of the ancient city (Fig. 2). Among these buildings, many face directly the openings of Renmin Road, where vehicles pass through frequently and have a severe impact on traffic going straight. Especially, in the rush hour vehicles turning left to get in or out have particularly tremendous bad effect on north–south traffic on Renmin Road, and even cause severe congestion in passage-intensive sections and rather poor traffic benefit in the whole line. Besides many units, Renmin Road also links up with lots of alleys and lanes. For only the west side of Renmin Road, there are more than 20 bypasses, streets and alleys connected directly to it, which makes traffic situation more chaotic in this road (Fig. 2).

2.1.2 Problems Arise in the Communication of the Inside and Outside of the Ancient City

Suzhou City presents the layout of "Ancient city in the center, Park in the east and District in the west, Wu in the south and Xiang in the north, and five districts cluster as a group". Renmin Road goes to Suzhou Railway Station and Xiangcheng District in the north and connects Suzhou South Bus Station and Wuzhong District in the south. Every day the average flow rate of motor vehicles entering the ancient city through Renmin Road reaches up to 563.8/h, and this number approximates 600/h in rush hours in the morning and evening. Due to the severe traffic jam in Renmin Road, average speed of vehicles is restricted to 16–20 km/h, thus causing serious detention and delay for vehicle getting inside and outside the ancient city.

2.1.3 Passageway for Bus Is Affected

On both sides of motorway, bus lanes are set aside in most sections of Renmin Road. At present, Renmin Road is the leading passageway for bus in the ancient city and there are six bus routes running through the ancient city. However, because the flow undertaken by Renmin Road is increasing day by day and many other vehicles turn left to get in and out of kinds of locations through the openings of bus lanes, normal travel for buses is greatly affected and the original advantage of bus passageway is increasingly weakened.

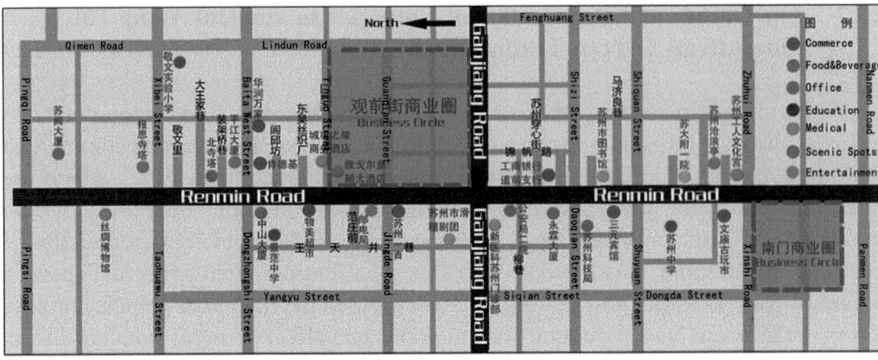

Fig. 2 The distribution of public facilities at Renmin Road (Source: author complete)

2.2 Scheme Configuration

Under the background that the problem of traffic jams in Renmin Road becomes serious increasingly, learning from the successful cases in other cities and combining the characteristics of Renmin Road in Suzhou City, relative departments in Sunzhou City officially begin carrying out the scheme of "No Left Turn in the Rush Hour of Public Utilities" in Renmin Road on March 15th, 2010 (Fig. 3). Following goals are supposed to be achieved by adopting this scheme:

(a) Use "No Left Turn" to thoroughly eliminate the problem of "too many turning left vehicles and too much delay time" in the rush hour of Renmin Road, thus reducing delay time for stop in the whole course of Renmin Road, elevating average speed of vehicles, improving traffic environment in Renmin Road and keeping traffic stream fluent in the rush hour.

(b) Use "Rush Hour" to control vehicles turning left, break the whole tremendous traffic pressure brought by tidal transportation in the rush hour of Renmin Road into parts and rationally disperse it into other time, and rule out unnecessary left turning flow of the rush hour so as to reduce the time of rush hour in Renmin Road and control the quantity of flow in rush hour of Renmin Road.

(c) Use no left turn control of entrance and exit of each public utility along Renmin Road to decrease the bad effects of various facilities on the traffic of Renmin Road, avoid congestion in the whole course of Renmin Road caused by vehicles turning left into kinds of locations in the rush hour and make sections which are congested before unobstructed.

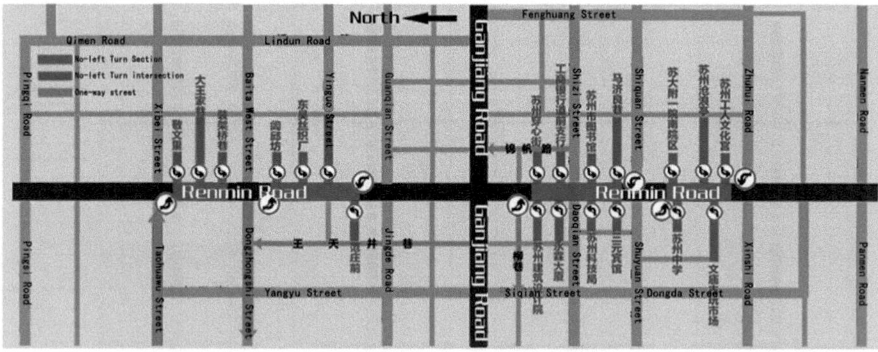

Fig. 3 The scheme of "No Left Turn in the Rush Hour of Public Utilities" in Renmin Road (Source: author complete)

3 Investigation of Implementations

3.1 Analysis of Traffic Benefit

In order to know specific implementations of "no left turn in rush hours of public utilities" in Renmin Road and the influence on its traffic benefit, interviews are made with relative transport agency to get data before "No Left Turn", and field observation and methods like following cars and counting cars are adopted to collect and process data. By the comparison of delay time caused by stopping, motor vehicle flow, and average speed before and after carrying out "No Left Turn" in Renmin Road as well as analysis and evaluation of factor model (Fig. 4), this paper tries to scientifically find out the effects of the scheme "No Left Turn in the Rush Hour of Public Utilities" on traffic benefit of Renmin Road:

(a) Delay for stopping: after carrying out "No Left Turn", situation in most sections where delay for stopping is comparatively long in Renmin Road has been improved to some extent and delay for stopping in the whole course has been even reduced from the previous 20–30 min to 10–14 min. But delay for stopping in some period for sections between Ganjiang Road and Daoqian Street is still rather long and no obvious improvement has been observed (Chart 1).
(b) Quantity of flow: after implementing "No Left Turn", quantity of flow does not show obvious change in sections of Renmin Road. But in general, quantity of flow in rush hours of the morning and evening in Renmin Road has decreased slightly, while it increases to some extent in normal time. As a result of this, rush hours of morning and evening in Renmin Road has been reduced by 15–30 min, in the meanwhile small peak traffic occurs during 11:00–13:00 (Chart 1).
(c) Average speed of vehicle: after putting "No Left Turn" into effect, average speed in most sections of Renmin Road has increased to relatively large extent

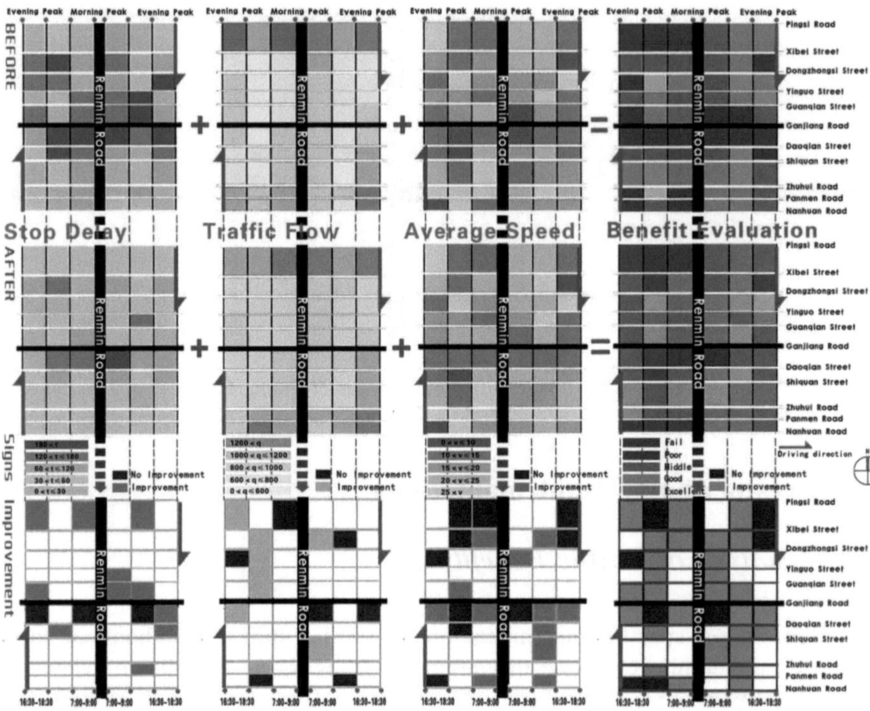

Fig. 4 Before and after "No-left Turn" implementing factor analysis model (Source: author complete)

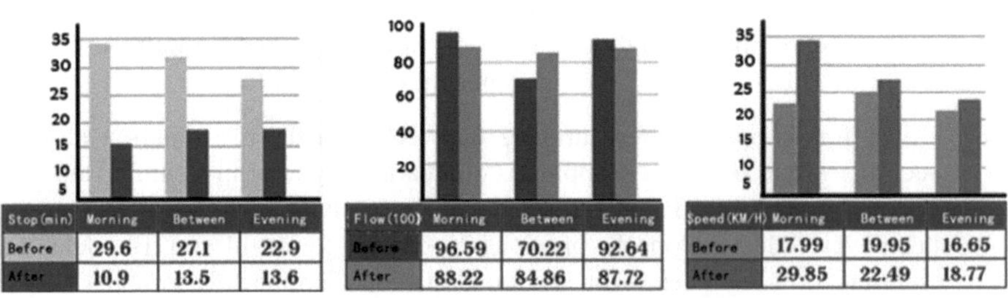

Stop (min)	Morning	Between	Evening
Before	29.6	27.1	22.9
After	10.9	13.5	13.6

Flow (100)	Morning	Between	Evening
Before	96.59	70.22	92.64
After	88.22	84.86	87.72

Speed (KM/H)	Morning	Between	Evening
Before	17.99	19.95	16.65
After	29.85	22.49	18.77

Chart 1 Compare before and after "No-left Turn" (Source: author complete)

with general average speed increasing from previous 16–20 km/h to 18–30 km/h, which improves the function that Remmin Road communicates the inside and outside of the ancient city in some degree (Chart 1).

(d) Evaluation of traffic benefit: Integrate three factors which affect traffic benefit, which are delay for stopping, quantity of flow and average speed of vehicle, to come up with comprehensive evaluation of traffic benefit before and after carrying out "No Left Turn". It can be drawn From the table that traffic benefit of Renmin Road has major improvement in general after carrying out the scheme, while problems still exist for traffic benefit of the south and north end of Renmin Road and sections between Ganjiang Road and Daoqian Street.

3.2 *Empirical Investigation*

After the study and analysis of traffic benefit in whole line of Renmin Road, aiming at overall analyzing the implementations of the scheme "No Left Turn in Rush Hours of Public Utilities" and its effects on civil life, further work of questionnaire and interview has been carried out. After the collation and statistics of questionnaire and interviews with citizens, it is found out that after carrying out "No Left Turn in Entrance and Exit during Rush Hour":

(a) Public transportation has improved: according to field observation and citizens' common reflection, because of eliminating left turning vehicles' interference with bus lanes on both sides of the road, operating speed of bus has improved, delay reduces and punctuality rate increases to a large extent, thus attraction of city bus has been distinctly elevated and the traffic function of Renmin Road as the passageway for bus is ensured.
(b) Means of transportation has changed in some way: the questionnaires and interviews show that, 40 % of citizens use private cars as the usual means of travelling before carrying out "No Left Turn". However, with the implementation of "No Left Turn in Entrance and Exit during Rush Hour" and the improvement of efficiency of bus, almost 10 % of these citizens commute by public transportation instead (Chart 2).
(c) Citizens working in enterprises and institutions along this line feel inconvenient for them to commute: because the immediate objective of "No Left Turn" is some public facilities on Renmin Road, directly certain inconvenience has been caused for vehicle getting in and out of these facilities. According to the statistics of the questionnaires, 60 % of citizens think that, due to the lack of certain publicity and guidance after "No Left Turn", many citizens are troubled with problems like "not clear about how to make a detour" and "have no idea where to turn around" (Chart 3).

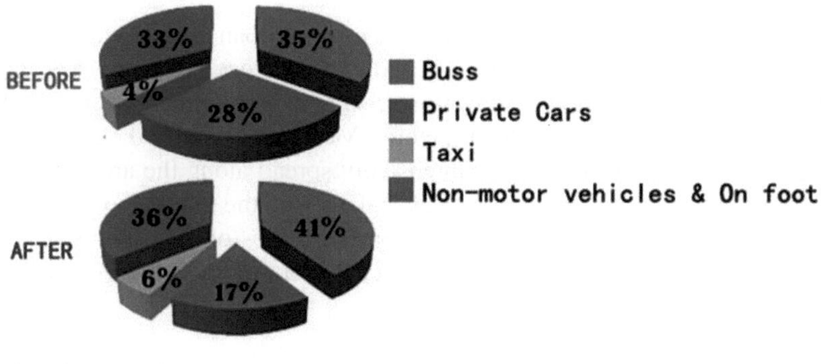

Chart 2 Trip mode change (Source: author complete)

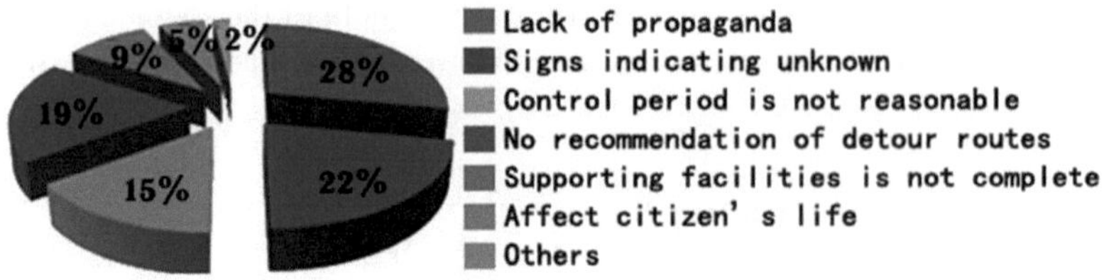

Chart 3 Citizen's attitudes towards "No-left Turn" (Source: author complete)

Through a series of factor analysis before and after carrying out "No Left Turn" and combining the empirical investigation made, it is discovered that overall traffic benefit of Renmin Road has increased after carrying out "No Left Turn" and traffic jams in rush hour has been eased in some way. But major congestion problem in some sections still exists and needs an in-depth study and rational dispersion. Besides, the implementation of "No Left Turn" has major effects on citizens who work in facilities along the road when they go on and off duty, so it's essential to execute follow-up management and improve surroundings for detour. Hence, a typical section of this road is chose to be further studied.

4 Suggestion and Prospect

4.1 Suggestion

Through the investigation of the implementation of "No Left Turn in the Rush Hour of Public Utilities", and in-depth study of traffic benefit, public opinions and responses, effects of "No Left Turn" on various facilities, bypass and management, the following recommendations for improvement are put forward.

4.1.1 Adjust Land Use and Reduce Openings in Arteries

Because urban traffic conflicts concentrate in central urban area, the first thing to do is to make use of land differential effects and move factories, warehouses and lands which fail to fit the function of downtown, such as government sectors, schools, public institutions to the outskirts of the city. Moreover, modify the distribution of commercial facilities which are too intensively spread along the arteries, properly disperse activities in city and seek new balance in the traffic volume. Related departments also should reduce openings on both sides of the arteries in order to protect traffic in the arteries from too much interference and keep traffic more smooth.

4.1.2 Optimize Public Transportation and Enrich Traffic Patterns

From the view of long-term development of city, it is indispensable to develop public transportation and multi-mode transportation. Related departments are called for to intensify the research on urban multi-mode transportation and develop the preliminary work of scheming non-motor vehicle system and pedestrian system which suit for regional characteristics. By short-term, medium-term and long-term overall traffic strategic researches, construct the modern public transport system of "railway transit and bus rapid transit as backbones, regular bus transit as main body and other public transit as supplement", thus transforming trip mode and enhancing traffic quality.

4.1.3 Adjust the Setting of One-Way Street and Perfect Traffic Microcirculatory System

It is recommended to make reasonable traffic scheme based on the original road system and after carrying out field visit. For example, add more and more reasonable one-way street and set tidal lanes on some roads, meanly, practice one-way traffic in rush hours of the morning and evening in some crowded roads and return to normal in ordinary hours to assure good traffic, and rationally set parking places on both sides of the road. Besides, further comb traffic in branches, consummate bypass in branches and traffic microcirculatory system, rationally disperse traffic flow, dig into traffic capacity, give full play to the comprehensive efficiency of urban road network, improve traffic order and alleviate traffic jams.

4.1.4 Improve Supporting Facilities and Follow up Management of "No Left Turn"

When carrying out "No Left Turn", related traffic administrative department called for to further improve and adjust supporting facilities, such as the perfecting road signs, t introducing changeable signposts, adopting road regulation and dispatching system, and promoting vertical parking lots. Besides, management, publicity and guiding work which supplement "No Left Turn" need to be strengthened so as to coordinate the hardware with the software and better ease traffic problems in Renmin Road.

4.2 Prospect

In the process of this investigation, from the current situation of implementation of "No Left Turn in the Rush Hour of Public Facilities", this paper makes scientific evaluation of the implementations in terms of traffic benefit, traffic function, and social effects of Renmin Road after executing the scheme, and inquires into existing problems. Then it investigates typical section of this road to find out the crux and collect public responses, trying to come up with effective advices and perfect the implementation of the scheme. In such an investigation, it is clearly comprehended that traffic jams is an inevitable problem with the development of the city. From the eyes of a designer, a different way to study this kind of problems is discovered. Traffic flow is like water flow, thus faced with the flow of motor vehicles which surges like flood, other than regulate rivers and watercourses as it is, more management and dispersion should be done to improve the circulating environment. Only by doing so, can traffic flow gallop in the "blood vein" of city and bring city vitality and vigor.

References

1. Cheng Kuanmin, Luo Zhizhong (2006) Characteristic analysis of left turning traffic flow in intersections and countermeasure research [J]. Road Traffic Technol (2)
2. Nan Chunli, Yan Baojie, Chen Hong (2006) Simulation model of the turning around behavior of vehicle under no left turn [J]. Comput Eng Appl (20)
3. Liu Xuefeng, Xu Menghua (2004) Discuss on traffic organization, planning and design for roads in Taiyuan City – introspection on canalized intersection and no left turn [J]. Taiyuan Technol (2)
4. Long Kejun, Chen Qun (2011) No left turn in intersections of road network and traffic organization optimization [J]. Coll J Beijing Univ Technol (6)
5. Li Yanyan (2011) Brief analysis of no left turn theory of traffic [J]. Traffic Sci Technol Econ (6)
6. Huang Weihua (2011) Design and reflection of no left turn regulation works in South Road around Kunming City [J]. Traffic Sci Technol Econ (6)
7. Wang Hui, Tong Lu, Wang Bo, Guo Zhong (2008) Impact assessment of traffic organization of no left turn in urban road intersections based on Visum [J]. Urban Public Util (6)
8. Zhou Le, Zhang Guohua, Dai Jifeng, Wang Jinqiu (2006) Traffic analysis of ancient city Suzhou and improvement strategy [J]. Urban Traffic (4)
9. Wen Guowei (2012) Analysis of traffic problem in famous historical cities [J]. Famous Cities China (1)

Problems and Redevelopment Strategy for Old Industrial Area

After an Accident in Nanjing's Mai-Gao-Qiao Area

Xin Yi

1 Problem Statement: A Specific Development Strategy for Old Industrial Area?

In China's rapid urbanization process, a great part of academic discussion concentrates on development strategy and tendencies in new development areas or in the urban regeneration of old city centers [1]. In contrast to such issues, this article tries to call attention to some other areas such as the former industrial districts. Usually located outside the old urban area, the industrial district is now becoming integrated into the urban area and is experiencing a deep transformative process.

In the author's perspective, such areas play an important role for the city's function as a whole but are not on most decision makers' agendas. Although there are several famous flagship projects in old industrial areas, such as Jiang-Nan Shipyard in Shanghai as an Expo site or Shou-Gang Steelworks' redevelopment in Beijing under influence of the Olympics [2]; the majority of old industrial areas in today's cities are under a series of restructuring processes without a systematic intervention.

As a result, the transformation of these industrial areas tends to be in a laissez-faire process [3], in which several influential local stakeholders dominate the redevelopment program for their own interests instead of a more comprehensive consideration on city level.

This article has chosen to examine an old industrial area in Nanjing, the Mao-Gao-Qiao (MGQ) area, which experienced a serious accident in 2010. The accident and the transformation process of this area will be illustrated in order to discuss the existing development mechanism in this type of old industrial area. The transformation pattern will be analyzed by reviewing today's urban development strategy as well as institutional frameworks. Additionally, the social and economic profile

X. Yi (✉)
School of Architecture, Southeast University, Sipailou 2#, Nanjing 210096, China
e-mail: xinyi.tum@gmail.com

X. Chen, Q. Pan (eds.), *Building Resilient Cities in China: The Nexus between Planning and Science*, GeoJournal Library 113, DOI 10.1007/978-3-319-14145-9_28

for different social groups will be delineated. Lastly, a specific development strategy for such kind of area will also be proposed in the conclusion.

2 Reorientation of an Industrial Area Against Rapid Urban Development

2.1 Site Profile

On July 28, 2010, there was a serious occurrence within the Mao-Gao-Qiao (MGQ) area in Nanjing. A former industrial site of 9 ha for the Nan-Su 4th Factory (Nanjing Plastic Industry Group) was undergoing the process land consolidation. This site was formerly used for chemical production and left intact a large amount of old industrial facilities and other remenants of the former use. Production had recently stopped, and the site was already transferred to a local government backed developer. Because of mistakes in the physical land consolidation process, there was leak from a propylene pipeline underground, leading to a serious explosion. Thirteen people were killed and at least 120 people were brought to local hospitals. This accident helped to focus the public's attention to the government's handling of safety issues [4] (Internet: http://baike.baidu.com.cn/view/4029213.htm).

As with many other former industrial areas, the MGQ area has been experiencing an extensive transformation against the urbanization process on the city level as well as a deindustrialization process. The chemical industry is seen as one of pillar industry in Nanjing but also has brought many accidents in its development to the city, especially in today's rapid urbanization process (Fig. 1). Such factories were

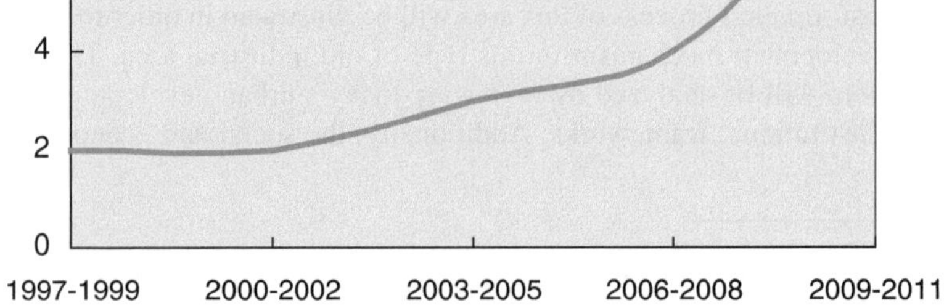

Fig. 1 Chemical industry accident in recent years (Source: MGQ investigation report)

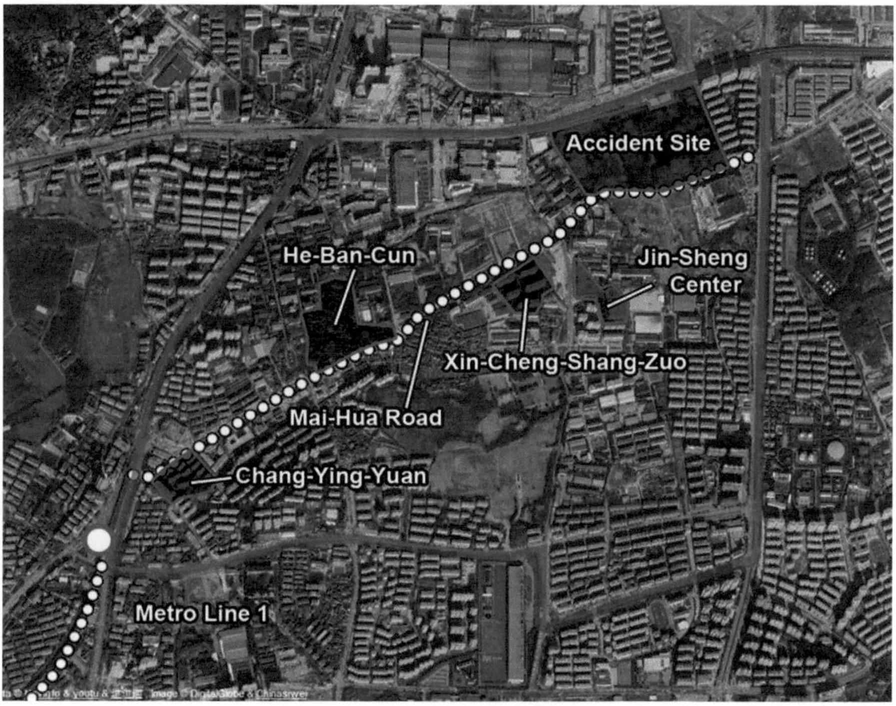

Fig. 2 MGQ area and investigated sites (Source: drawn by author)

located in the suburban area before but have been integrated into the large urbanized area in recent years (Fig. 2).

The investigated MGQ area is located in the northern part of Nanjing's central city area, the Qi-Xia district. It covers 290.1 ha [5]. Physically, an old industrial area is identifiable with four main streets serving as its boundary. The northern boundary is a highway that connects the eastern part of Nanjing's central city to the new Xin-Gang industrial area.

As one of the four former large chemical industrial areas in Nanjing, this area is home to many factories. In addition to the industrial facilities, there are also residential apartments built by the factory owners for their workers, newly developed commercial residential buildings, as well as several old urban villages, which remain their own administrational organizations for the rural area [6].

2.2 *Transformation of General Condition*

It is necessary to retrospect Nanjing's urbanization process over last year. In the former master plan (1991–2010), the central concept for urban development concentrated on a balanced relationship between the inner city's regeneration and an

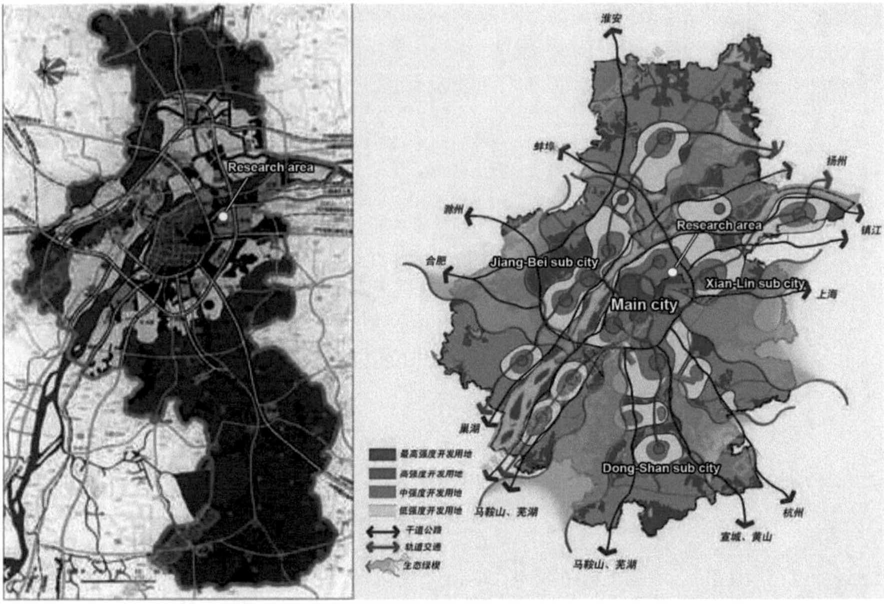

Fig. 3 Master plan (1991–2010) on the *left* and Master plan (2007–2030) on the *right* (Source: Nanjing's Master plan (2007–2030))

outward urban extenpansion (Fig. 3). In contrast, the most recently established master plan (2007–2030) transfers its focus into cultivating a polycentric structure across the whole region. According to this plan, in the whole 6,582 km^2 administration area of Nanjing, 4,388 km^2 is designated as urban area and 834 km^2 as central urban area. The latter area assumes the essential urban functions for the city and region. This central urban area is composed of one main city and three sub cities, Dong-Shan, Xian-Lin and Jiang-Bei (Fig. 3). The related spatial development strategy is embodied in following aspects:

- To enlarge the urban built area by pursuing the aimed spatial structure with emphasis on the three new sub cities.
- To accelerate the development of the new urban area in main city and to enhance the urban quality in the old city.
- To implement the development strategy in the urbanization process for surrounding area in the whole city.
- To accelerate the implementation of a public transportation-oriented strategy and to improve travelling conditions.
- To improve the urban environment and to strengthen the protection of the historic city.
- To enhance the regional service capacity and the city's comprehensive competitiveness.

(Source: Master plan (2007–2030), city of Nanjing)

In comparison to the comprehensive master plan, this spatial development policy leaves great deficiencies for the city's further development. In the face of greater urban expansion, the demographic development is much slower. The overall population of all of Nanjing has increased from 6.13 to 7.41 million from 2000 to 2007, about 3 % growth per year on average. Such a speed shows a limited capacity to meet the great leap of the rapid urbanization process. Correspondingly, a series of large projects with great land demand have been initiated and planned in the newly planned sub cities. For example, the Xian-Lin area is filled with many new campuses for different universities like Nanjing University; and there are enormous new industrial areas such as Xin-Gang industrial area along the Yangtze River. As a result, vast resources have been invested in the new sub cities, especially technical infrastructure in the newly developed urban area and in connecting utilities with the city center. Furthermore, the currently low population density in new urban areas leads to the shortcomings in social infrastructure and other service facilities.

2.3 *Role Transformation: From Urban Periphery to City Corridor*

The subject of this investigation, the MGQ area, is located on the main city's periphery and is also connected to the Xian-Lin sub city to the east. As a result, the master plan shows a secondary role for the MGQ area. The aforementioned six aspects will have few direct influences or systematic intervention from city level on the research site. The master plan only mentions briefly that urban renewal measures for the social and economic transformation should be developed (and offers no guidance on how to develop these measures). Additionally, a district level public facility sub-centre has been planned in the northwestern part of the MGQ area because of its high population density; but the related implementation measure has not been discussed. There is also no strategic project in this area. In the master plan, the existing spatial structure of research area has been overlooked; and the existing diagonal main street in the site is removed (Fig. 4).

Such types of final outcome oriented blueprint master plans provide very limited guidelines for the actual implementation process. However, the economic and functional relationship with other surrounding urban areas has stimulated the further transformation of this area. In general, MGQ area encounters great changes in the general condition, such as location and accessibility. With the completion of Metro Line 1 and development of Xian-Lin district in the east, the MGQ area has been converted to a corridor between the main city and the sub city and has been further integrated into the whole urban structure.

Within the study area, there is a permanent residential population of approximately 8,000 and a temporary population of 9,000 (MGQ report), the latter of which is composed of a large amount of domestic immigrants from outside Nanjing. The

Fig. 4 Land use and main street in Nanjing's master plan (2007–2030)

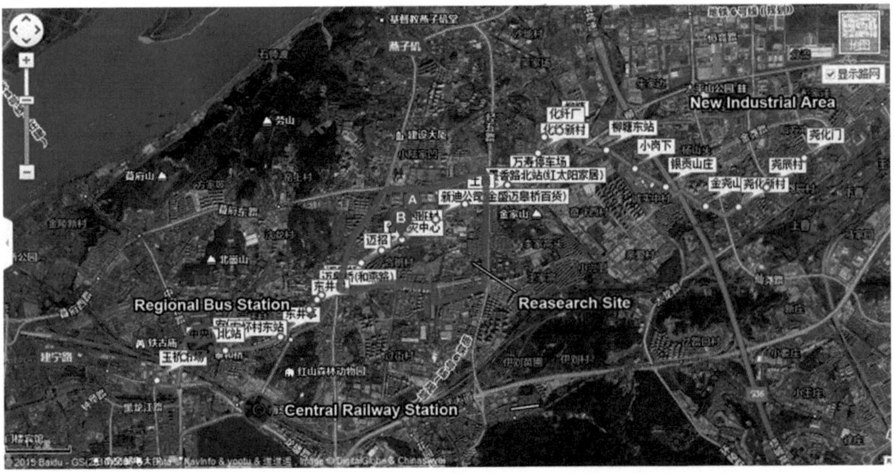

Fig. 5 Bus line 76 through the diagonal street across MGQ area (Source: internet in Baidu)

eastern new Xin-Gang industrial area has provided enough work for these immigrants, but there is limited residential facilities and social infrastructure for them.

The MGQ area appears to be strongly supported by the government as it is crucial in supporting the industrial area in the east. By observing the path of the public bus line 76 (Fig. 5), as it travels from the regional bus station and central railway station to the newly developed industrial area, it is clear that the MGQ area serves as a corridor between the main city and new industrial area in the east. Additionally, the central railway station in the west brings great inflows of immigrants into this area [7].

However, even with such an important transportation function, MGQ has not been identified as significant in the master plan's land use and traffic structure. As a result, there is also no incremental planning concept. The temporal dimension in the implementation has received very limited consideration. In this meaning, the spatial transformation of MGQ encounters an obvious unregulated development process. No specific regulation can be carried out.

At the same time, MGQ's profitable location between main city and the new industrial area, (based on a functional interrelationship in working position and housing) has attracted great attentions from local developers. As a result, a series of encroaching activities have occurred in MGQ based on the coalition between private developers and other local government backed institutions. In actuality, the accident on July 28, 2010 was caused under such kind of coalition. After the accident, a local community director and accomplices were charged with corruption and thrown into jail. Of course, there are many other factors contributing to this accident; but it would be reasonable to say that the very limited planning regulation should also take responsibility to some extent.

3 Site Investigation for Specific Development Strategy

3.1 Special Institutional Framework

It is necessary to investigate the existing institutional conditions and stakeholder relationships of MGQ for the strategy discussion in following text.

Actually, the institutional problem is due to the coexistence of a market economy and a centrally planned economy [8]. As it is well known, such industrial areas were initially organized by the state-owned enterprises. Based on the centrally planned principle, these government-owned companies were appointed to run specific sectors of the economy. Belonging to diversified industrial sectors of different scales, these enterprises are put under supervision entities with respective administrational levels, such as community, district, city, province or even central government level. In this principle, these enterprises are also authorized to self-organize in their designated industrial area. To a large extent, they will be more responsible for their own supervision entity rather than the local government, in which they are located, especially if the supervision entity has a higher administrative level (Fig. 6).

In other words, the systematic intervention on such former industrial area requires the leading organization allied with a relatively higher administrative level. If this doesn't take place, the coalition between the local government and such state-owned enterprises will still dominate the site's development, but there will be no comprehensive improvements. It would not be very likely to initiate an artificial cooperation among different enterprises, if they have different supervision entities. As a result, this general condition leads to a corresponding nibbling way encroachment of the old industrial area.

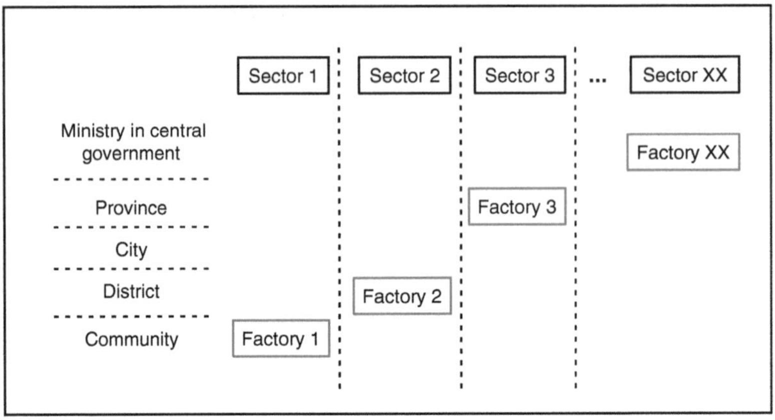

Fig. 6 State-owned industrial enterprise of different sectors under supervision entities of different administrational level (Source: drawn by author)

3.2 Example: Stakeholder Structure in the Accident Site

The accident site is a typical industrial area, which experienced different periods of industrial development. The following is an analysis on the site's development history and also the stakeholder relationship with respective roles in the accident site. (http://2009.nbd.com.cn/newshtml/20100804/20100804023704679.html).

Nan-Su 4th Factory The former industrial enterprise "Nan-Su 4th Factory" operated this area for industrial production until the end of the 1990s. The factory was then closed, and their workers lost their jobs and given only a very small compensation payment. After that, this area had been used as storage functions for building material or logistic firms with short-termed contracts; but overall, there were no changes in its physical structure. In 2005, this factory was integrated into the "Nanjing Textile Industry Group". There were a number of ideas to develop this area from industrial use to commercial and residential functions. The substantial procedure was started at the beginning of 2010. The land consolidation work was also carried out. The required land use transformation document was also approved.

Mai-Yan Construction As a local government backed real estate developer, Mai-Yan Construction was founded on February 20, 2009. The three shareholders with respective shares are the following: Qi-Xia district state-owned investments center (50 %), Nanjing Kang-Qiao Ltd. Corporation (44 %) and Nanjing Yang-Zi-Jiang Economic & Technology Development Ltd. Corporation (6 %). The second shareholder – Kang-Qiao – developed itself from an enterprise of collective ownership from the original local rural community. Its corporate representative is also an officer in the community commission. Mai-Yan Construction authorized land consolidation assignment to "Hong-Yun Corporation", which then directly caused the accident.

Qi-Xia Construction As a listed company, this corporation (51.04 %) has founded a "Mai-Yan Real Estate Corporation" with Mai-Yan Construction (48.96 %). Then as transferee, Mai-Yan Construction has set up transfer contract to purchase all other share of "Mai-Yan Real Estate" from Qi-Xia Construction in 2013.

Nanjing Textile Industry Group This group was founded in 2005. The former Textile Industry Bureau, City of Nanjing, was merged together with Textile Industry Corporation to set up Nanjing Textile Industry Group. Considering the higher administrational level for this group, the land was under supervision of Textile Industry Bureau before and then subsequently belonged to Nanjing Textile Industry Group. Based on the above mentioned administrational hierarchy principle, the Qi-Xia district level has no direct authority to influence this group. As a result, Nanjing Textile Industry Group would be both owner and developer to convert this former industrial area to commercial and residential function in the coalition with "Mai-Yan Construction".

3.3 Zone of Transition with a Community Development

The research area is composed of an extensive spectrum of social groups, which relate to the transformation process of this area. During the deindustrialization process, a series of interventions have been implemented by the different stakeholders and social groups. The site information will be summarized in following categories.

3.3.1 Residential Facilities

Physically, there are two types of building forms, multistory residential towers and office buildings for the middle class and hastily constructed urban villages for low income newcomers. Outside Nanjing Clusters, these buildings coexist together in the whole research area. Primarily, multi-story buildings are located along the boundary and within villages in the site's interior. Some multi-story buildings have been built by former state-owned enterprises, but units within these building have been already sold to the long-term residents. Other multi-story buildings have been built in recent years by local developers. In contrast, temporary residents with limited economic capacity have accommodated themselves within the urban villages. Other residential facilities include public housing for temporary residents, but these provide limited space because of shortage of investment (Fig. 2).

In the relatively early developed residential area "Chang-Ying-Yuan", the composition of residents is elderly retirees (49 %), teachers from public schools in Qi-Xia district (33 %), individual purchasers (11 %) and others (7 %) (Source: sample survey by the author). The retired residents are from surrounding areas, whose houses have been torn down; so the owners were compensated by the developer.

These people then purchased the new apartments with this compensation. Teachers receive the housing by public allocation from local government or have paid for it but at a discounted price. Other residents have paid for the apartment by themselves. The average price there is slightly over 10,000 Yuan per square meter. At the moment, this price is relatively lower than that in city center.

In the other urban village "He-Ban-Cun", a large minority of residents are from other provinces such as An-Hui; but usually, the homeowners are original village residents with a large part of their income from the tenants' rent from the floating people. Some of the tenants have lived in these conditions for more than 8 years, and have developed networks with each other. Many new comers have found this area by means of their own personal networks. As a result, they have seen this area as a small community a self-consciousness. When the author spoke with the tenants, they always mentioned the work of "we" and emphasized the regular security issues in what they believe is a spatially chaotic village. According to the author's observations, situations such as physical structure and hygienic conditions are actually quite acceptable. In this meaning, the social control by the local community seems effective, since the security issues are always a threatening issues for such kind of area, which accommodate a large percentage of new immigrants.

3.3.2 Working Facilities

As mentioned before, there are still several currently functioning factories but with limited positions. On the diagonal main street "Mai-Hua Road", there are a large amount of facilities e.g. tea markets, office buildings with renovated industrial structure, economy hotels, and large shopping centers.

However, most residents in the area have their places of employment beyond MGQ. From the information of one employment agency in the He-Ban-Cun, the author is informed that 80 % of the residents in the urban village work mainly in the Xin-Gang industrial area (Source: sample survey by the author). In addition to for the public bus line in connection with the industrial area, the large enterprises also provide shuttle buses for their worker on the Mai-Hua road. Other residents there work in the service sectors, (bars, restaurants, and café houses) or in other industrial areas, which can be reached by public transportation.

Some businessmen working in the markets or shopping centers have bought their own apartments in the commercial residential buildings. Some new commercial residential buildings have taken such people as important goal market (Xin-Cheng-Shang-Zuo) (Fig. 2). The other residents living in the multi-story buildings have their jobs all over the city, since the Metro Line 1 provides a good accessibility to other parts of city.

3.3.3 Service Facilities

As one of the most important facilities for the residents, many commercial facilities are located on the Mai-Hua Road; and they exist as the important interface between the residents from multi-story buildings and those from the urban villages. Except for the above mentioned commercial facilities, there is lots of space for leisure and recreational activities also on this road. Some facilities like the "Jin-Sheng Center" provide such space. On the Mai-Hua road, there are also many people who take walks by themselves or with dogs. The wide street is used as common living room for the whole quarter.

In the chess rooms, half of people playing mahjong are from urban village and the other half from the multi-story buildings. The recreational space in Jin-Sheng center is open from 5:30 p.m. to 2 a.m. because most workers will visit there after work hours. One disadvantage is the lack of facilities for elderly people or green space. Elderly people have limited choices and usually stay at home for housework or take care of children in the inner yard of commercial residential buildings.

Based on the investigation, it would be summarized that the residents in the village and in the multi-story building still hold very different preferences and social background on one hand; but in public space such as Mai-Hua road and within many leisure facilities, there is the chance for different social groups to intermix.

3.3.4 Technical Facilities

As one of four large industrial areas, MGQ area embraced several large industry facilities as well as many other small and middle factories. The pipes go through the whole area, especially in the northern part with a clear proximity to the residential and commercial areas. Considering the factories and their direct sphere of influence for safety, almost 40 % of whole area are under the threat of potential industrial accidents [5] (Source: MGQ investigation report).

In addition to the industrial facilities, there are also other supply units: one gas station for old residential area because there is still no gas pipe connected and two petrol stations for the passing traffic flows and also for the requirements from industry and local residents. During the explosion accident, an adjacent liquefied gases station exploded and lead to a greater explosion accident.

The other social infrastructures (e.g. hospital and fire station) could not be developed in accompanying with the extension of new residential and commercial facilities, since there is no systematic intervention for the whole area to provide the required space. Existing measures concentrate in maintaining the existing facilities to avoid the security issues (Fig. 7).

Fig. 7 Industrial facilities in the MGQ area (Source: MGQ investigation report)

4 Conclusion

Based on the above discussion, a specific development strategy will be proposed. It is admitted that other than several large industrial areas comprehensively redeveloped for the occasion of mega-events, former industrial area in today's Chinese cities under a transformation process are encountering a dilemma between little direct support for ambitious planning from city level and piecemeal development by coalition between local government and state-owned industry enterprises. The master plan alone cannot solve such complicated problems with its outcome oriented blueprint measure on one hand; and the piecemeal development can only lead to the deterioration of such chaotic status but even brings the serious security issues like explosion accident in MGQ area.

On the other hand, many other informal relationships in economic, social, as well as spatial aspects show the important meanings of such areas in the city as a whole. In this meaning, a specific planning method with consideration of such relationships of the urban economy should be developed.

It is required to develop a specific planning approach for the development of old industrial areas. With the consideration that there is no strategic project, and the city government has no further resources to lead an overall intervention in the area in near future; it can be concluded that a third road among ambitious master plan and encroaching activities by local developer should be pursued.

In comparison to the profound transformation outside, there is limited systematic intervention from the city level. To maintain the functional relationship between the MGQ area and the new industrial area in the east, it is necessary to see the research areas as heterogenic and dynamic objects as a whole and cultivate corresponding social, economic and spatial relationships correspondingly. The development path for this area should also be considered.

For city government, they have authority to give restriction of further development of old industrial structure and, with legislation, to stop such coalition between local community director and industry enterprises. Generally, the efforts should focus on controlling the further piecemeal development. Limited control and regulation to the informal relationships should be utilized, so that the security issues would be addressed and the low income groups' demands will be considered. Such functions are also important for the new industrial area in the east with residential and service facilities. In long-term, the whole area should be redeveloped by the systematic intervention from city level. At the same time, it is also necessary to improve the living conditions in the newly developed area, so that the people could be better accommodated there rather than always stay in an informal urban village.

References

1. Broudehoux A-M (2004) The making and selling of post-Mao Beijing. Routledge, London/New York
2. Xin Y (2011) Urban transition via Olympics. Fakultät für Architektur, Technische Universität München, München
3. Park RE, Burgess EW, Mackenzie RD (1925) The city: suggestions for the investigation of human behavior in the urban environment. University of Chicago Press, Chicago
4. Internet source: http://baike.baidu.com.cn/view/4029213.htm
5. Xiong G, Zhu Y, Wu X (2011) MGQ investigation report – investigation for chemical industry security in Nanjing Mao-Gao-Qiao area (not published report). School of architecture, Southeast University
6. Guan W (2013) Political economic analysis on urban space redevelopment plan: a case study of Yan Ziji old industrial district in Nanjing. International Urban Planning, Beijing
7. Internet source: http://2009.nbd.com.cn/newshtml/20100804/20100804023704679.html
8. Wu F, Ma LJC (eds) (2005) Restructuring the Chinese city: changing society, economy and space. Routledge, London

Analysis of the Planning Mode of Resilient Urban Space Structure

Zhou Min, Lin Kaixuan, and Huang Yaping

1 Introduction

In the information society, as a complex giant system, the city has become more and more powerful, while also becoming increasingly vulnerable. A series of natural and man-made disasters, such as floods, extreme weather, disease infection, terrorist attacks, and so on, have been difficult for individuals to cope with. In the face of crises, the problems that arise can show how resistant cities are to all kinds of change and disaster, and their ability to maintain vitality and development. In recent years, the concept of flexibility in cities has been given more attention. The concept of the resilient city, on the one hand, stresses the buffering capacity of external shocks. On the other hand, it reinforces the enhancement of the city's ability to learn and seize opportunities to sustain its vitality. From a research standpoint, domestic and overseas research has extended to a multi-dimensional perspective, which consists of ecological, technological, social, and economic principles. In the face of urban sprawl resulting from China's rapid urbanization, there is need for a new spatial development concept to guide urban development. Based on the resilient city model in contemporary China, this paper presents the elastic urban space planning mode, which provides a reference for the sustainable development of cities in China.

Z. Min (✉) • H. Yaping
School of Architecture and Urban Planning, Huazhong University of Science and Technology, Wuhan 430074, China
e-mail: zhouminhust@126.com; hust_hyp@sina.com

L. Kaixuan
Jiangsu Institute of Urban Planning and Design, Nanjing 210000, China
e-mail: lkxzd@126.com

X. Chen, Q. Pan (eds.), *Building Resilient Cities in China: The Nexus between Planning and Science*, GeoJournal Library 113, DOI 10.1007/978-3-319-14145-9_29

2 Resilient Space Structure Is Proposed

Regionalization has caused the relationship between the city and the outside world to exceed the scope of the administrative jurisdiction. With it has emerged mixed urban land use, the multi-center urban spatial structure, and the integration of ecological space. In traditional urban planning, once a city's structure or size has been determined, the quantitative measurements remain relatively static compared to a city that is in the dynamic process of continuous development. How is it possible to meet the dynamic requirements of urban development? How might a city exceed the established spatial structure model in order to meet the needs of the city's dynamic growth? Based on this line of thought, this paper presents the concept of resilient space structures. The basic idea is that in understanding the extent of city conditions, it is possible to enhance the city's public bearing capacity by adjusting the structure of urban space, thus recognizing the resilient demand for urban development.

3 The Basic Attribute of Resilient Urban Spatial Structure Mode

I believe that, from the concept of urban resiliency, a resilient urban spatial structure mode should have four basic attributes, including ecology, growth, economy, and coordination (Fig. 1).

3.1 The Ecology of the Resilient Urban Spatial Structure Mode

In the 1970s, UNESCO initiated the "Man and the Biosphere" (MAB) program and put forward the concept of the "eco-city." Today, the concept of the eco-city has triggered widespread awareness. From a specific physical design level, the main objective of the eco-city is the construction of ecological sustainability harmonious with intensive human settlements. From a practical planning point of view, the concept of ecology has gradually been explored [1]. The green belt, or green wedge, in the planning of major cities is proof of this. Undeniably, the arrival of the information society and the concept of ecological sustainability will significantly impact urban spatial structures.

Ecology is the most basic attribute of the resilient spatial structure. It is also the basic element of protection that can enhance the ability to respond to disasters, as well as the capability of self-recovery. Based on spatial elements, an eco-city embodies a space that is able to provide residents with ample open spaces, including city and forest parks, scenic areas, rivers, and lakes. An intrinsic characteristic of a

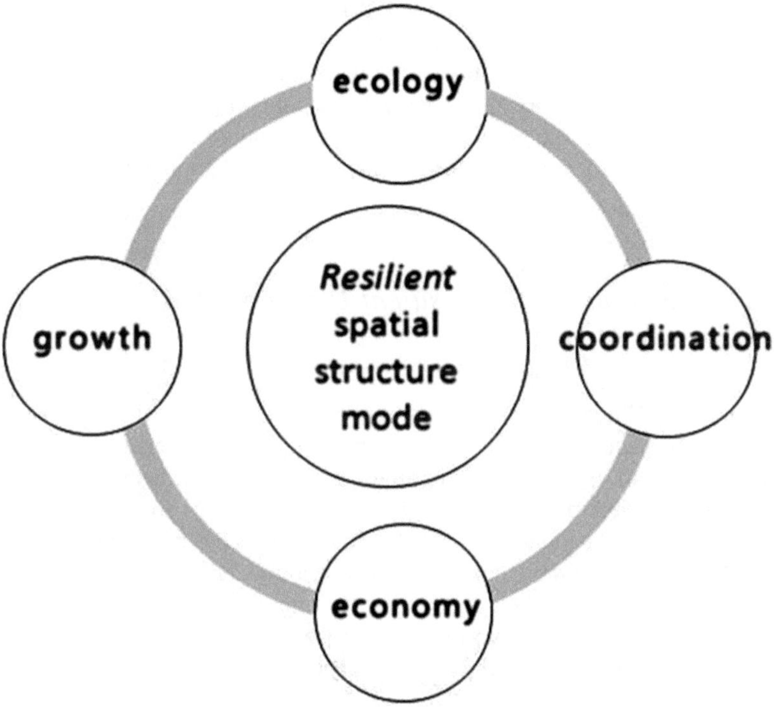

Fig. 1 The connotation of resilient urban spatial structure (Source: Lin Kaixuan [6])

city is its mode of urban spatial structure. The ecology of this spatial structure pattern directly affects a city's ability to retain flexible growth. In other words, if a city wants to maintain its flexibility for growth, it first needs to have an ecological spatial structure. The dotted green parks, green belts, green hearts, green wedge areas, and ecological networks will directly impact the ecology of the urban spatial structure. Therefore, the city's ecological space structure pattern should integrate these into an organic green space system.

3.2 The Growth of the Resilient Urban Spatial Structure Mode

During the rapid development of the urbanization stage, due to the law of differential rent of land, the centralized, single-center urban growth model was seen to be a relatively high-efficiency mode of development. But results from this model being put into practice have shown that these cities have poor resilience. The uncertainty of the market economy has facilitated the sprawl of urban spaces, similar to the planning practices of Greater London and the Tokyo metropolitan area.

Copenhagen, Stockholm, Shenzhen, and other cities, however, have adopted the axis growth mode in the stage of rapid development, and they have shown great resilience and strong vitality in face of the market economy. This suggests that the spatial structures of the city mode have levels of vitality based on the varying growth characteristics. The growth of the urban spatial structure mode will also be an effective method of promoting urban sustainable development.

3.3 The Economy of the Resilient Urban Spatial Structure Mode

In the context of China's "space-time compression," China's cities have developed rapidly over the past 30 years through extensive land use patterns of development. This urban growth mode results in high-growth, high consumption, and high pollution, thus has led to the unsustainable city. A consensus about the construction of a compact city and its intensive use of land resources has been gradually formed, which reflects the urban development of the economy. The sustainable urban space structure mode should, through planning, reflect the character of the economy with a functionally scientific layout of the space and economic and efficient use of land. The outcome will be a city that is more compact and concentrated [2].

3.4 The Coordination of the Resilient Urban Spatial Structure Mode

Coordinated development between urban and rural areas must be emphasized. Urban and rural areas are the two most important spaces for human survival. A complete, sustainable spatial structure model should take into account the integrated model of development for both the urban and rural areas. The 1944 London Plan suggested that the integration of both urban and rural planning ideas have penetrated the planning practice. The proposal, which was created to establish the development of the outlying areas of the city, suggested that the model of urban spatial structure should have coordination between urban and rural areas. In other words, a successful urban spatial structure model should demonstrate this form of coordinated development.

4 Resilient Overall Urban Spatial Structure Mode Is Proposed

The overall urban spatial structure is part of urban area structure research. As part of resilient and concept-oriented development, combined with the evolving trends of the urban spatial structure, the author proposes a "multi-center, organizing network" resilient urban spatial structure model. The point of this structure model is to use large, contemporary cities or mega-cities as the research object, combining the city's natural conditions with their socio-economic development and other factors. The author will then divide the city into integrated planning data and build a "sub-structure of the fragmented local balance." The "sub-structure of the fragmented local balance" is not only an integral and important part of the urban continuum, but also the relative individuality of the city unit. These may be massive, linear, or cyclic forms developed by the group mode between areas. Through the shared contact of the transportation network, the ecological space is split and forms the network of geographical spatial structure (Figs. 2 and 3).

5 The Elements Resolution of Resilient Overall Urban Spatial Structure Mode

In view of the resilient urban spatial structure, the "multi-center, organizing network" proposed by the author will combine cases from the four levels involved in the public centers structure. These include the functional land use structure, the traffic structure, the eco-spatial structure, and a further analysis of the elastic spatial structure planning mode.

5.1 Public Center: "Ownerless-Center and Balanced" Mode or "Main-Center and Hierarchical" Mode

With the increasing integration of the urban function, the urban public center has transformed from a closed-end, single-center, and centralized mode to a multi-center network mode. The open multi-center network is less concerned with the overly concentrated central area, but is effective in containing the excessive fragmentation of the space structure pattern. In view of this, the author believes that the future of the public center structure in China's major cities will be optimized in the following two ways.

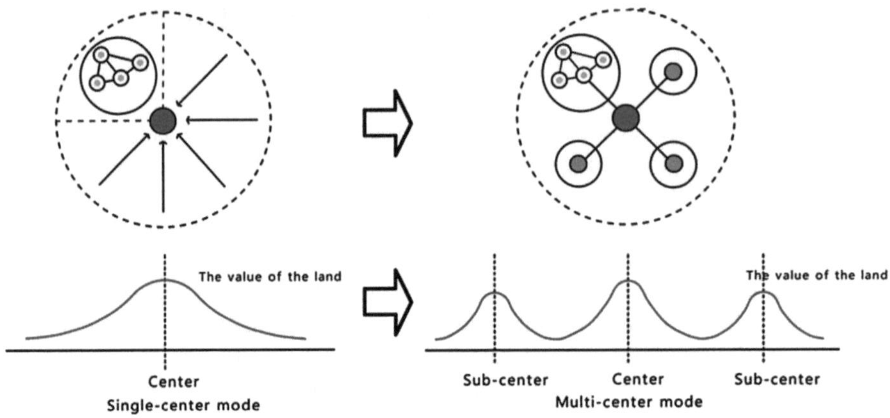

Fig. 2 The diagram of resilient overall urban spatial structure mode (Source: Lin Kaixuan [6])

Fig. 3 The mode of "ownerless center balanced" multi-center structure (Source: Lin Kaixuan [6])

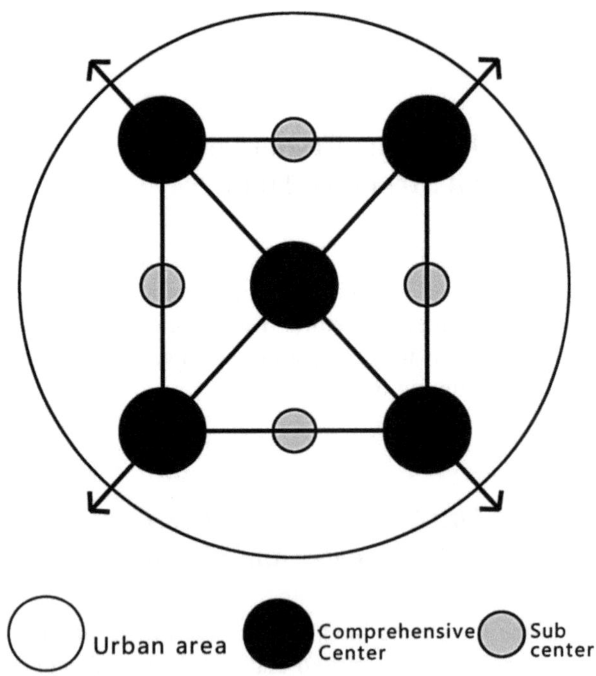

5.1.1 Building the "Ownerless-Center and Balanced" Multi-center Structure Mode

The "ownerless-center and balanced" structure mode means that there is no significant main center. At the same time, there are a number of equivalently sized area centers. With the emergence of the rapid transit system and information network, various centers have formed an orderly division of labor and collaborative structure.

The mode characterized by that layout of the city is more flexible. This mode is adapted for the cities that have formed the "strip-space" and "modular-space" structure modes, such as Shenzhen, Lanzhou, Xining, Shaoxing, Ningbo, and Suzhou.

5.1.2 Building the "Main-Center and Hierarchical" Multi-center Structure Mode

The "main-center and hierarchical" structure mode recommends a clear, strong main center within the city along with varying sizes, scales, and groups of sub-center land forms. The mode is characterized by a highly concentrated layout of the city. The boundaries of the city center are more obvious, and show a centralized layout of the land. This structure mode is adapted for cities that have formed a cluster spatial structure planning mode, or cluster-type space structure planning mode, such as Beijing, Kunming, Xian, Wuhan, Nanjing, and Changsha.

The radiating effect of the city's traditional center is still strong. On one hand, there is a need to keep the traditional hub. On the other hand, there is a need for appropriate evacuation to relieve the pressure of the central area (Fig. 4).

5.2 Functional Space: Compact-Group Structure and Diversity-Composite Layout

5.2.1 The Structure of Compact and Group-Style

Compact-group structure can effectively inhibit the sprawl of the city. The concept of compact urban areas is a sustainable design philosophy and an important idea found in New Urbanism planning. Compact development emphasizes a reasonable density of urban construction and efficient, economical use of land resources. Practice has proven that "the spread city" is economically inefficient because it increases the cost of travel. Compact land structure can save land resources, as well as promote social interaction. At the same time, the development of compact urban space should be controlled. There can be green belts established and urban growth boundaries set in order to control the development of urban space.

The concept of the group-style means that a relatively independent path is formed between the various functional areas of the city. The growth of this centralized pattern will only exacerbate the sprawl of urban space. Of course, the group-style does not mean to automatically divide the layout as mixed land use is one of the future trends. The group-style layout mode is meant to emphasize the principle of "large partition, small mixed," which means that the overall city needs to be divided into different functional areas and each area needs to achieve diversity in order to operate independently.

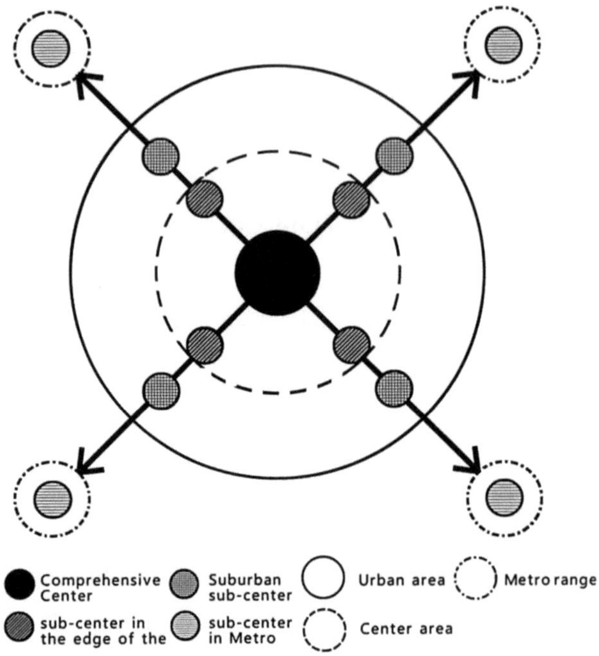

Fig. 4 The "main-center and hierarchical" multi-center structure mode (Source: Lin Kaixuan [6])

5.2.2 The Layout of Diversity and Composite

In the 1930s, the functional area in the "Athens Charter" occupied the mainstream thinking of the planning profession. Until Jane Jacobs strongly criticized the city's features as single and rigid, people gradually started to pay attention to the principle of diversity as a functional space. The planning concepts of new urbanism and smart growth both stress land diversity and mixed-use development. Speaking from the macro level, the diversity network layout is a mix of uses that form the overall function of a city. It reflects the basic properties that make up a city into a gathering place of integrated services. Speaking from the micro level, it emphasizes the diversity of the urban space, including architectural style, which is comprised of form, color, and, landscape design. The core of the city is made up of the individuals that occupy it, and in order to ensure its continued existence, the city must first meet the needs of the people. Also, the diverse layout of the network should further promote compact development of the city through people-oriented, well designed, and efficient-use of the land (Fig. 5).

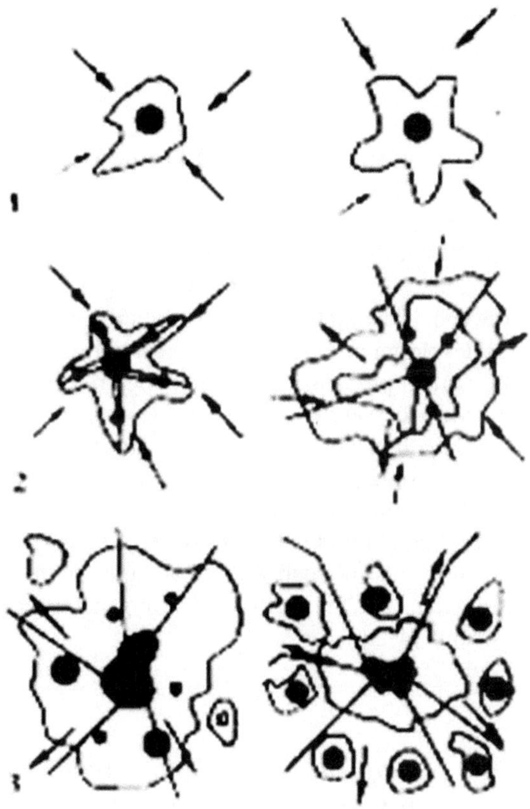

Fig. 5 The growing of the metropolitan areas' group layout (Source: Xie Shouhong [7])

5.3 *Road Traffic: Composite Network and Public Transport-Oriented*

5.3.1 Building the Fast Complex Network of Road Traffic Structure

The trend of urban regionalization and spatial diffusion determines the future direction of large city road traffic structures, which is transitioning to a fast, open network. From a regional perspective, the fast transport network from the inner city will form the main artery of the regional urban transport connections, accommodating inner city rail, high-speed railways, and highways [3]. From the inner city perspective, with the continuous expansion of city size, the transport connections between the various functional groups of the inner city are organized by fast public transport, such as subway, light rail, BRT, and fast roads. Whether it is a radial road network or grid-based road network, the road traffic structure model must be coupled with the functional layout of the land. Road traffic is serviced for guiding the rational distribution of urban functions, so traditional closed road traffic structure should gradually transform to an open network of road traffic. This coincides with the background of the market economy under the conditions of urban regionalization, the proliferation of urban space, and the restructuring of urban functions.

5.3.2 Persisting in the Road Traffic Mode of Public Transport-Oriented

The traditional mode of development for transportation showed cyclical characteristics of "traffic demand increases – an increase in traffic supply – increased traffic demand." Such a model will eventually be a vicious circle of "crowded – eased – crowded." Taking the pattern of suburbanization in the United States, for example, this development leads to a loose spatial structure that brings traffic congestion and other urban problems. With the popularity of the car in China's major cities, it has become necessary to ease the traffic congestion problems through the development of public transport-oriented road traffic patterns. Compared to other modes of transportation, public transport has a large carrying capacity, low energy consumption, low pollution, and other advantages, which are especially true of the subway, light rail, and rapid public transport. At the same time the public transport network should be incorporated with urban land use in order to create a reasonable arrangement of urban functions, generating a fast and convenient way of life for the residents [4, 5].

5.4 Ecological Space: The Combination of Artificial Landscape, Natural Landscape, and Multi-factor Integration Development

5.4.1 The Combination of the Artificial and Natural Environments

The artificial and natural environments are the two most basic elements of the urban space. The development of the urban space is focused on the relationship between those environments. An ecological city must be an organic combination of the artificial and natural environments. The development of the urban space corresponds with the natural environment evolution process, which guides the urban land layout and development. This layout design, combined with the construction of an artificial environment, should be reinforced by the natural environment. The focus should not be on the needs of the new development and construction, which might destroy the natural environment. This combination of the natural environment with the artificial environment can evoke a perception of charm and an individual identity, and may also be an expression of sustainable urban development. There are individuals, who desire to see a return to nature and a pursuit of high quality of life, that advocate for this planning structure because it introduces nature into city.

5.4.2 The "Green Center, Green Belt, Green Wedge" Multi-factor Integration Development, Etc.

From the macro level of planning, the ecological space structure of future urban areas should be formed by the "green center, green belt, green wedge," and a variety of ecological domain surfaces. For the mass type planning model of the urban spatial structure, found in cities such as Beijing, Kunming, Xi'an, Zhengzhou, and others, the process can be optimized and restructured by incorporating parks into the interior city, constructing a green belt around the city, and repairing the damaged ecologically sensitive areas. The environment may be transformed through the building of a sustainable ecological space structure in the inner city by constructing a "green way." The construction of a "green belt" emphasizes an element of constraint, which affects the scale of city space from the natural ecological environment, and reflects an extreme development concept. For the cities that use "strip-," "ring-," "cluster-," or "combined-" type space structure planning modes, such as Shenzhen, Qingdao, Wuhan, Chengdu, Xiamen, and others, the green center and green wedge are the main elements of an ecological space structure, and these cities all have clear boundaries between their urban groups. A green open area formed between groups should provide access to the center city as much as possible. This would meet the needs of both residents in the inner city as well as those in the periphery groups to be in close proximity to nature. The green center (blue core) in "combined-" and "ring-" type cities is common. A green center is often the ecological core of these cities, which are usually placed in a central area of a city. The layout of the urban space with a green center is based on the ecological protection and restoration of the green heart, along with moderate development. Together, with the green wedge and green way, this forms a complex network of urban ecological space structures (Fig. 6).

6 Summary

This paper proposes the concept of "elasticity of the urban spatial structure," and suggests that, under the condition of a certain size, we can foster urban public capacity by adjusting the urban space structure and give way to the elasticity of demand of urban development. Secondly, the basic principles of the urban space structure are optimized, namely through ecological and economic growth and coordination of cities. Finally, a sustainable planning mode of urban space structure is proposed; namely the "multi-center, groups network" for contemporary cities in China, with the guidance of an elastic growth concept. The optimization strategy that was put forward, which comprises four aspects of development (the structure of the public center, land layout, the structure of the road traffic function, and the ecological space structure), serves as a reference for the sustainable development of Chinese cities.

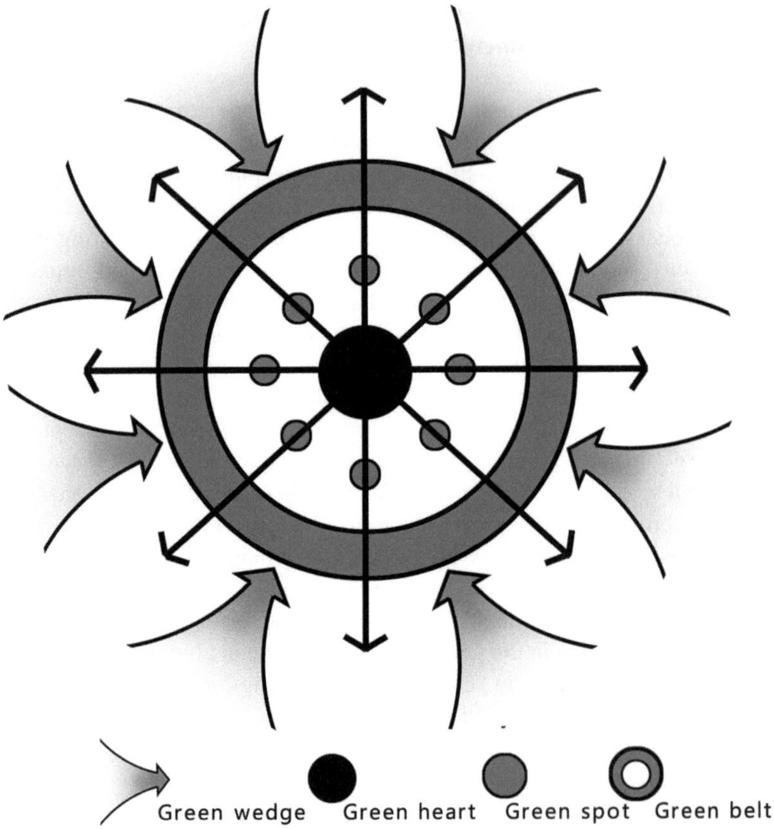

Fig. 6 The diagram of ecological space optimization structure (Source: Lin Kaixuan [6])

References

1. Guo Lijun (2008) Urban spatial structure in the knowledge economy [M]. Tianjin University Press, Tianjin, pp 116–117
2. Deng Qinghua (2002) Ecological urban spatial structure [D]. South China Normal University
3. Lin Zhen, Yang Haoji (2002) Urban transport planning in sustainable development [J]. Comprehensive Transport, pp 53–56
4. Huang Yaping (2002) Theory of urban space and spatial analysis [M]. Publishing House of Southeast University, Nanjing
5. Huang Yaping (2011) Urban land development and spatial development: the case of Wuhan [M]. Huazhong University of Science and Technology Publishing, Wuhan
6. Lin Kaixuan (2013) From chaos to order: the restructuring and reorganization of the spatial structure of contemporary metropolitan urban area [D]. Huazhong University of Science and Technology
7. Xie Shouhong (2004) Spatial organization of metropolitan area. Press of Science, Beijing

The Impact of Building Control on Urban Planning and Building Management in Hong Kong

Han Zou and Charlie Q.L. Xue

1 Introduction

Hong Kong had advanced by leaps and bounds over less than 200 years from a fishing village to a well-known international city with an image of impressive skyscrapers. At the same time, the building industry and market in Hong Kong had developed rapidly. There are lots of experiences in the field of urban planning and architectural design worth referencing.

Because of the imbalance between land shortage and dense population, Hong Kong has to utilize its land resources very efficiently by adopting high-density mode. The density of public housing reaches at least 2,500 residents per ha, which is twice the density of the most crowded residential areas in mainland China [1]. For the same reason, the price of land is so high that architecture design usually hangs about the borderline of legal limitation. Cities often respond to development pressures by setting targets for increased urban densities; therefore, the establishment of high-rise cityscape and compact urban settings is unavoidable [2].

A review of the evolution of the building industry, especially the aspects of urban planning and architectural design reveals that the supervision and control under the continuously revised legal system by a number of government departments is remarkable. The general policy and regulation, while defining the building standards, inevitably lead to certain uniformity in design. The evolution of law relevant to architecture directly affects the development of architectural design. The

H. Zou (✉)
Institute of Civil Engineering and Architecture, Hubei University of Technology (HBUT), Wuhan 430068, China
e-mail: zouhangogo@hotmail.com

C.Q.L. Xue
Division of Building Science and Technology, City University of Hong Kong (CITYU), Kowloon, Hong Kong

X. Chen, Q. Pan (eds.), *Building Resilient Cities in China: The Nexus between Planning and Science*, GeoJournal Library 113, DOI 10.1007/978-3-319-14145-9_30

institutional framework of legal system is composed of statutory laws and regulations, contractual constraints and public administration.

During the period of 1841–1997 as British colony, a full set of building control method was formed under the influence and management of suzerain. Although the current system of building control spans over the entire history of Hong Kong, the requirements of minimum health and safety standards for the planning and design of new buildings existed at the beginning of colonization.

This article will review of the building control system prevailing in Hong Kong, and analyze its influence of urban planning, architectural design, and relevant industry.

2 A Brief Review of Building Control Evolution in Hong Kong

2.1 Division of Time Phases of Building Control Evolution History

To promote a healthy and safe built environment, Hong Kong's Government enacts statutory standards in the form of ordinances, regulations and guidelines.

At the first century the Building Control policies in Hong Kong were mainly concerned about public health and dangerous buildings that were affected by environmental factors. Although the legislation didn't receive favorable reaction from both the judiciary and the public because of the unsettling social conditions, it did lay down the basic principles and mechanisms for maintaining public safety.

After the World War II, the Building Control System, a set of prescriptive or performance-based requirements of planning, design and construction, was established and was gradually improved over time. The Building Control System also controls the registration authority personnel and contractors, the vetting and approval of building plans, issuances of occupation permits, monitoring of site works, and authority to cease works, demolish or alter works. The Building Ordinance and the Building Regulations (Administration, Planning and Construction) were statutory parts. The Codes of Practice issued by Building Authority is lower grade of control and non-statutory. At the end of 1970s, the Building Control System was basically finalized. Since then more details of ordinance and regulation were added or replaced, such as offences, appeal, Authorized and Registered Persons in Buildings Ordinance, the height of stories, staircases and fireman's lifts, firefighting and rescue stairway, supplement of site coverage and plot ratio, and the use of verandahs or balconies in Buildings (Planning) Regulations.

The building control mechanism has been frequently reviewed and amended to meet the changing need in time for the purpose of social and economic lever. The evolution history of Building Control in Hong Kong and the experiences of legislation and its implement may be useful for other urban centers in mainland China.

The history of Hong Kong's Building Control System evolution that spans nearly 200 years can be divided into various phases that are shown in Table 1.

2.2 *Abstracts and Characteristics of Buildings Control in Each Phase*

2.2.1 Building Control Closely Relate to Public Health (1841–1934)

For nearly one century, the building control in Hong Kong was closely related to public health.

The first ordinance relevant to building control was The Building and Nuisances Ordinances 1856. The Public Health Ordinance 1883 was enacted for the needs of sanitation by controlling buildings following the suggestion of Chadwick Report 1882 that the building condition should be improved to avoid a cholera epidemic. In 1887, the Amendment of Public Health Ordinance required that back yard (open space), definition of overcrowding, and occupation permit be authorized by the Sanitary Board. Two years later, the Building Ordinance 1889 was enacted departing from the Health Ordinance which, for the first time, defined precisely the buildings and work. In 1894, the Closed Houses and Insanitary Dwellings Ordinance was enacted for sanitation needs. Further, the Building Amendment Ordinance 1901 was amended to permit the use of poor quality blue bricks only on the top story of buildings, and the heights of buildings were limited to no more than four floors. All the Ordinances above were not comprehensive until the Public Health and Building Ordinance 1903 was enacted. This Ordinance established the fundamental concept of building control that served as source for administrative or detailed amendments in the next 30 years.

2.2.2 Building Control Basically Self-Contained (1935–1954)

Since the Chinese Revolution broken out in 1911, large population has flooded into Hong Kong. The flood of refugees didn't follow the Colonial legislation that made it difficulty to control. The Building Ordinance 1935 was enacted to deal with the buildings and building work as well as the Sanitary Services with the Urban Council in charge of medical and health services. Higher standards of lighting, ventilation and fire safety were stipulated (Section 43–45) which shows the need of vertical dimension. It provided the Building Authority with more power to deal with building issues (Section 119) which made the amendment fairly satisfactory in

Table 1 Phases and evolutions of building control

Phases	Enacted time		Name of ordinance
I. Buildings control closely relate to public health	1856 No. 8[a]		Buildings and Nuisances Ordinance
	1883 No. 7		Order and Cleanliness Amendment Ordinance
	1887 No. 24		Amendment of Public Health Ordinance
	1889 No. 15		Building Ordinance
	1891 No. 25		Building (Amendment) Ordinance
	1894 No. 15		The Closed Houses and Insanitary Dwellings Ordinance
	1895 No. 7		Buildings (Amendment) Ordinance
	1899 No. 34		Insanitary Properties Ordinance
	1901 No. 30		Buildings Amendment Ordinance
	1903 No. 1[a]		Public Health and Building Ordinance
	1908 No. 14		Public Health and Building Amendment Ordinance
	1927 No. 6		Public Health and Building Amendment Ordinance
II. Buildings control basically self-contained	1935 No. 18[a]		Buildings Ordinance
	1936 No. 2		Buildings Amendment Ordinance
	1941 No. 12		Buildings Amendment Ordinance
III. Buildings control overhaul and supplement	1955 No. 68 (operated1956 G.N.A 45)[a]		Buildings Ordinance (3 Regulations: Administration, Planning, Construction)
	1959 No. 44		Buildings Amendment Ordinance
	1962 G.N.A. 97[a]		Buildings (Planning) (Amendment) Regulations
	1967 L.N. 91		Buildings Ordinance (Application to the New Territories)
IV. Buildings control system mainly finalized	1969 L.N. 54		Buildings Amendment Ordinance
	1972	No. 71	Buildings Amendment Ordinance
		L.N. 235	Buildings (Amendment) Ordinance (Application to the New Territories)
	1979 No. 24		Buildings Amendment Ordinance
	1980 No. 72		Buildings Amendment Ordinance
	1982 No. 41		Buildings Amendment Ordinance
	1987	No. 57	Buildings (Amendment) Ordinance
		No. 60	Buildings Amendment Ordinance (Application to the New Territories)
V. Buildings control system gradually amended	1990 No. 91		Buildings Amendment Ordinance
	1992 No. 42		Buildings Amendment Ordinance
	1994 No. 77		Buildings Amendment Ordinance
	1996 No. 54		Buildings Amendment Ordinance
	1997 No. 36		Buildings Amendment Ordinance

Source: Hong Kong Government Gazette
[a]Important points of buildings control system evolution

the next 20 years [3]. Further amendment was made in 1936 which provided provisions for evacuation of occupants. Because of the occupation by Japan in 1941, most operation was ceased until 1945.

2.2.3 Building Control Overhaul and Supplement (1955–1971)

Since Hong Kong population grew very rapidly during the postwar periods, the Building Ordinance was relaxed in 1956, allowing the development of high-rise blocks on land [4]. It was the urgent demand for housing and safety that gave impetus to government housing organizations [5].

After the World War II, the Building Ordinance has been overhauled [6]. The Building Ordinance 1955 together with three subsidiary regulations, the Building (Administration) Regulations, the Building (Construction) Regulations and the Building (Planning) Regulations, were enacted and operated in 1956 for building development, design and maintaining [7].

> To provide for the planning, design and construction of buildings and associated work; to make provision for the rendering safe of dangerous buildings and land; and to make provision for matters connected therewith.–Buildings Ordinance (CAP 123), 1956.

At the same time, the "Practices Notes for Authorized Persons (PNAPs)" and the "Code of Practice and Design Manuals" were published to detail how to implement the control.

Remarkably, the concepts of 'site coverage', 'plot ratio' and 'floor space' as more complex control approach to intense development were first introduced in Buildings (Planning) (Amendment) Regulations 1962.

2.2.4 Building Control System Mainly Finalized (1969–1989)

The Building Ordinance (Application to the New Territories) was enacted in 1967 to exempt the buildings, site formation and drainage work in New Territories from the Building Ordinance. Since 1969, the building ordinance (regulation) was reviewed and amended nearly every two years. It can be found from the evolution of ordinance relevant to buildings that the Building Control system was basically finalized at the end of 1970s.

> Projecting windows will not be regarded as GFA and will be accepted as not counting for site coverage and plot ratio, if they satisfy all the following criteria within the storey...... (PNAP68 in 1980)

As one of the green initiatives to end-users, bay windows with less than 500 mm depth have been exempted from GFA calculation since the 1980s. Bay windows increase the flow of natural light into a residential unit as well as provide special views of the outside [8].

Codes of Practice of fire safety were improved in Oct 1989 such as for Means of Access, Means of Escape and for Fire Resisting Construction since fire safety first mentioned in 1955 Building Ordinance.

2.2.5 Building Control System Gradually Amended (1990–1997)

Since 1990s, every two years the Building Ordinance was amended; however, all the amendments of Building legislation were piecemeal and fragmented.

Codes of Practice for Means of Access code was revised in May 1995. Codes of Practice for Fire Resisting Construction and for Means of Escape were revised in 1996.

The Fire Safety (Commercial Premises) Ordinance (Cap. 502) was enacted in 1997. Draft code on Site Safety Supervision (and Technical Memorandum for Supervision Plans) and Design Guide for Barrier Free Access (BFA) were issued in 1997.

3 The Framework and Supervision Institution of Building Control System

3.1 The Framework of Building Control System

The Building Ordinance is the highest statute law norms for standardizing the construction activities in Hong Kong which is enacted by the top legislature—the legislative council.

As mentioned above, the framework of Building Control System was gradually formed by two parts. The statutory part is consisted of Building Ordinance and its subsidiary legislation to ensure the implement of the Buildings Ordinance—Building Regulations, such as Building (Administration) Regulation, Building (Construction) Regulation, Building (Planning) Regulation, Building (Demolition Works) Regulations, Building (Private Streets and Access Roads) Regulations, etc [9].

The Building Ordinance includes the examination and approval of the project drawings, the supervision of implementation phase of the construction, the maintenance after the final acceptance of construction, and appeal system for violation behavior. It also includes the maintenance of old buildings and technical standards of construction, etc [10]. And the Regulations cover the administrative management, construction and demolition, planning, private roads, pathways, health equipment engineering, etc [11].

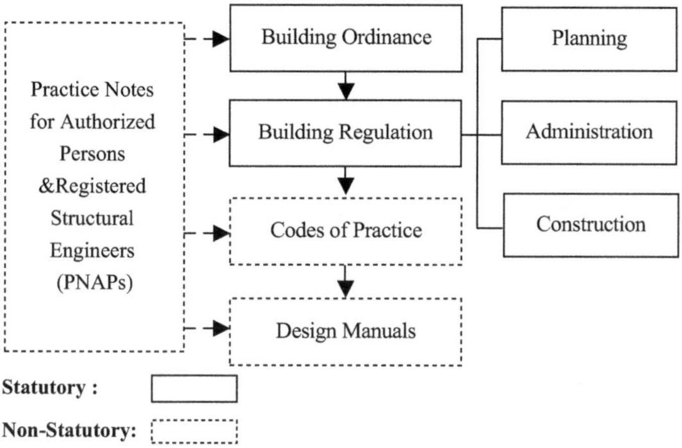

Fig. 1 Framework of building control system in HK

The non-statutory part is consisted of Practice Notes for Authorized Persons & Registered Structural Engineers (PNAPs), Codes of Practice and Design Manuals. The Building Control System is shown in Fig. 1.

3.2 The Supervision Institution of Building Control System

The building industry in Hong Kong can be divided into public construction led by government and private construction led by developer. The former is on behalf of government by Architectural Service Department and Housing Department. The latter is under the supervision of the Building Department following the control of Building Ordinance and the subsidiary Regulations. With regard to new private buildings, the Buildings department scrutinizes and approves building plans, carries out audit checks on construction work and site safety, and issues occupation permits upon completion of new buildings.

The Hong Kong Housing Authority is exempted from the Building Ordinance for the construction of public housing estates (Fig. 2).

4 The Impact of Building Control on Urban Planning

As described in the previous chapters, the Building Control was in charge of urban development control for nearly more than one hundred years until the Government started to prepare the Colony Outline Plan in 1965. It was revised to become the Hong Kong Outline Plan and consisted of two parts: Part I relating to planning

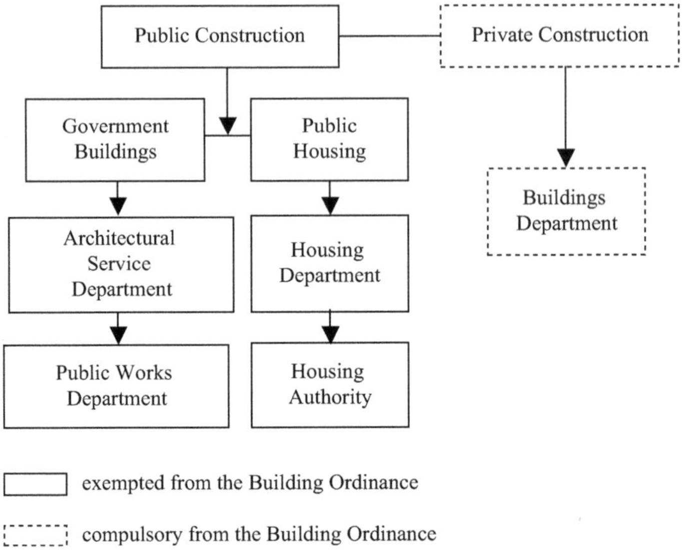

Fig. 2 Framework of supervision institution of control system

standards (known as the Hong Kong Planning Standards and Guidelines) and Part II on development strategy (known as the Territorial Development Strategy).

At present, it is indicated in the Hong Kong Planning Standards and Guidelines that The Buildings Ordinance, including its associated Building (Planning) Regulations, is the only statute that explicitly sets density limits and provides for their enforcement. The Building (Planning) Regulations specifies the maximum plot ratios and site coverage permitted for domestic and non-domestic buildings in relation to building height.

All land in Hong Kong is leasehold; all private land developments are controlled by the government through the statutory control of planning, building and land lease [12]. It is reasonable to take the evolution of Building Control to study the urban planning history of Hong Kong.

In addition, it is a visual and objective research method to establish the contact of architecture types and urban morphology in the same time and space. It is necessary to identify the urban morphological unit and architectural typological process [13]. The concept of urban morphology is concerned with the physical urban form and associated socio-cultural representations.

Let's look back at the socio-cultural history. Hong Kong was captured for the main function as transit port; therefore the public buildings such as industrial, commercial and business buildings were firstly built in Hong Kong and they play the most important role of the Hong Kong's economy [14]. It is, thus, inevitable to pay attention to the evolution of public buildings. On the other hand, the long-time problem of urgent need of housing, the residential zoning, takes another important role of urban planning and urban form in Hong Kong.

Therefore, this paper takes public buildings and private housing as examples to present the impact of building control on urban planning by analyzing the historical urban morphological unit and typological buildings.

4.1 Evolution of Commercial Buildings Zoning: Urban Density of Central Praya in Hong Kong

Central, the heart of Victoria City, is the central business district of Hong Kong. The Central Praya area, as the earliest developed land in Hong Kong, with its proximity to Victoria Harbour, has served as the centre of trade and financial activities since the earliest days of the British colonial era in 1841.

The urban density evolution of Central Praya can be presented by the building density and the skyline which rest with building height (see Table 2).

The evolution can be divided into five phases that are closely associated with the five phases of Building Control System Evolution (see Table 1).

In 1860s, Government House and Various barracks, naval base and residence of Commander, Flagstaff House were built at the east end of Government Hill. Between 1860 and 1880, the construction of City Hall, Theatre Royal and other financial structures made Central the heart of Hong Kong.

The Amendment of Public Health Ordinance 1887 required at least 300 ft^3 of living space per person. It also stated a 4-ft gap between buildings and a backyard of new buildings. Quantitative Order was thus started by defining and controlling density relating to the volume of building [5].

The standard of overcrowding was defined in a new Public Health and Buildings Ordinance 1903 that "in excess of a proportion of one adult for every fifty square feet of habitable floor space or superficial area and five hundred and fifty cubic feet of clear and unobstructed internal air space" in the urban area, and "more than one person to every one thousand cubic feet of clear internal space" in the European Reservation or the Hill District.

Although the Buildings Ordinance was basically self-contained from Public Health since 1935, it can be seen from the images in I and II phases that the building height and urban density did not change much in the first 100 years.

But the situation was greatly changed since 1956. The urban density increased a lot in just 20 years. Under the pressure of population explosion after World War II, the Building Ordinance was made to high-density urban development. The building bulk is controlled by a much more complex system including building heights, volumes, open spaces and lanes. The height of the building could not exceed twice the width of the street. Setbacks at an angle of 76° are permitted to allow natural lighting to streets (Regulation 17 & 18). The Volume of a building is set in a formula, depending on the location of the site and use of the building (Regulation 20):

Volume = Factor (F) times the width of street times the area of the site.

Table 2 Historic images of Central Praya in Hong Kong

Phases/time		Images of Central Praya
I	1870s	
II	1940s	
III	1960s	
VI	1980s	
V	2000s	

Source: Hong Kong museum of history

This volume control system was to be blamed for too crowded urban space and large buildings. The everywhere huge building bulk can be seen in the image of phrase III.

The volume system was amended in 1962 and replaced by Development Intensity Control consisting of plot ratio and site coverage. Three types of site were designated to improve the control system which made the new generation of building form—smaller site and taller height—called "pencil-like" buildings. The impact is obvious as can be seen in the image of phrases IV and V.

The high-density way of urban development is inevitable in Hong Kong due to shortage of land [15]. Each improvement of building control is made to cater to these needs [16].

4.2 Comparison of Private Housing Zoning: Urban Morphological Unit in Different Historical Period

Urban morphology focuses on the form, physical structure, plan, layout and their formation and transformation process in functional areas of cities. The town plan is itself subdivided into three constituent parts or elements: streets and their arrangement in a street system, plots and their aggregation in street blocks, and the block-plans of buildings (Conzen [13], p. 5). The first level of street system seems to be the steadiest element in the evolution of urban form.

This chapter seeks to find the impact of Building Control on the urban grain (street pattern, plot pattern and building pattern) and its character by examining and comparing the patterns of ten private housing zoning as urban morphological unit in different historical periods. The cartographic sources are chosen involving the historical urban development background with corresponding Building Control above. They are compared at same scales (See Table 3). These zonings are all influenced a lot by the residential building form that follows the Building Control.

Sai Ying Pun was built to the west of Central Praya as Chinese residential area in the early years of Hong Kong. The existing urban form of Central and Sai Ying Pun are different mainly because of the early racial segregation policy which lingered to 1940s– Europeans were assigned areas above High Street where Chinese were excluded from living. It was stated in the Public Health and Building Ordinance 1903 that 'It shall not be lawful to erect any Chinese domestic building other than quarters occupled by servants, within the European Reservation or the Hill Distriet' (Article 200). The urgent need of land and Chinese traditional shop-house building type made the urban form so crowded. Meanwhile, the ordinance regulated the type and height of buildings as well as open space for private streets, such as 'open space extending across the entire width and in the rear of such building and of a minimum depth of 8 feet' (Article 13). The Building Height to Street Width ratio in 1903 was nearly 1:1. According to the old land lease, the building height was restricted 35 in..

Table 3 Plan of private housing zoning in Hong Kong

Phase/location		Plan of private housing zoning
I	**1900s** Sai Ying Pun/Sheung Wan	
II	**1930s** Wan Chai/Causeway Bay	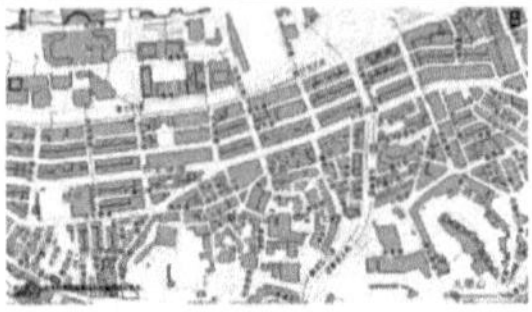
III	**1960s** To Kwa Wan (13 Streets)	
IV	**1970s** Jordan (Man Wah Sun Chuen)	
V	**1980s** Lai Chi Kok (Mei Foo Sun Chuen)	
VI	**1990s** TaiKoo (TaiKoo Shing, Kornhill)	
	1990s Shatin (City One)	
	1990s Hung Hom (Whampoa Garden)	
VII	**2000s** Tin Shui Wai (Kingswood Villas)	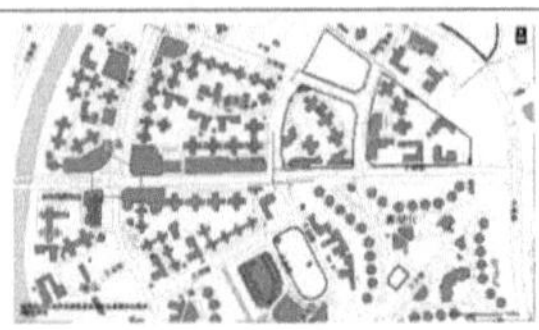

Source: The Government of HKSAR Building Department

That's why most buildings are two or three stories and that the streets are narrow in Sai Ying Pun (Phase I).

This ratio was amended to 1:1.25 in Building Ordinance 1935 which required 'a clear intervening space or area of a width of not less than one-fourth of the height of the cutting shall be left between such building, along its whole extent' (Article 79). It can be found in phase II of Table 3 that the street width of private housing zoning in Wan Chai and Causeway Bay is wider than 1900s. By the 1850s this area was already becoming a Chinese residential area and the reclamation of Praya East had an effect on Wan Chai from 1922 to 1931. The residential zoning between Queen's Road East and Johnston Road was built here before the World War II (Phase II).

As mentioned above, the Building Ordinance was not improved much until 1955. The Building Height to Street Width ratio in 1955 was nearly 1:3 by this turn of amendment since the structure technology was improved such as reinforced concrete and structural steel. The To Kwa Wan in Kowloon peninsula was not developed much in 1950s until 1960s. Most residential dwellings in the area are mid-rise flats of 10 or less floors, built in the mid-twentieth century. There are 83 buildings which could be the oldest large-scale residential zoning existing in the 13 Streets area, built between 1958 and 1960. It is clear that the Building-block Plan is similar to the Wan Chai formed in 1930s (Phase III).

In 1962, the intensity of development was first introduced in Buildings (Planning) (Amendment) Regulations. Higher density of construction was allowed under the pressure of population explosion. "The erection of buildings containing 20 or more storey is rapidly becoming more common, and it seems clear that skyscrapers, at any rate in the business centre, will be the future form of development" (Hong Kong Annual Report 1957, p. 199). The buildings were mainly controlled by 'site coverage', 'plot ratio' and 'floor space'. For example, the Man Wah Sun Chuen of 14 stories is built in 1965 which has 8 blocks. The surrounding urban morphological unit seems to be integrated until now (Phase IV).

At the end of 1970s, the building type changed dramatically in Hong Kong. More and more "cross type" appeared and replaced the former flat type. Take the large-scale residential estate Mei Foo Sun Chuen built during 1968–1978 in Lai Chi Kok as example, the new-coming urban morphological unit appeared and pointed to its elder neighbour urban form. This "cross type" of residential zoning has been commonly built since 1980s (Phase V).

It can be seen from the plan of Taikoo, Shatin and Hung Hom in 1990s (Phase VI) that the "cross type" residential zoning became common in Hong Kong. The well-known large-scale private housing, such as TaiKoo Shing, Kornhill, City One and Whampoa Garden, is all this kind of building form. Compare with the Mei Foo Sun Chuen, there seems to have more public open space, but in actuality, it results in much higher density.

This situation spread to declining industrial district and New Territories such as Kwun Tong, Lam Tin and Tin Shui Wai in 2000s (Phase VII).

5 Conclusions

There is always contradiction between land resource shortage and comfortable human environment in Hong Kong. This extreme condition made the unique case study of high-density city in China [17].

Looking back at the close relationship between the evolution history of Building Control and the development of urban planning and building management especially in the field of commercial and private housing, it is obvious that the monopolistic control of land sales and the development control mechanisms of Hong Kong government have played an important role in guiding and monitoring the development of commercial and private housing market.

Moreover, the Building Control System is reviewed and amended frequently over time to address urgent needs, for the purpose of public health, social safety or economic consideration.

On the other hand, the estate developer and constructor pursue the economic benefit in their nature. Facing the high value of land, they have to explore the right way of game with government and client.

Therefore, it is worth adopting the scientific and suitable theory and technology to direct the urban development and living environment. Though there is a full-order legislation system of Building Control and Urban Planning scheme with quite a number of details, the present guidelines should be improved up to date, not only the theory and technology (e.g., the energy-efficient building, the public communication space, and the multi-purpose space), but also the government institution. It's also important to have public participation in the process of establishing or amending the Building Control for better living environment.

Otherwise, residential buildings played a major role in forming the physical structure of Hong Kong. Beside, in the view of social economic factor, the residential issues always contribute to the problem of class distinction in Hong Kong. Public housing and private housing are developed under different control system and follow different standards which made different urban forms. These two systems could learn from each other in order to build more harmonious community.

Fortunately, up to now, less than 30 % of the land is developed in Hong Kong. It is inevitable to pay attention to the land development potential. In addition, there are lot of old urban districts that can be redeveloped. The urban form and building density did not change much in the first 100 years, but the development has accelerated since 1960s. There is no sign of shrinking city appearing in Hong Kong now. It is time for thinking over the present problems and working on more efficient management of building control.

Acknowledgment This is part of a study supported by the Doctor Scientific Research Startup Found of Hubei University of Technology, No. BSQD12149 and the National Natural Science Foundation of China (NSFC), No. 51278438 and Strategic Research Fund, City University of Hong Kong, No. 7002844.

References

1. Xue CQL, Manuel K, Chung R (2001) Public space in the derelict old city area, a case study of Mongkok, Hong Kong, vol 1, 6th edn, Urban design international. Palgrave-MacMillan, London, pp 15–31
2. Hui SCM (2001) Low energy building design in high density urban cities. Renew Energy 24(3):627–640(14)
3. Hong Kong Government Buildings ordinance(1935–1997). Government Printer, Hong Kong
4. Tse RYC (2001) Impact of comprehensive development zoning on real estate development in Hong Kong. Land Use Policy 18:321–328
5. Yuen B, Yeh AGO (eds) (2011) High-rise living in Asian cities. Springer, Netherlands, p 90
6. Hong Kong Government Hong Kong annual report (1962–1997). Government Printer, Hong Kong
7. History Law of Hong Kong Online: http://oelawhk.lib.hku.hk/
8. Ng E (2003) Studies on daylight design and regulation of high-density residential housing in Hong Kong. Lighting Res Technol 35(2):127–139
9. McInnis A (2000) Building law handbook. Butterworth Press, Hong Kong
10. Booth P (1996) Controlling development: certainty and discretion in Europe, the USA and Hong Kong. UCL Press, London/Bristol
11. Chan EH (1998) Building control in connection with real estate development. In: Poon NT, Chan EH (eds) Real estate development in Hong Kong. Pace Publishing, Hong Kong, pp 66–81
12. La LW-C (2000) Planning buildings for a high-rise environment in Hong Kong: a review of building appeal decisions. Hong Kong University Press, Hong Kong
13. Conzen MRG (1960) Alnwick, Northumberland: a study in town-plan analysis, vol 27, Institute of British Geographers Publication. George Philip, London
14. Yiu CY (2008) The emergence of a performance-based building control system, contractual and regulatory innovations in building and real estate. Pace Publishing, Hong Kong
15. Tang B-S, Choy LHT, Wat JKF (2000) Certainty and discretion in planning control: a case study of office development in Hong Kong. Urban Stud 37(13):2465–2483
16. Edwin HWC, Tang B-S, Wong W-S (2002) Density control and the quality of living space: a case study of private housing development in Hong Kong. Habitat Int 26:159–175
17. Charlie Q.L. Xue, Han Zou, Baihao Li, Ka Chuen Hui (2012) The shaping of early Hong Kong: transplantation and adaptation by the British professionals, 1841–1941. Plan Perspect 27(4):549–568

MIX
Papier aus verantwortungsvollen Quellen
Paper from responsible sources
FSC® C105338

If you have any concerns about our products,
you can contact us on
ProductSafety@springernature.com

In case Publisher is established outside the EU,
the EU authorized representative is:
Springer Nature Customer Service Center GmbH
Europaplatz 3, 69115 Heidelberg, Germany

Printed by Libri Plureos GmbH
in Hamburg, Germany